D1047142

JOE BIDEN

ALSO BY JULES WITCOVER

85 Days: The Last Campaign of Robert Kennedy

The Resurrection of Richard Nixon

White Knight: The Rise of Spiro Agnew

A Heartbeat Away: The Investigation and Resignation of Vice President Spiro T. Agnew (with Richard M. Cohen)

Marathon: The Pursuit of the Presidency 1972–1976

The Main Chance (a novel)

Blue Smoke & Mirrors: How Reagan Won and Why Carter Lost the Election of 1980 (with Jack W. Germond)

Wake Us When It's Over: Presidential Politics of 1984 (with Jack W. Germond)

Sabotage at Black Tom: Imperial Germany's Secret War in America, 1914–1917

Whose Broad Stripes and Bright Stars? The Trivial Pursuit of the Presidency 1988 (with Jack W. Germond)

Crapshoot: Rolling the Dice on the Vice Presidency

Mad as Hell: Revolt at the Ballot Box 1992 (with Jack W. Germond)

The Year the Dream Died: Revisiting 1968 in America

No Way to Pick a President: How Money and Hired Guns Have Debased American Elections

Party of the People: A History of the Democrats

The Making of an Ink-Stained Wretch: Half a Century Pounding the Political Beat

Very Strange Bedfellows: The Short and Unhappy Marriage of Richard Nixon and Spiro Agnew

JOE BIDEN

A Life of Trial and Redemption

JULES WITCOVER

wm

WILLIAM MORROW

An Imprint of HarperCollins*Publishers*

All insert photographs, unless otherwise indicated, are printed with grateful acknowledgment to Valerie Biden Owens: p. 2, bottom, and p. 3, top: Associated Press; p. 3, bottom, and p. 5, bottom: photographs courtesy of Senator Ted Kaufman; p. 4, top: photograph courtesy of Jim Harrison; p. 4, center: Terry Ashe / Getty Images; p. 4, bottom: White House photograph; p. 5, top: Evan Vucci / Associated Press; p. 5, center: John Duricka / Associated Press; p. 6, center and bottom: photographs by Bradford Glazier; p. 7, center: photograph by Suzanne Rubel; p. 7, bottom: *Washington Post* / Getty Images; p. 8, top: Jae C. Hong / Associated Press; p. 8, bottom: Khalid Mohammed, Pool/Associated Press.

HarperCollins books may be purchased for educational, business, or sales promotional use. For information please write: Special Markets Department, HarperCollins Publishers, 10 East 53rd Street, New York, NY 10022.

FIRST EDITION

Designed by Lisa Stokes

Library of Congress Cataloging-in-Publication Data

Witcover, Jules.
Joe Biden : a life of trial and redemption / Jules Witcover. — 1st ed.
p. cm.
Includes bibliographical references and index.
ISBN 978-0-06-179198-7
1. Biden, Joseph R. 2. Legislators—United States—Biography. 3. Vice Presidents—United States—Biography. 4. United States. Congress. Senate—Biography. 5. United States—Politics and government—1945–1989. 6. United States—Politics and government—1989–. I. Title.
E840.8.B54W58 2010
328.73092—dc22
[B]

2010017723

10 11 12 13 14 OV/RRD 10 9 8 7 6 5 4 3 2 1

To Curtis, present at the creation, and remembering best friends Bob Healy and John Mashek, gone too soon

He's always been a guy for the heart. He wants to be the guy wearing a cape: Captain Good.

—Tom Lewis, *Joe Biden's close friend at Archmere Academy and roommate at the University of Delaware*

CONTENTS

PROLOGUE

THE LONG ROAD that Joseph Robinette Biden Jr. traveled to the American vice presidency—and, in the tired but accurate cliché, to within a heartbeat of the presidency—was one of triumph and tragedy. It was a journey of great hopes and expectations raised, then dashed, then resurrected, then dashed again and revived again, but always short of the man's dream of attaining the Oval Office.

Despite the Horatio Alger saga that has grown about Joe Biden over the years, the bright and ambitious Irish Catholic lad born in gritty Scranton, Pennsylvania, in 1942 did not exactly rise up from poverty. In his family roots and tribulations, however, he did experience the rigors of working-class urban America. His father, Joe Biden Sr., did encounter hard times, but not until after he had enjoyed the beneficence of a rich uncle who positioned him for a lucrative executive job that soon found him on the polo fields of Long Island and in the cockpit of his own private plane.

But the economic woes that eventually brought the elder Biden and his family to Wilmington, Delaware, another town of blue-collar profile, maintained the saga. At the same time, it provided the launching pad for the rapid political climb of Joe Biden Jr.—a cocky and aggressive young lawyer, elected to the New Castle County Council at the age of twenty-seven and to the U.S. Senate at twenty-nine, five weeks short of the required age to sit in the august body.

Blessed with a boyish charm, athletic agility, a dazzling smile, and the quick and overflowing gift of gab, young Joe tap-danced through an elite Catholic prep school, the University of Delaware, and the Syracuse Law School. He met the smart and beautiful girl of his dreams during a spring break in the Bahamas, married her, and quickly began a young family of two boys and a girl. Even before he was to be sworn into the Senate, Joe Biden had all the earmarks of a golden boy reaching for the stars and armed with all the personal, social, and intellectual accoutrements for grabbing his share of them.

Then came with shattering suddenness the family tragedy that was to be the darkest hour of Joe Biden's life, and yet a springboard to eventual personal recovery and political achievement. When, after fifteen years of impressive service in the Senate, he first reached for the presidency, a fateful run of carelessness and hubris brought the golden boy down, seemingly ending that high ambition. Next, a life-imperiling stroke that forced him to the sidelines ironically saved his life and in time set him, two decades later, back onto the same quest for the presidency.

Sustained through all this time by a restored and steadfast family life, Biden tried again in 2008. But not even his broader experience as chairman of the Senate Judiciary and Foreign Relations Committees could erase the public remembrance of his failings in his first presidential run, and his second bid met the same quick fate of the first.

In the course of that rejection, however, Joe Biden so impressively redeemed himself as a capable, responsible, informed political figure that the man who won his party's presidential nomination chose him to stand at his side in the historic election of 2008. In Barack Obama's emergence as the nation's first African American president, he brought Biden with him as his standby and partner in governance, as close to the achievement of Biden's lifelong goal as circumstances seemed to promise.

Embarking on the vice presidency at the age of sixty-seven, Biden was no longer considered likely to use the office as a stepping stone to the Oval Office, or to presidential nomination, as other standby predecessors had done, especially since the days of Richard Nixon. Assuming a second term for Obama, Biden would be seventy-four years old before he could stand for the presidency a third time. But in the vice presidency, he had the fortuitous opportunity to return public esteem to the office, after the eight years of his predecessor, Dick Cheney, whose arrogant embrace of greater power had diminished it in the eyes of many Americans.

Before assuming the vice presidency, Biden had chided Cheney by saying he hoped to restore "balance" in the office, an observation Cheney interpreted as Biden's intent to diminish it. The account that follows, chronicling the personal and political rise, fall, and rise again of Joe Biden, will also consider how in the first year and more of his vice presidency he measured up to his promise to repair the reputation and function of the office that in his later years unexpectedly came his way.

ONE

SCRANTON

ON NOVEMBER 20, 1942, lost amid the latest news of Allied gains on the battlefronts of World War II, the *Scranton Times* reported on page twenty-six that "a son was born this morning in St. Mary's Hospital to Mr. and Mrs. Joseph R. Biden of Baltimore, Md. Mrs. Biden is the former Jean Finnegan, daughter of Mr. and Mrs. Ambrose Finnegan of 2446 North Washington Avenue."[1] That was all the item conveyed.

The child, the couple's first, was named after his father, and from the start was lovingly called Joey. At the time, Joe Sr. was working out of town, so Tom Phillips, a neighbor in the middle-class Green Ridge section of the grimy old eastern Pennsylvania city, drove the mother-to-be to the hospital. It was not reported that young Joey had anything to say upon delivery, an observation about him that later in life would have been unlikely.

Elsewhere in the *Times*, the daily horoscope provided a not-quite-clairvoyant glimpse of the future vice president of the United

States. It foresaw "great joy and good fortune" for the child as well as "secret or unusual gains." It said he would be "progressive, forceful, rigid in his judgment, and fond of study." It predicted "a happy marriage" and declared that "remarkable traits, talents and abilities will be manifested in the child who is born on this date. Early fame and renown will be achieved, especially if the passions and emotions are controlled."[2]

The horoscope foresaw in a general way Joe Biden Jr.'s success, his liberal and determined manner, his marriages, and especially his "early fame and renown." It proved to be somewhat off the mark, however, regarding controlled passions and emotions.

Coinciding with Joey's arrival came triumphal news in the *Scranton Times* of U.S. and British forces engaging Nazi troops across North Africa and Eurasia, and confronting the Japanese naval armada in the Pacific. Front-page headlines in the local paper proclaimed:

> SEVEN MORE JAP SHIPS SUNK IN SOUTH PACIFIC
>
> GERMANS DRIVEN BACK BY U.S. FORCES
>
> NAZIS SUFFER SMASHING DEFEAT IN CAUCASUS[3]

But Scranton on that day otherwise seemed far removed from the war, without the excitement experienced to the south in Philadelphia, where a front-page story in the *Philadelphia Inquirer* offered a lively local angle. Just arrived at the Philadelphia Navy Yard, a critical home-front cog in the war effort, was USS *Boise*, previously reported lost in a great sea battle.

"As the cruiser steamed slowly up the Delaware, coming home from the war for overhauling, river craft blew whistles and sounded horns in greeting," the paper reported. "And at the Navy Yard the crews of other ships lined the decks and joined with thousands of workmen in one of the most rousing cheers ever heard in these parts."[4]

The "Inky," as the newspaper was known locally, noted six Japanese flags painted on the ship's bridge proclaiming its kills, and reported the crew's conquest of the City of Brotherly Love: "Browned by the tropic sun and scrappy over the great exploit, a crowd of victorious seamen from the cruiser Boise swarmed ashore in Philadelphia last night, to find the town belonged to them. . . . The Boise's men filled the USO centers, the Stage Door Canteen, movie houses and other places of amusement. They swung along the crowded streets arm in arm, and their long tour of sea duty was evidenced by their rolling gait."[5]

On the day of Joey Biden's birth, the Inky and the Scranton paper were full of other war news, and comic strips reflected the times as well. They featured Joe Palooka flying U.S. Army fighters against the Nazis, Superman capturing a German villain called "the Monocle," Tillie the Toiler dressed smartly in a Women's Army Auxiliary Corps uniform, and Barney Google as a recruit being dressed down by a fearsome drill sergeant. But thoughts of the war were temporarily ignored that day at the small Finnegan home in the Green Ridge neighborhood of Scranton. There, for the occupants, nothing could detract from the more momentous arrival of Joseph Robinette Biden Jr.

The boy's family was of humble origins but with fortuitous connections. His paternal grandfather, Joseph H. Biden, from Baltimore, married Mary Elizabeth Robinette, from an old colonial family of French origin in Maryland and West Virginia, and in 1915 a son was born to them, named Joseph R. Biden. The grandfather began in the oil business as a boy helping the family of a German immigrant, Louis Blaustein, deliver kerosene to Baltimore neighborhood families, and continued as a salesman as Blaustein's business became the American Oil Company, later known as Amoco. The young Biden family eventually was sent to York, Pennsylvania; Wilmington, Delaware; and then in the 1930s to Scranton, Pennsylvania, where the son graduated from Saint Thomas Academy.[6]

Joey's maternal grandfather, Ambrose Finnegan, called "Pop" by the grandkids, was born in Scranton to Irish immigrants who had been apple pickers in upstate New York. He went to college in California, where he was an acclaimed quarterback at Santa Clara, before returning to Scranton. There he first worked for coal and gas companies before becoming an advertising salesman for the local newspaper and marrying Geraldine Blewitt, daughter of a Lehigh graduate, engineer, and Pennsylvania state senator, the only previous politician in the family.[7]

The couple had four sons and a daughter, Catherine Eugenia, called Jean, who was a local school homecoming queen when the first Joseph Robinette Biden came on the scene from Wilmington. A tall, handsome, dashing, and refined figure, he courted and married Jean, a spunky Irish lass with a mind of her own, and the new couple moved in with the Finnegans on North Washington Street. In addition to Ambrose and Geraldine, also living there were Geraldine's unmarried sister Gertie Blewitt and brother Edward Blewitt, who had a bad stutter and was called "Boo-Boo" after the way he tried to say his own name.[8] A year later, Joey arrived.

With America's entry into World War II, a well-to-do uncle on his father's side, Bill Sheen, inventor of a product used originally to line cemetery vaults, had hit pay dirt. He expanded into the manufacture of a special watertight sealant required for merchant marine ships sailing from Eastern Seaboard ports. Joe Sr. and his cousin, Bill Sheen Jr., became close friends, and Bill Sr. set the two young men up in the booming wartime family business, Sheen Armor Company.

The cousins found themselves for a time in financial clover. With Bill Sr. lavishing them both with expensive gifts, they drove the fastest new cars and even piloted a small plane into the Adirondack Mountains on trips to hunt elk and bear. Bill Sr. regularly bought three new Cadillacs for himself, his wife, and his son, and a Buick roadster convertible for the "poor cousin."[9]

Subsequently Joe Sr. was transferred to run the firm's branch office in Boston, where Joe Jr.'s sister, Valerie, was born in 1944. The family was now ensconced in a four-bedroom Dutch colonial in a Boston suburb, with plenty of money and perks to fly back to Scranton to visit the Finnegans whenever they wanted. Joe Sr. was as generous to his family and friends as his benefactor was to him, as long as the money flowed.

But the good times did not last. After the war, Bill Jr. lost his bearings in high living and Joe Sr. had to rely on his own devices. First he planned with an old friend to open a furniture store—until the friend absconded with much of the seed money. Undeterred, he hitched up with Bill Jr. in a crop-dusting enterprise over Long Island and upstate New York farm country, with the family moving from Boston to Garden City. But that venture failed, too, with Bill Jr. running out on him. Jean, weary of warning her husband about him, decided to return to Scranton with the two small kids in 1948. Then, to make matters worse, Bill Sr., who had helped to underwrite the enterprise, bailed out.[10] It was back to Scranton for Joe Sr., where the Bidens found sanctuary again in the Finnegan homestead. It was quite a comedown for the previously high-flying aficionado of the polo fields and the airways, but he accepted it while still aspiring to better circumstances.

Harmony generally reigned within, though not without some friction at times in a household in which the Finnegan side was Irish, from County Derry, and the Biden side was English and French. Joe Jr. remembered his maiden aunt Gertie telling him that "your father is not a bad man. He's just English."[11] Indeed, according to Joe Jr., his grandfather Joseph H. Biden and a brother, Charles, emigrated from Liverpool to Baltimore in 1825, though there were some suspicions that the Bidens came from Germany. However, Joe Jr. recalled later, "My grandfather Biden, who died the year before I was born, used to always say when someone asked if they were English, 'No, we're Dutch,'" which only confused the matter.[12]

On top of that, Joe Sr.'s acquired tastes for the better side of the tracks had given him a certain ease and polish that did not always fit in the often raucous lifestyle of the street-smart Finnegans or, for that matter, of blue-collar Scranton. Uncle Boo-Boo didn't take to what he saw as Joe Sr.'s superior ways, Biden later wrote, and would needle him, complete with stutter. "He could not stand rich guys. When my dad was making money, during the war, he used to remind him he'd never been to college, that no Biden had. "B-b-b-b-Bidens have money, L-l-l-l-Lord Joseph, but the Finnegans have education."[13]

The city then was a bustling anthracite coal–mining town nestled in eastern Pennsylvania's Lehigh Valley. It was marked by the low-lying Allegheny Mountains and lower- and middle-income brick and frame houses on small patches of ground, along with occasional large, impressive homes occupied by coal and other industrial and business executives. The Green Ridge neighborhood had both, with Ambrose Finnegan's modest shingled abode in the first category. The town was peppered with small Catholic churches, local groceries, schools, and taverns, and a distinctly hardscrabble, blue-collar working-class aura, largely of Irish, Italian, and Polish roots. A strong sense of togetherness prevailed that was particularly represented in the Biden clan.

In time, Joey as a healthy and happy five-year-old and Valerie as an adoring sister of three were joined in the family circle by two brothers, James and Frank. Together, they comprised an unusually close and mutually affectionate sibling quartet destined to become the core of the eldest brother's successes and tribulations in all the years ahead. Their grandparents were doting presences, and the clan always returned to the family hearth from wherever the father's business fortunes might take it.

Joey was now starting at Saint Paul's—a Catholic school of course, in heavily Catholic Scranton. The driving forces in his life were already in place as they would remain throughout—family and faith. Neither anchor, however, prevented young Joey from being a freewheeling kid in the rough Green Ridge neighborhood of the old coal-dominated town. He and his buddies Charlie Roth, Tommy Bell, and Larry Orr hung out in an old city dump and used the streets and vacant lots of the neighborhood for baseball and football. Joey, though small for his age, had the spirit and athletic quickness that made him a leader and a daredevil.

Many years later, when their old childhood pal became vice president, Tommy Bell, now an insurance agent in Scranton, remembered the young Biden as "an aggressive guy, he was a leader, he was a risk-taker. He was a good guy, a friend, with a sense of fairness. He was one of those people who was just fun to be with." Orr agreed: "He stuck out in the crowd. He was kind of the go-to guy." Bell added: "We weren't thinking it was at all odd that Joe Biden achieved greatness in life. That was not something we were startled about. We just assumed it would happen, whatever it was going to be."

Politics, however, never came up in those days, Bell and Orr agreed. "We didn't talk about those things," Bell said. "We talked about the normal things—girls, sports. We loved sports. Charlie was, I'd say, the glue. He kept us together as a group. Joe was the leader but Charlie was very worldly. He knew what girls were for long before the rest of us. He had a girlfriend before anyone of us." Orr chimed in: "We were still playing baseball." But it was Joe who kept the same gang together well into their adult years, with repeated trips back to Scranton after the family had moved away. He invited Charlie, Tommy, and Larry to every Biden family and political event wherever it might be, including all the milestones marking Joe's rise to national prominence, and they attended many of them.[14]

Roth was the closest of the gang to Joe, Bell and Orr said, and the chief instigator of pranks, egging on young Biden. Bell remembered a Roth challenge from their high school years that became part of the Joe Biden lore. "When you deep-mine coal, not strip-mine it," Bell explained later, "you bring out of the ground coal, dirt, rock, and slate, and you separate the coal from the debris. And in the process of cleaning this material that comes out of the ground, it's supposed to be mostly coal but it's not."

In a large structure known as a "breaker," he said, "the coal would get separated from the debris, and over many years of these mines, the debris piles became huge, and this area was dotted with them. Nearly every neighborhood had a coal breaker—106 coal breakers, I believe, in the city of Scranton over a period of time.

"In our neighborhood was the large Marvine Coal Company, and they were in business many, many years. They built up a 25- or 30-million-ton pile of debris that spontaneous combustion could set on fire. It didn't burn with a heavy flame. It smoldered, as coal would, because whatever was left in the debris, it was low-quality, and it gave off a very sulfurous smell to it, like a fart. But the wind blew it up the valley, so we never noticed it. Being hard coal, anthracite, it burned very slowly and would set off a blue flame."

The steep hills, known locally as culm banks, could be like quicksand and treacherous to the sneakered foot. "These things would burn," Bell said, "and sometimes they would crater. Underneath they would burn more rapidly than on top. If you stood there you could fall through, twenty or thirty feet." One Sunday afternoon, Bell recalled, he and Joe were accompanying Charlie as he was collecting for his newspaper delivery route when they came upon one of the piles. "It was a couple of hundred feet high, like a pyramid," Bell remembered, and Charlie told Joey he would give him five dollars if he would climb to the top.

Without hesitation Joey took off, picking his way until he reached the peak. Five bucks was a lot of money in those days, and no challenge could go unaccepted by Joey. Roth eventually paid up, and Biden framed the bill and later put it on the wall in his Senate office. Bell said Roth challenged him, too, and he followed young Biden up, but Roth never paid him. "Charlie would never stiff Joe, but he stiffed me," Bell said, laughing.[15]

Larry Orr remembered another summer Sunday morning when he, Roth, and Bell had just left Mass and were standing on a low overhang having a water-balloon fight. "Here comes this guy driving up the street in his convertible with his arm around this girl," Orr said, "probably going to take her for a ride somewhere. We were about twelve or thirteen at the time, and Joe says, 'Let 'em have it!' The first one hit the hood of the car and sounded like a bomb going off. I thought the guy was going to have a heart attack. The second one hit the front seat with this girl in it, and this guy comes to a screeching halt, put on the brakes, and he chased us. He was in good shape. Me and Joe ran up to Dunmore, which was the next town, and we hid behind tombstones. This guy would not give up. He was fast, but we were faster, and we knew the territory, too. He finally gave up. He couldn't find us. We waited about fifteen or twenty minutes before we came out of the cemetery."

Joe apparently could not resist a moving target. One snowy winter morning outside the Finnegan house on North Washington, Orr remembered, "this guy in a coal truck was just turning the corner with his window down when Joe hit him on [the] right . . . side of his head with a snowball. This guy was so mad he got out of the truck and Joe and I ran. We ran up to his house and ran in. His aunt Gertie was there, and when the guy was going up after us, she beat him down the steps with a broom. 'Get out of here! Get out of here!' We were guilty as hell, but we could do no wrong with Aunt Gertie."[16]

Another young Green Ridge neighbor, Jimmy Kennedy, a few years older than the others, lived across Dimmick Avenue, a dirt alley separating the backyards of the Biden and Kennedy homes. The first time he laid eyes on Joey, he remembered long afterward, this scrawny kid in short pants with a distinct stutter called across the alley and defiantly challenged him: "You ca-ca-can't catch me." Kennedy couldn't, and Joey with his speed, shiftiness, and toughness played himself into tackle football games with the older boys.

Being two or three years younger made no difference, Kennedy recalled later, "because of his determination. He was gonna give you a hundred percent. He was gonna get dirty. He was gonna give whatever it took. Outweighed, outmanned, he was gonna play."

Kennedy, now a Scranton city magistrate, remembered the scrappy Joey, bleeding badly from falling on a broken soda bottle, being carried up the street to a doctor by his shaken father—and the same day appearing back on the street, ready to resume the game.[17]

On another occasion in the Green Ridge neighborhood, Joey and Jimmy Kennedy were fooling around in a field where construction for Marywood College was going on. A huge dump truck was at work, slowly digging holes and carrying the dirt away. As the driver drove it back and forth, Kennedy suddenly dared his young friend to run under the moving truck. Joey, eight or nine at the time, with no hesitation dashed under it, between the front and rear wheels and out the other side, unscathed.[18]

"Joe was just a daring guy who wasn't frightened by anything," Bell said. "He wasn't at all mean or cruel; in fact he was exactly the opposite." Orr added: "You could say Joe had guts, and he wasn't afraid to take a chance."

Once when construction workers were putting up steel girders for a new Marywood auditorium, Orr recalled, Joe climbed up to the top, grabbed a heavy rope secured to one of the girders, and swung far out and back in his own Tarzan imitation for the edification of his buddies—and without, like Johnny Weismuller, the film Tarzan of the day, a safety net.[19]

After a Saturday afternoon of watching a jungle or Western movie and a filmed race of crazy characters at the Roosevelt Theater, known to them as the Roosie, the four musketeers would wend their way home. Joey and Tommy would climb onto the roof of one of a string of adjacent garages and hop from one to another, reenacting scenes they had just seen on the silver screen. Larry and Charlie would trail them up the Green Ridge alleys.

Another popular prank by the Biden gang was "rapping," performed in the late 1940s when electric streetcars still rolled between downtown Scranton and near-in subdivisions like Green Ridge. "Joe would rap streetcars," Bell said. "You would run behind it and pull the cord"—the long pole in the center of the roof running up to the power line strung above, causing the streetcar to halt. "Joe did it with gusto," Bell said, and he and Orr "followed the leader." But again, they said, Charlie Roth was the instigator. "He was the snowball maker, was the best way to put it," Larry Orr said.[20]

In the early years in Scranton, Joey's best pals paid no attention to his stutter and never ribbed him about it, they said later, but the same was not true in school. Bell and Orr remembered that one of the teaching nuns at Saint Paul's School, Sister Eunice, nicknamed him Bye Bye Blackbird because he stuttered his own name as Bu-Bu-Biden. It was the same cruel inspiration that had earned Joe's uncle his name of Boo-Boo.[21]

The uncle was an inveterate newspaper reader who, though Valerie was his favorite, inculcated in Joey the habit of reading the editorial and op-ed pages in the local paper. Along with their grandfather, Pop Finnegan, somewhat of a political junkie, Boo-Boo encouraged young Joe's interest in politics. But he had a negative attitude born of career disappointments and lost opportunities of his own, Valerie said, which somehow seemed to spur an optimism in Joe. Also, she said, "my brother Joe is the direct product of my mother saying, 'When bad things happen, if you look hard enough something good will come.' " As for the stutter, she said, "my brother was never going to let an impediment get in the way of what he wanted to do. He was going to do it anyway."[22]

When Valerie was old enough, Joe simply took her along with him as a pal. "From the time I opened my eyes," she remembered long afterward, "he was there, he had his hand out and said, 'Come on, let's go.' There weren't any girls in the neighborhood when I was growing up, and I was a full-fledged tomboy. I hung with him; he was my best friend. He taught me everything. He was an athlete. He taught me how to throw a baseball, he taught me how to tackle, he taught me how to jump for the basketball net, how to ski; there was nothing I would do without him.

"He had a bicycle, and the boy's bike had a bar across the front,"

Valerie recalled of those early days. "I used to run next to him when he was going up a hill, and as soon as he got up he'd say, 'Okay, hop on,' and I could jump on that bike, backwards, forwards, on the handlebars, or the back fender. He just took me everyplace. He said if you're gonna play with the guys then you're gonna learn how to throw a ball. So I throw a baseball like a boy.

"He always told me that whatever he did, he told me I could do it better: 'I can throw this ball. Come on, you can throw it farther. Throw it. Hit me. Go ahead and hit me.' He always told me I could do anything. I'm a scrapper. I was the tiniest kid in the class, and I was scrappy. I wanted to keep up with them. I wanted to beat them."

As for the little sister insinuating herself into the gang, Valerie said, "his concept was, 'If you like me, you like my dog; if you like me, you like my sister.' Wherever he went I was there and the guys just accepted me. He was a wonderful, wonderful brother, and what he did through all my years growing up was, he always watched out for me, but he never acted like he was doing me a favor. On the other hand, I always watched out for him."[23]

That description of the relationship between Joe and Valerie Biden pretty well described how it continued to be through childhood, adolescence, and well into adulthood and Joe's life in politics. From the start, the vivacious Valerie was her oldest brother's best friend, confidante, and adviser, and eventually his devoted, enthusiastic, and innovative campaign manager, in a particularly tight Irish Catholic family that put loyalty, along with religion, above all other considerations.

Later, when the two other boys, Jimmy and Frankie, arrived, it was more of the same. "My mom and dad always told us that we were responsible for each other," Valerie said. "There were four kids, and once we walked outside that door and that door shut, we were in the outside world and we were the Bidens. If we had something to say, if you wanted to punch somebody [among the four] in the nose, you walked inside the house and did it, but you didn't get in a fight outside. There was family, and then there was *family*. It was never a burden, it was just natural."

The four Biden kids had a deal with their parents to settle their disagreements on their own. Each sibling had the right to call a meeting among themselves, Valerie said. "We would close the door and sit at the table and resolve it," with their parents excluded.

As for the senior Bidens, she said, "they were not shy. Each had a backbone of steel. They were principled people. My parents tried to teach about basic decency and basic justice, and sometimes we got it and sometimes we didn't. My parents never hit us. The one thing that would have driven them to it would be if we had been deliberately mean to somebody. That's very different than if you went and attacked a person where they were most vulnerable, just because it felt good. That would have been the biggest disappointment to my parents. All they had to do was say, 'Oh, I'm so disappointed,' and that was like a knife through you."[24]

In Scranton, Sunday was always family church day, with the Finnegan clan attending Mass together at Saint Paul's and then returning to Ambrose Finnegan's for a long dinner. Afterward the men gathered in the kitchen to discuss and debate sports and politics, and here Joey was able to listen in and pick up wisdom in both fields.

No retelling of the Sunday scene at the Finnegan household in Scranton can come close to Joe Biden's own account in his excellent memoir, *Promises to Keep*. In this tight Democratic circle, Harry Truman was lauded as a stand-up guy with "no artifice," but Adlai Stevenson was seen as "too soft" while General Eisenhower, a Republican, was given "the benefit of the doubt" because "he was a hero of the war, after all." Joey kept his ears open and his mouth shut most of the time, but the political bug was already nipping at him.[25]

After Joe Sr.'s run of good fortune ended and harder times befell the Bidens as work in Scranton became scarce, he was required to take odd jobs to keep food on the table. The situation was a far cry from his elegant polo-playing and hunting days as a younger man cavorting on the Long Island estate of his cousin Bill Sheen. Long before his marriage, Joe Sr.'s family had lived in Wilmington, Delaware, and now he found work there cleaning boilers for a heating and air-conditioning company, commuting back to Scranton, a distance of more than 140 miles. Finally, when Joey was ten the family pulled up stakes and moved to Wilmington for good.[26]

But the Bidens' association with Scranton never ended. Their friendship endured, Tommy and Larry said later, because even after the Bidens moved to Delaware, they constantly returned to Scranton. "One of the things about the Biden family," Tommy recalled, "they came back every holiday—Easter, Christmas, Thanksgiving—they were here all the time. Weekends, long periods of time in the summer. We as kids didn't even know Joe had moved away. He was always here." Larry added: "And when he did come back it was like he never left. The ball kept on rolling."[27]

When Charlie Roth died in 2000, not only Joe but the whole Biden family traveled back to Scranton for his funeral, and Joe delivered the eulogy. It was a tribute that the then U.S. senator repeated regularly at the death of old friends and their relatives in both Scranton and Delaware. With the Bidens, family and friends were uncommonly bound together. When Joe's mother, Jean, called Mom-Mom in the family, had her ninetieth birthday in Wilmington, Tommy Bell, Larry Orr, and their wives attended the family celebration, and later were invited to Joe's inauguration as vice president in Washington as well.[28]

In later years Joe Biden was to become so closely identified with Delaware that campaign signs for him, rather than imploring voters to support him when he sought another Senate term, simply said: DELAWARE'S JOE BIDEN. But 140 or so miles to the north, he remained Scranton's Joe Biden, who never forgot where he came from.

TWO

WILMINGTON

THE MOVE TO Wilmington wasn't easy for the Biden kids, leaving their school and neighborhood pals in Scranton. But Joe Sr. had found a job in Wilmington a bit more suited to his dignified demeanor—selling cars—and the family rented a small but new garden apartment in the suburb of Claymont. It was directly across from an impressive boys' private Catholic high school called Archmere Academy. The sight of it eased the transition for Joey and set him to dreaming of a future there, as improbable as it seemed for the financially struggling Bidens. His mother made the adjustment to Wilmington easier for him by having him repeat the third grade when they got there; he had missed much of it with absences due to having his tonsils and adenoids removed, a common practice in those days. So he got off to a fresh start at another Catholic school, Holy Rosary, in Claymont.

The presence of the nuns, according to young Joe later, also made the new school experience more comforting, and he soon took up with new friends in the neighborhood. Joe's first entrée of a sort to Archmere was Catholic Youth Organization football, coached by a man whose sons attended there and was permitted to hold his team's practices on the school's field. Joey was small but fast with a ball tucked under his arm, and early on he showed the promise of future success, which he dreamed would be achieved on the Archmere squad. But until then, immediately ahead for him was finishing grade school at Holy Rosary, where he faced and addressed the most challenging demon of his young life.[1]

Joe Biden, for whatever reason causes such things, continued to be afflicted with stuttering, which seemed to come on him when he was in unfamiliar circumstances or under new pressures. The sisters at school tried to help him deal with it or to avoid situations that brought it on. For a short time his parents sent him to a speech therapist but it didn't do much good. The problem didn't seem to hinder him as he addressed his major interests of the time—especially sports, in which verbal expression mattered little, and in which he excelled with an agility and daring that impressed his circle of friends.

When Joey was twelve and the family moved to a new house in a better neighborhood called Belfont in 1955, he transferred to the seventh grade at Saint Helena's, again a Catholic school. Like many Irish kids educated in Catholic schools, his sister recalled, "everybody wanted to be a priest or a nun, and Joe talked about it."[2] Biden himself recalled a Trappist priest coming to the school and telling the boys about a training high school for would-be priests. He went home and told his mother about it, but she discouraged him. "My mother said, 'If you still feel that way when you're older, then you can go, but you're much too young to make that decision.'"[3]

At Saint Helena's, young Joe encountered a nun who may have caused him to think twice. As he related in his memoir in what he called "the Sir Walter Raleigh incident," he had by this time developed a way to cope with his speech problem. He would anticipate meeting a neighbor and prepare himself with what to say and how to say it. If he had to read in school, he would memorize passages and at the suggestion of one of the sisters he developed a cadence that helped ease his delivery. In class, where students were seated alphabetically, he would always be in the first row and could calculate in a reading exercise which paragraph would fall to him to read aloud.

On this particular day, he wrote, the phrase he knew was coming to him was: "Sir Walter Raleigh was at gentleman. He took off his cloak and laid it over the mud so the lady would not dirty her shoes." As he launched into the first sentence, the teacher interrupted, asking him to repeat a word. He began to stutter, at which point, he wrote, she began to mock him, repeating his name in a stutter: "Mr. Bu-bu-bu-bu-Biden . . ." As she said this, he wrote, "I could feel a white heat come up through my legs and back of my neck. It was pure rage. I got up from my desk and walked out of the classroom, right past the nun."

He left Saint Helena's and hiked the two miles home, where his mother was waiting in anger. The school had called to report he had left. She ordered him into the car, along with brother Frankie, a toddler, and headed for Saint Helena. As Joe related it: "I could tell she was mad. I knew I was in trouble. 'Joey, what happened?' she asked. 'Mom, she made fun of me. She called me Mr. Bu-bu-bu-bu-Biden.'"

Jean Finnegan Biden took the two kids into the principal's office, left them in the anteroom, and demanded that the offending nun be called in. When she arrived, Joe's mother asked her what she had said. The sister insisted she hadn't really said anything, but when pressed whether she had mocked her son's stutter, the nun answered: "Yes, Mrs. Biden, I was making a point."

As Joe noted in his memoir: "I could see my mother pull herself up to her full height, five foot one. My mother, who was so timid, so respectful of the church, stood up, walked over in front of the nun, and said, 'If you ever speak to my son like that again, I'll come back and rip that bonnet off your head. Do you understand me?' Then the door flipped open and my mom grabbed Frankie from my lap. 'Joey,' she said as she left, 'get back to class.'"[4]

One of his first and best neighborhood friends was Tom Lewis, who went to Saint Helena grade school and played football and other sports in local playgrounds with him. All the boys were highly competitive. "We all kind of tried to best each other, like walking down the street, who could walk faster," Lewis recalled years later. "It wasn't done in a humiliating way, it was never cutting. You get a laugh when you can get a laugh. It was never done to be superior, always to get a laugh. We were raised to be extremely respectful. We never said anything to take advantage of someone's physical disability or unfortunate position in life."[5]

This latter comment came in the context of young Joe's speech problem, and it clung to him through the lower grades and into high school.

All the while, he continued to pin his hopes on attending exclusive Archmere Academy. When his father hesitated over the high tuition, Joe applied for and got into a work-study program that put him on the school's grounds crew over the first summer. He was taken by the collegiate atmosphere and especially the sports program, which he joined though he was the second smallest boy in the freshman class, in both height and weight.

Public speaking was a firm requirement but Joe for a time was excused because of his stutter, which did not go unnoticed by chiding classmates. They called him "Joe Impedimenta," a word they picked up in their two-a-day Latin classes. Either that or "Dash," not for his speed afoot but because his stutter apparently reminded his buddies of the Morse code, as in dot-dot-dash-dash. In the end, he mused later, "carrying it strengthened me and, I hoped, made me a better person." Practicing long passages and reciting them before a mirror at home, Joe watched his own facial expressions and strove to control them.[6] Uncle Boo-Boo, also a stutterer, gave Joe moral support and became an inspiration for him to conquer the problem. In other ways, though, he vowed he would not wind up like his uncle, the subject of ridicule mired in a dead-end job peddling mattresses in a five-state area.[7]

Later, at Catholic Archmere, Joe thought again about the priesthood. "Even though [by that time] I dated a lot of girls," he remembered, "I still thought I might want to be a priest. I remember thinking, 'Well, I'll go to Saint Norbert's College [in Wisconsin], run by a French order called the Norbertines." But the headmaster at Archmere, Father Justin Duny, told him: "No, Joe, why don't you go off to college and then decide whether you want to be a priest?"[8] Valerie remembered their mother saying the same: "Fine, Joey, if you want to be a priest. But first you're going to go to high school and going to go to college. And then when you get out of there and if you want to be a priest, that would be a great blessing, that would be a wonderful thing. But you're not going anyplace until you've lived a bit of life."[9]

Archmere worked magic on Joe. He grew a foot in height by his junior year, and as a senior led the academy's undefeated and untied football team in scoring with ten touchdowns. The team had a new coach that year, John Walsh, a world history teacher, and he had inherited a small and dismal squad. In the previous twelve years the team then known as the Archers had won a total of only ten games, and Walsh turned it around with an impressive exercise in confidence-building. A University of Delaware player and graduate, Walsh recalled that in 1960 at age twenty-four he was just getting married and hunting for a coaching job. When his prospective bride, a local girl, told him: "Well, Archmere is open," he replied, "Do they have a football team?"[10]

Hired and arriving in the late spring, Walsh called the team together—only seventeen players at the time—and after looking them over he told his wife, "I think we can go all the way." And they did. In a small grouping of five area private schools called the Quindependent Conference, Archmere went undefeated in eight games and won the title.

On the third or fourth day of practice, Walsh scheduled a scrimmage with the Delaware area's football powerhouse, Salesianum Prep, on the understanding it would be against its third team. Instead, the rival coach said he wanted to play his first team, for openers at least, and Walsh reluctantly agreed. He returned to the school bus where his players were waiting, watching with awe the three Salesianum squads warming up, and he broke the news. "I think our kids thought I was insane to scrimmage them," Walsh recalled, "but we end up losing the scrimmage on the last play, and at that time I think they realized that they could be good."

In the opening game of the season, Walsh said, they played a team "that absolutely slaughtered Archmere the year before. We won that game, 6–0, and we kept on getting better all year. But I definitely think the thing that turned it around was the Salesianum scrimmage."[11]

During spring practice, one of the best halfbacks quit the squad and Walsh switched Biden from end to that position, to take advantage of his quickness coming out the backfield for long passes from the star quarterback, Bill Peterman. Joe, the coach said, "caught my attention right away because of his outstanding ability to catch a football." Archmere under Walsh began playing out of Delaware's winged-T formation with Joe as the wingback, a formation that usually emphasized running. "But when we did throw, Joe was the main recipient," Walsh said. "He had very good speed, not a 'burner' but above average, and excellent hands" as well as being an above-average runner. Joe got a new nickname, "Hands," which was much preferable to having been ribbed earlier by some schoolmates as "Stut," referring to his still frequent stutter.[12]

Tom Lewis, who got his own inoffensive nickname, "Spike," from Joe, recalled that his friend "had incredible hands, and he was very competitive. If you played against him you'd almost kill each other. He always had confidence, or made everybody believe he had confidence. Mr. Cocky. He'd never shy from a confrontation. Joe was the money man."[13]

Coach Walsh recalled that Biden wanted to be team captain, but when the squad chose its leading overall scorer, the other halfback, Mike Fay, there was no resentment on Joe's part; "He was still a leader, he had a lot of enthusiasm, he was upbeat, he was a team player; he never was a stitch of trouble. He really got along with people." In those days, most players played both ways; he started on offense and was a spot player on defense.

The Archmere team of 1960 remained very close through the years. In 1985 they held a twenty-fifth reunion at Joe's home and were planning a fiftieth when one starting player who was terminally ill, Bob Stewart, requested that they have a fortieth. Joe hosted that one as well, and plans went forward for a fiftieth in 2010, possibly this time in Washington, also hosted by the man who then would be vice president of the United States.[14]

Fay, who also attended the University of Delaware with Biden, remembered: "Halfbacks were the glory boys, so in a way we were competing with each other." Archmere had won only one game out of seven in their junior year, Fay recalled, "but the team in '60 had the best glue holding it together. It really was a team. There were a lot of individuals who didn't mind being the center of attention, but the whole team worked very well together on the goal of winning games." Coach Walsh and the scrimmage against Salesianum, which had won thirty-two straight games, were the key, Fay said.[15]

Joe also played baseball and basketball at Archmere, was very popular, and had a semisteady girlfriend named Maureen Masterson, who was Valerie's best friend. "She was my high school sweetheart," he recalled much later. "We dated until, I guess, my sophomore year in college. I went out with other girls but she was my girlfriend and a great person." In fact, he said, he introduced her to her eventual husband, Jim Greco, who was on Joe's floor at Harter Hall at the University of Delaware. They all remained close friends, with the Grecos invited to Biden family celebrations thereafter, including the 2008 presidential and vice presidential inauguration.[16]

Joe, Maureen Greco says now, "was pretty much the way he is

today. He was a good person, he had a value system and he didn't veer from it. He had great parents. He was never going to drink or smoke; it was just something he didn't want to do; it never appealed to him." They met on an elementary school playground where she was playing basketball and he was playing baseball. She was a year behind Valerie Biden at Ursuline Academy.[17]

Every year, Valerie, Maureen, and two other high school and Delaware pals went off together for a retreat, and later Valerie and Jill Biden organized what they called a "Senior Prom" of old friends and their spouses from around the country; they rented a firehouse and a sixties band to renew old times.[18]

At Archmere, where Joe was also an outfielder on the baseball team, "he was a good player but it wasn't his favorite sport," teammate Lou Bartosheshky recalled. "He ran well and played good defense, but he could never hit a curve ball. He played a lot as a junior and started as a senior." Off the field, he said, Joe got along with everybody and got elected to a lot of class offices.[19]

David Walsh, Biden's best friend at the school, said that before Joe arrived, "Archmere had always been the doormat" in the private school football league in which it played. "He was a natural athlete at whatever he did," Walsh said. But Joe was not your average jock either, he said. "He was an early environmentalist, and in high school in the sixties he was more concerned about desegregation in Delaware than I was. I for one was mindless about it, but he led some protests." Walsh remembered that later, at the University of Delaware, "Joe would talk about it. At the Charcoal Pit, a hangout for everybody else, blacks weren't allowed. He noticed; I never did."[20]

High school kids including Joe and the rest of the Archmere football team would often go to the Charcoal Pit, a hamburger joint that looked (and still looks) like an old diner, with coin machines at each booth for playing records (now CDs). The team included the only black student at the school at the time, Frank Hutchins, and legend has it that Joe led the whole team out on one occasion when the proprietor declined to serve him with his white teammates in the main dining area.

"As I remember it," Fay said, "there was a bunch of us who went there. They had a take-out window, and Frank was told he could get something there and eat it in the car or something. They wouldn't let him come inside, and the whole bunch of us left. Joe was part of the group. I don't remember any one person saying, 'Let's leave,' but I'm sure somebody said that, because everybody decided, 'Well, we'll either eat together or we're not going to eat at all.' "[21] Hutchins much later said: "There were a number of these kinds of incidents, and it is clear to me Senator Biden could be referring to any one and be perfectly accurate."[22]

At Delaware, Fay remembered, the campus was very conservative, and Joe sported a relatively high flattop haircut in the fashion of the day. Most of the football players belonged to the Theta Chi fraternity, but Joe did not join any. He did go to fraternity parties but Fay said he never saw him smoke or take a drink.

Above all, especially after he shook his affliction, Joe was a talker. Dave Walsh recalled: "Friends and detractors back in high school would say about Joe that he never saw a soapbox he didn't want to get up on. He was very knowledgeable about history and politics. I guess we were probably seventeen, in the backyard of my house, and my father asked him: 'Joe, what do you want to do?' and he said: "Mr. Walsh, I want to be president of the United States.' And when my dad heard it and I heard it, it wasn't like anybody would laugh at that because, yeah, he was so talented. They'd say, 'Yeah, he could do that.' It was unlikely, somebody from Delaware, but you'd talk to anybody right through high school and college, yeah, he was special. But nobody [really] thought he could do this.' "[23]

Valerie for one had no doubts: "My brother was a natural leader. People just gravitated to him. From the time Joe was a little boy, he always appealed to the better instincts of human nature; he looked for the things that could bring kids together. He was always the president of his club or the leader of the fort, not because he pounded on his chest and said, 'I'm the leader, come follow me.' "[24]

"He was one of those guys that you follow, that you'd listen to," Fred Sears, another childhood friend, recalled much later. "He had a pack of friends who did what he wanted to do." With good behavior drummed into him at home, Sears recalled, Joe mostly avoided occasions of trouble, going to canteen dances on Friday nights at the Du Pont Country Club or at Archmere's rival Salesianum Prep. "I have to say," he said, "I can't remember seeing him with a drink, not that any of us drank in high school."[25]

Joe's grades were only satisfactory B's, but he was elected class president in his junior and senior years. His chief achievement, however, was fulfilling as a sophomore the school's public-speaking requirement before its morning assembly. As a senior, he gave the class welcome at his graduation in 1961 with nary a stutter.[26]

Dave Walsh, recalling his classmate's trial, said, "Kids can be really mean and nasty to other kids, but he was such a leader that he took it and laughed about it, and we would kid other kids about whatever nicknames they had. The really nasty one was 'Stutterhead.' "

As for "Impedimenta," Walsh said, "We all had to take Latin at Archmere, so that's where the kids heard the definition and nailed him with it. Most of the nicknames were not derogatory, but that was one that certainly could have been taken that way. At least once a year you had to get up in front of the whole school in the auditorium and give a speech about something. Every once in a while you'd get that little stutter, but in the course of those four years, he overcame it."[27]

Much later, Father Joseph McLaughlin, headmaster at Archmere, recalled the then senator Biden addressing the students and telling them one of the most important courses he had taken at the school was public speaking. "Now there's no evidence of a stutter at all," the headmaster said. "It's such a victory for him."[28]

Biden's sister observed later that even his verbal affliction seemed in the end to have had a positive character-shaping effect on him.

"Because he knew what it was to be different or to be the brunt of a joke," she said, "when you know what it feels like to be laughed at. There's very few things that are more humiliating than being laughed at for that stuttering. So he always attempted to bring people in."[29]

Indeed, talking to people, which only recently had been such a nightmare for young Joe Biden, was soon to emerge as his strength. Freed of the impedimenta that once had been the subject of such ridicule, his inheritance of the Irish gift of gab held out great promise in his stated ambition to become a trial lawyer—and his harbored visions of much more. Most of the boys his age had dreams, but his friends said Joe was a great one to see in his mind's eye how they would play out, whether it was on the football field, where he would deftly evade tacklers on a long touchdown run or, later, in his career ambitions. If he could see it, they said, he believed he could do it, and so from his earliest days he reached high.

THREE

BUILDING A DREAM

IN THE SPRING of 1960, as another young Irish Catholic, named John F. Kennedy, from Massachusetts, was pursuing the Democratic presidential nomination in the West Virginia primary, Joe Biden sat in the library at Archmere Academy musing on his own future. He entertained no such lofty ambitions seriously then, but the notion of the first Catholic in the White House stirred him and set his mind to the prospects of entering politics one day.

Subsequently, Kennedy's inaugural address, with its emphasis on public service, squared with the ideals Joe had learned in Catholic schools since childhood, and were reinforced by the examples and preachings of his parents at home. Young Biden had other dreams as well, such as becoming a college and professional football player. But he weighed only 140 pounds at the time; politics may have seemed the more reasonable option.

So he began to look into how a young fellow could make a start, especially if he lacked the family connections and wealth of a Kennedy.

He found a copy of the Congressional Directory in the Archmere library and proceeded to skim the short biographical sketches of the senators. He saw that many of them were lawyers and concluded that the best course for him was to go to law school. One thing he indisputably had was a way with words, now that he had pretty much conquered his stuttering.[1]

In the fall of 1961, upon entering the freshman class at the University of Delaware with law school and then politics in mind, Biden selected political science and history as his study majors. At the same time he threw himself into freshman football and generally into the life of the campus, getting involved in a lot of extracurricular activities, including the dating scene.

The next summer at nineteen, Joe got a good taste of existing racial tensions as the only white lifeguard at a public swimming pool near a heavily black-populated housing project called Prices Run. When a tough black gang member kept bouncing on a high board in violation of club rules, the hot-tempered Joe called him "Esther Williams," the white movie swim star of the day, and ordered him out of the pool. A confrontation occurred in the lot outside the pool and young Biden stared the tough down, then softened the situation by apologizing for calling him by a woman's name, while also telling him he'd be thrown out if he broke the pool rules again.[2]

At the same time, Biden said later, the lifeguard experience was a window for him on a world he had not known or thought much about. Meeting and talking with the other lifeguards, all black, "was a real epiphany for me," he said. "It was the first time I got to know well and became good friends with all of these inner-city black guys, all of whom were college students at historically black colleges. I actually got pulled into their lives a little bit. . . . I'd play on the same team with them and be the only white guy of all ten guys on the basketball court. . . . And I came to realize I was the only white guy they knew. They'd ask me questions like 'What do white girls do? Where do you live?' It was almost like we were exchange students. It was a real awakening for me. I was always the kid in high school to get into arguments about civil rights. I didn't do any big deal, but I marched a couple of times to desegregate the movie theaters in downtown Wilmington [that were] still segregated." But, he acknowledged, "I wasn't part of any great movement," yet he made many black friends who later were part of one of his strongest political constituencies.[3]

Meanwhile, with his gregarious way, Joe was elected president of the predominantly white freshman class, and he stood out as a sharp dresser. Fred Sears, now also at Delaware, remembered him as "very clean, very neat—you'd never catch him with his shirttail out or walking around in sneakers. Those of us from that [private school] area of Wilmington would be wearing Weejuns—you know, penny loafers. I wouldn't say we'd wear sport coats to class, but a nice sweater and a shirt. I don't even remember him ever wearing blue jeans. But I wouldn't call him preppy, just always well-dressed and always took care of himself. He never joined a fraternity. I'm not sure whether it was the money or the drinking side of it, but he went to a lot of fraternity parties and certainly was very social."

Drugs were not part of the scene at that time, Sears said. "We did not even know what marijuana was," and in any event Joe would not have had any part of it. "I'm sure he went out with girls who drank," he said. "One girl I do know told me, 'To tell you the truth, I had one date with him in college, and when I lit up a cigarette, he told me to throw it out the window right away or this date was going to be over. I knew this relationship wasn't going to go very far.' That was pretty gutsy on Joe's part," Sears said. "She was a cutie. I could have lived with that cigarette."[4]

Joe and his sister were both teetotalers all through high school and college, and he remains one. Valerie said later it was not by accident, because there were cases of alcoholism in the family. "We saw the heartbreak. There were enough people around us [who drank], and there was nothing good that came from it," she said. "I believe, as does my brother, that you are genetically disposed to it; genes will out, so I didn't want to take the chance. If I didn't drink, I'd never be chemically dependent on it, so that's why I didn't drink." In school, she recalled, "Joe was always the designated driver. All the parents [thought], as long as Joey Biden was there that was fine, because they always knew he was responsible. Joe would do wild and crazy things but he was always sober. You couldn't blame it [on] because he had too much to drink."

At the University of Delaware, Joe sometimes was called out by campus police for being around the women's dormitory. But part of it at least, Valerie remembered, was that her room was on the ground floor, and he would hand her a soda or a snack through an open window. As an eye-catching homecoming queen, she'd regularly have male bees buzzing around her nest. "They'd grab him and tell him he was not supposed to be there," Valerie said. "He'd say, 'She's my sister,' and they'd say, 'Sure, buddy,' and chase him off."[5]

Sears remembered that Joe, with his flashing smile, "dated a lot of good-looking girls," and once he got out of Archmere "where the priests were on your butt to get your work done, he may have had a little more freedom than he needed right off the bat."[6] Valerie, two years behind him at Delaware, remembered: "He always had girlfriends—he was handsome, he was an athlete, he was smart, and, because my father sold cars, he always had a car on the weekend. He was the catch in town. He was always a one-girl man at the time; he was a practitioner of serial monogamy. We always doubled. He dated my girlfriends and I dated his boyfriends. I went out with his friends for every prom. The only thing I ever did by myself was get married."[7]

As for interest in politics then, Sears recalled that Joe "looked like he might have had something else long-term in mind and not just being a typical college BS rah-rah guy," but he never talked about seeking public office. "We didn't think he was going to law school, that's for sure," he said much later. "I don't know what his grades were, but I know they were not a hell of a lot better than mine were. We were all in that range where B's looked pretty good, and a C was okay."[8]

In any event, when young Biden posted disappointing first-semester grades, his father made him drop football. No Biden had ever graduated from college and Joe Sr. was determined that his son would do so. But by his junior year his grades had slipped further and, as he put it in his memoir, "I was no longer sure, given the state of my academic transcript, that I could talk my way into a good law school."[9]

For help, young Joe turned to a favorite political science professor named David Ingersoll, who talked turkey to him. He told him he'd never get into law school unless he knuckled down in the remaining three semesters left to him, took on a heavier academic load, and excelled. The talk worked; he decided to move into an off-campus apartment with Tom Lewis. "There was nothing to do but study," Lewis remembered. "The TV worked only about half the time. We would just listen to the news at eleven and we would go home on weekends."[10] Not only did he lift his grades but he also tried out for football again in the spring. He impressed the coaches and, as he wrote in his memoir later, "it looked like I had a shot to start at defensive back [in the fall]. I couldn't wait for next September; I could almost see the fall season unfold in my head. So I was feeling pretty good when I headed to Florida for spring break after our last practice. That trip changed everything."[11]

Sears, who accompanied Joe along with two other Delaware undergraduates, recalled how they had talked about going but didn't know how they were going to get there. "There was a bulletin board in the student center," he remembered, "and one day some gal put up a card that she was driving to Fort Lauderdale and looking for people to share the gas. We all signed up. We were really only going to Fort Lauderdale and didn't decide to go to Nassau until we got to Lauderdale. After a day we realized there were about ten thousand guys for every girl, and this was not worth our time and effort. I remember seeing some ads or maybe one of these planes flying with a tail on the beach: '$28 round-trip ticket to Nassau.' [I said] That's way in our price range, let's go over."

So over they went, to Paradise Island in the Bahamas, which was Joe's first plane ride. "We didn't have a place to stay," Sears said. "I remember that one of the first things we did when we got there, we found some guys who had a room over the top of a drugstore or some retail store on one of the main streets of downtown Nassau—two bedrooms and a kitchen. They already had six guys staying there, but they said if you guys want to chip in on the rent, it was going to be less than five dollars or something ridiculous like that. Of course it was even money you were going to end up sleeping on the floor. In those days it didn't matter."[12]

As Joe recounted in his memoir, three of the Delawareans headed right for the public beach, where they found they were not the only college kids there. To one side, beyond a high chain-link fence, was the private beach of a swanky hotel. "Through the fence," Joe wrote, "we could see dozens of beautiful young college girls sunning themselves on the British Colonial beach. You had to be a hotel guest to go to that beach, but we were determined to get in. The three of us liberated three towels that guests had hung on the fence to dry, wrapped them around our waists and walked past the guards at the main entrance. We walked like we belonged, and it worked."

The invading threesome sauntered through an atrium and out on the deck of a swimming pool facing the beach. "There were college kids roaming all over the place," Joe wrote, "but we spotted two girls sitting poolside on the chaise lounges nearest the beach. One was a brunette, the other blonde. 'I've got the blonde,' I said. 'No,' said my buddy. 'I've got the blonde.'"

At this point, according to Biden, Fred Sears said, "I've got a coin. You guys can flip for it." But as Joe wrote, "I didn't wait to see who won. I approached the blonde. My other buddy took the brunette.

She was very attractive and wore a tight leopard-skin bathing suit, so he didn't seem to be really disappointed. I walked up and sat down on the edge of the blonde's chaise. Over my shoulder I heard my friend hit the wall: 'I'm Mike.' 'You're in my sun, Mike.'" Then it was Joe's turn: "Hi, I'm Joe Biden." "Hi, Joe. I'm Neilia Hunter."[13]

Sears later described the scene from his vantage point: "We were on that beach ten minutes and I can remember seeing her, Neilia. It was almost like the movie *10* [with Bo Derek]. We're looking around the beach, there's some groups of girls sitting next to each other, and we saw this group of two or three girls and she stood out. Her hair and maybe the way the sun was hitting her, it was, like, spectacular. And we're all saying, 'Oh my God, who's going to talk to her? Are we going to pull fingers here, or one-potato, two-potato? What are we gonna do?' And by the time I looked up, Joe had a fifty-yard dash on the rest of us. He wasn't going to wait to see who was going to talk to her. It was Joe.

"Of course, we're all saying, 'Oh, well, he'll be back. This will last about two minutes. He's got the gift of gab, but we know Joe. He can't fool everybody all the time.' I don't remember him ever coming back until we left on the plane at the end of the trip. He was with her all the time. I don't remember he ever brought her back, God forbid, to where we were staying. I just remember when we got on the plane coming back, he said, 'I'm in love and I'm going to . . .' And I say, 'Well, no wonder. For God's sake, Joe, she's outstanding.' And he says, "And I'm going to Syracuse Law School.' [I said] 'What? Who the hell picks Syracuse? For God's sake, if you're going to law school, go south!' He said, 'That's where she lives.' I don't remember ever seeing him with anyone else after that."[14]

Biden in his memoir confirmed Sears's account. Of that first meeting, he wrote, "I fell ass over tin cup in love—at first sight. And she was so easy to talk to. She was in school at Syracuse University, right up the road from her hometown, Skaneateles, New York. She was just two months from graduation and hoped she'd be teaching at a junior high school in Syracuse that September."

He then told what happened next: she broke a date with another guy and they had dinner that night. He only had $17 left and she had to slip him some money under the table to cover the tab. For the last four days of spring break, Joe saw Neilia every day and night and arranged to go see her in Syracuse the next weekend. Whereupon he told her: "You know we're going to get married." To which, according to Joe, she whispered: "I think so."[15]

When he got back home, the first person Joe told about Neilia was his sister. At the time, she was only a freshman at Delaware and wasn't allowed by her parents to go away on spring break. When Joe came into the house, Valerie recalled, "He opened the door to our dining room and he yelled: 'Vaaaal!' I was upstairs throwing myself across Mom and Dad's bed talking on the phone. I hung up and he ran up the stairs and he said: 'I met the girl I'm going to marry!' And that was it."[16]

Meanwhile, in Skaneateles at the home of her parents, Neilia phoned her old roommate at Syracuse, Bobbie Greene (now McCarthy), and excitedly gave her version of the encounter at Paradise Island. "She called me so excited," her friend recalled years later. "She told me all about him. It was very clear that this was something different. Neilia asked me, 'Do you know what he's going to be?' 'No, what?' I asked. She said, 'He's going to be a senator by age thirty and president of the United States!' And she never had any doubt."[17]

From then on, Joe made a weekly 320-mile commute to Neilia's home in a car from the lot then being managed by Joe Sr. On the first weekend, Bobbie recalled, she was at her small cottage on campus with her hair in curlers when Neilia came over with him. "She couldn't wait for me to see him," she said, "so I took out the curlers and came down the stairs. He was sitting in a chair and he was gorgeous. I thought, 'That's the best-looking man I'd seen in my life,' well-dressed, with that shock of hair and that dazzling smile. I had no doubt that if he wanted to be a senator he could be. It was clear they would be together from the beginning. She was completely over the moon."

Neilia and Joe and Bobbie and her boyfriend, later her husband, Dan Greene, double-dated that weekend, with Joe driving a beautiful turquoise convertible that made an impression. "Nobody in our crowd had cars like that," Bobbie remembered. It was, of course, a loan from his father's car lot.

Neilia was living with her parents at the time, her friend said, although she was a member of Kappa Alpha Theta, a highbrow sorority at the university. "It was very snooty and Neilia wasn't comfortable with the sorority rules," Bobbie recalled. "They complained for example when she wore slacks to a football game." As for Joe, Bobbie said, with Neilia living at the family house in Skaneateles, "he was very proper. He always stayed at a rooming house nearby. Joe was a very straight arrow."[18] Sometimes he brought Tom Lewis with him from Delaware.

On Joe's weekend drives to Syracuse, he would park outside Neilia's school and leave notes for her on the car's windshield. On one occasion she left him a note suggesting that while he was waiting he ought to take a look at the town's law school. According to one report, besides Syracuse Law, he was considering the prestigious Cornell Law School about an hour's drive to the south.[19] He was impressed with what he saw at Syracuse Law and vowed to apply there.

The love affair developed rapidly, and one weekend Neilia went to Wilmington to meet the Biden family members, all of whom approved. Her father at one point made a feeble effort to dissuade Neilia because of Joe's religion, but soon relented. A confirmation of Joe's own determination came when University of Delaware football practice was to begin two weeks before the start of the fall semester. Despite his burning desire to play and the coach's encouragement that he had a good chance to make the team, Joe told him he wouldn't be there because playing would have meant not seeing Neilia all through the fall. On the long weekend drives up to Skaneateles, Joe wrote later, he planned their future—law school for him in Syracuse, marriage, a large family, becoming a trial lawyer with his own firm, and then running for public office.[20]

With this scenario set in his mind's eye, Joe in his final semesters at Delaware was able to submit better grades along with letters of recommendation from his professors to the Syracuse Law School. One such letter described him as having promise to be "a 'natural' lawyer, able and argumentative, ready to take a position and defend it with strong and well-presented arguments."[21]

For all of young Biden's more serious demeanor, as he approached graduation at Delaware one incident suggested that his outward confidence may have shielded some uncertainty. One classmate, John DiEleuterio, recalled much later: "We were talking one day in the student center, and he seemed a bit agitated. I said, 'What's the matter?' And he said, 'I just found out something very interesting. Do you realize that after four years at this institution as a political science major, you have to take an oral exam?' I said, 'Well, what does that mean?' He said, 'It means you go before a panel and they ask you any number of questions, and if you fail your oral exam, you don't graduate!"

DiEleuterio, who much later became director of Biden's Senate office in Wilmington, went on: "Well, if there ever was a person who didn't have to worry about passing an oral exam, then or now, it would be Joe Biden. But he said to me, 'You know what? I think I've found a way that we can avoid taking the oral exam.' " Biden explained that if a senior was taking eighteen credits in his last year, and he took three more for a double major in something like history, you could graduate without the exam. "So he did that and I did that," his friend said, "and we both graduated with a bachelor of arts degree in political science and history, and didn't have to take the oral exam. So he saved my life that day. He would have sailed through it with flying colors." DiEleuterio speculated that Biden might not have wanted to take any chances on flubbing the oral exam, so intent was he to go on to law school—in Syracuse especially.[22]

An indication of Biden's aspirations came, he said later, in writing a paper in his senior year about the Republican congressman from Neilia's district, John Taber, an old lion who was chairman of the House Appropriations Committee. "The reason I picked it," he said, "was I had fallen ass over tin cups in love with Neilia." He wrote the paper, Biden said later, because he was already thinking of running for public office in upstate New York "and I wanted know about the area to see if you could win."[23]

In the spring of 1965 Biden was admitted to the Syracuse Law School. On his first day, he met another first-year student named Jack Owens who would quickly become his best friend at Syracuse, and years later Valerie's husband and his brother-in-law. Owens recalled later how they had struck up a conversation at their hall lockers and stood there talking for three hours. "During that long and ranging conversation between two strangers," Owens remembered with some wonderment, "he told me how he was going to run for the Senate from Delaware. I grew up in New York, so it was kind of a shock to think on that grand a scale. It was very indicative of Joe that he was able to think that big, that realistically. It was an eye-opener for me." His reaction, Owens said, was, "'Hey, what are you talking about, buddy?' He was only twenty-one, twenty-two years old. But he projected a kind of confidence, without being obnoxious about it."[24]

With alumni and state financial aid and a job as a resident adviser in an undergraduate dormitory for room and board, Biden got by. A fellow resident adviser, Roger Harrison, remembered Biden as having "a certain self-assertion but not in any way arrogant. He was very likable, engaging."[25] At the same time, young Biden was hardly an honor student. Another classmate, Bill Kissell, reflected: "You knew Joe was different. If I had studied [only] as much as Joe did, I would have flunked out. There was some resentment [that he got by] because he was not a student, but he was doing other things. He had other projects, and Neilia clearly was one of them."[26] Others remembered him hanging around the law school library, jawing with other students, but seldom burying his nose in a weighty volume for hours as others in his class did.

Still another classmate, Clayton Hale, said of the first day he met Biden: "The professor called on Joe, calling him 'Mr. Bidden.' Joe answered: 'I'm unprepared, and by the way, my name is Biden—B-i-d-e-n.' He was not bashful at all." But, Hale recalled, he had not yet shaken the stuttering problem entirely. "I know he wanted to get into politics, and he spent time talking to high school classes as therapy," Hale said.[27] Joe also befriended another first-year classmate with a stutter and showed him how to break it before a mirror as he himself had done.[28]

On the same matter of Biden's casual approach to his academic challenges, two other classmates, Bob Osgood and Don MacNaughton, also remembered the time the professor in one of their classes called on Biden and asked him to discuss a certain case that he hadn't read. Whereupon Joe stood up in his coat and tie and discoursed in great detail, making it up as he went along. Others in the class quickly grasped what was going on, and even Joe himself began grinning as he spun out his answer. When he finished and sat down, Osgood remembered, the class broke out in appreciative applause.[29]

Meanwhile, having missed varsity football at Delaware, Biden formed a law school touch football team in a Syracuse club league and worked off excess energy as its quarterback and playing rugby as well. Again he won a reputation as a natural leader. To this day, Owens said, "he has a way of causing people to do a thing his way because they want to do it his way." But, Owens said, "his off-hours at the law school were mainly with Neilia."[30] Harrison recalled accompanying Biden to Skaneateles for dinner at the Hunter home on the lake, and remembered they always returned to the dorms afterward.[31]

"He readily admitted he didn't take law school all that seriously," Owens remembered. "He wanted to get out into the real world. I would say he was not what you would call a devoted student. He did not study hard, and he did not do well, but Neilia helped him with last-minute cramming for exams."[32]

Between peaks of concentration, however, Biden would fall back into lazy and careless academic habits, missing classes and often relying on the lecture notes of compassionate classmate Clayton Hale. In those days, Hale said later, a more casual attitude prevailed at the school and as a good note-taker he willingly passed his labors on to other classmates. "Joe and I had an arrangement," he said, "where Neilia would take my notes and copy them over for him." Years later, for a class reunion Joe could not attend, he sent a video in which he thanked Hale for getting him through the law school.[33]

At one point, however, young Biden slid into a situation that much later would have damaging consequences to his career. As he described it in the memoir: "About six weeks into the first term I botched a paper in a technical writing course so badly that one of my classmates accused me of lifting passages from a *Fordham Law Review* article; I had cited the article, but not properly. The truth was, I hadn't been to class enough to know how to do citations in a legal brief. The faculty put my case on the agenda of one of their regular meetings, and I had to go in and explain myself. The deans and professors were satisfied that I had not intentionally cheated, but they told me I'd have to retake the course the next year. They meant to put the fear of God in me; the basic message was that I had better show some discipline or I'd never get through the first year. But the dean of the law school wrote a note to the dean who oversaw my work as a resident advisor: 'In spite of what happened, I am of the opinion that this is a perfectly sound young man.'"[34]

An associate law professor, James K. Weeks, for whom the paper in question was written, wrote in 1967 that Biden was "far from distinguished scholastically," but that his grades on written exams "do not present a totally accurate picture. He knows what he is doing and appears to possess good judgment and a highly developed sense of responsibility. He is the type of individual one is more than willing to take a chance on, for he is unlikely to sell short your expectations."[35]

John Covino, a close friend in his law school class who said he shared Biden's casual approach to his studies, remembered Joe coming up to him one day and declaring: "You're not going to believe what just happened. They accused me of plagiarism!" And, Covino said, "that was the last time I heard of it. Nobody heard about it at the time."[36]

Some others in the class, however, did learn about the matter at the time or later. Several, including Covino, Kissell, and Hale, named another first-year student, Arthur Cooper, as the villain in the piece. School policy assigned students to review the papers of classmates, and Cooper spotted and reported Biden's failure to cite the *Fordham Law Review* material properly. "Everybody hated the course," Kissell recalled, "and I think it was sour grapes" by the studious Cooper, who was a contrast from the easy-going Biden.[37] Cooper, Hale said, "certainly wasn't very popular in the class,"[38] and Covino observed that if somebody "was going to nail Joe," Cooper would have been a prime candidate.[39] The whole incident was soon forgotten after Biden retook the course in question and passed it the second time around—forgotten, that is, until it resurfaced years later in a manner that shook Biden's political ambitions to the core.

By cramming before finals with Neilia's help and prodding, Joe completed that first year and they were married in a large family wedding in Skaneateles a few weeks later, with all the Bidens present along with friends from Archmere, college, and the law school, as well as his old Scranton schoolboy buddies as ushers. With the acquiescence of Neilia's father and to Joe's great relief, the vows were administered by a priest in a Catholic church.

Joe and Neilia Biden moved into a small apartment in the Strathmore residential section of Syracuse, a homey neighborhood of small frame houses reminiscent of the Green Ridge section of Scranton where Joe had spent his early childhood days. Bobbie and Dan Greene were frequent visitors. "Joe was very focused on his going to have a political career," she remembered, "and he was definitely a Democrat. We were all against the Vietnam War. It took him a while to come around on the war, but he did come around."[40] Another neighbor, Joseph Fahey, later a county judge, said of Biden at the time: "He was pretty cautious. He was a middle-of-the-road liberal. He always favored working within the political process to make change [rather than engaging in street protest]."[41]

The young Biden couple plunged into neighborhood friendships. Neilia taught at the Bellevue Elementary School, within walking distance of their apartment in a two-family house at 608 Stinard Avenue. Joe meanwhile would take refuge from his legal studies by rounding up local boys to play stickball, football, and other street games by the hour. His flashy light green sports car was a magnet for boys and girls alike.

A young neighbor, Jane Fahey (now Suddaby), and her friend Ann Keough, each about twelve years old, were standing on the corner when, as Jane later put it, "an extremely handsome young guy was driving down the street in a '67 Corvette with the top down. Not many people in our neighborhood drove a Corvette, so Ann and I as a joke stuck our thumbs out, being wise, as if hitchhiking, and he pulled over! We were horrified and delighted at the same time. He chastised us for hitchhiking in a good-natured way, and then introduced himself and explained that he and his wife, Neilia, had just moved in on Stinard."

The car was a wedding gift from Joe Sr., who took their old vehicles as a trade-in on the new eye-catching sports model. The girls and other young kids in the neighborhood soon became good friends with Joe and Neilia. While she corrected students' papers, Joe would take a break from his law books and come out onto Stinard to throw a football or to organize games in which he of course would be the star quarterback. Jane and Ann and other young girls would stand and chat with Neilia over the fence at their backyard and sometimes were invited in for dinner with the young couple.

One hot summer day, Jane recalled later, "Joe was so fried from studying he came out to the kitchen, indicated he wanted to go swimming, and then wished for a pool in their backyard. He then climbed up on the counter, pretending he was going to dive into a swimming pool, and begged Neilia and me to catch him. It was hilarious."

On other hot days, Joe and Neilia would take neighborhood kids to the local Marble Farms store for ice cream, "cramming two or three of us in the Corvette with the top down. . . . It was such a great treat not just for the ice cream or the Corvette ride, but because they really seemed to enjoy the activity and also had a genuine interest in you as well. For people in their twenties, they were tremendously generous of their time and resources, not at all self-absorbed as you might expect someone of that age to be."

Neilia, she recalled, "was a beautiful woman, and she would lay out in the backyard sunbathing, and boys who were my age would be staring from the bushes." She and Joe would have neighborhood kids in for pizza, and when they got a young German shepherd from the neighboring Del Vecchio family, "it was a great excuse for us to knock on their door to see the puppy, though it was Joe and Neilia that we really wanted to see." Interestingly, the pup was named Senator, called "Tor" for short, which may have been an indication of where Joe's career thoughts were.[42]

On another day, Joe was in his apartment on Stinard Avenue when he heard some older boys outside making fun of a neighborhood boy, Kevin Coyne, who had a speech impediment. "He jumped over the fence and read them the riot act," Coyne later recalled. "He scared the daylights out of them," and the two became good friends. (Years later, when I told Coyne in passing that Biden also stuttered as a child, he said surprisingly that he had never heard it before.) Kevin regularly walked the puppy for the Bidens and he became a kind of guide for them in the neighborhood, showing Neilia where the best stores were and sometimes shopping for her.[43]

A young girl at Bellevue named Pat Cowin (now Wojenski)

became friends with Neilia as her student in the eighth grade. Young Pat's mother had a series of medical problems and Neilia took the girl under her wing after school. "She probably realized I needed somebody like her in my life," Pat said later. They continued to correspond after the Bidens moved to Delaware and Pat was planning to visit them in Wilmington until fate intervened.[44]

When the Bidens eventually left Stinard Avenue, they held a big barbecue to which everyone in the neighborhood was invited, kids and parents. Kevin Coyne helped them move. "I realized it was coming to an end," he recalled of the moment. "When they were pulling out, I was looking out my window, crying," he said. "My mother said, 'Why don't you go out and say good-bye?' I was too crushed to go."

One of the strange items Kevin spotted in the move was an old soapbox with a pair of Joe's sneakers fastened on top of it. "That's my friends thinking they're funny," Joe told him, a gag by way of chiding him as Mr. Soapbox.[45] Indeed, by his final year in law school, he had demonstrated an increasing proficiency and style in oral argument. "What had terrified me in grade school and high school was turning out to be my strength," he said in his memoir. "I found out I liked public speaking. With all that practice quoting Emerson, I could memorize long passages; I never had to look down and read a text. So when I talked, I could watch an audience for their reaction. If I felt myself losing them, I would extemporize, tell a joke, focus in on a single person who wasn't paying attention and call him out. I fell in love with the idea of being able to sway a jury—and being able to see it happen right before my eyes."[46]

Law professor Thomas Maroney, who arrived at Syracuse in Biden's last year and had him in his class on writing and interpreting legislation, recalled that he thought he "might not wind up being the top student in his class but he was going to wind up as somebody someday. He had a natural bent, a presence." Maroney remembered: "Joe got an A in Legislation, the only A he got in law school. And I was not known as Santa Claus in giving grades."[47]

Over summers during law school, Biden did a number of jobs around Skaneateles Lake or on the campus, working at a marina or as a hotel night clerk, driving a school bus, and working for the Schaefer Beer Company, stacking beer bottles and cans in local groceries.

Joe's only plunge into politics while at Syracuse Law was to run for first-year class president. He lost by a single vote in a runoff to a New Yorker named Bill Brodsky, who apparently won with heavy support from his brothers in Phi Epsilon Pi, a predominantly Jewish fraternity on campus. Biden never forgot a fraternity brother named Charlie Goldman coming up to him, waving his fraternity ring in his face and declaring: "You see this ring? This ring beat you!" They became great friends, despite the fact that Goldman's father, a strict Hassidic Jew, would write to his son and daughter weekly admonishing them, Biden said, "not to hang with that goy!"[48] Brodsky, years later the chairman and chief executive officer of the Chicago Board of Options Exchange, remembered Biden telling him "that I broke his record—he had never lost another election."[49] Joe had no fraternity membership of his own to rally around him.

Despite his election defeat, Joe Biden's vision for his future was becoming true as he had laid it out in his head. He had met, wooed, won, and wed the girl of his dreams, and with some bumps along the way was headed for law school graduation with his goal in sight of a career as a trial lawyer before the bar of justice. Nothing, it seemed, could stop this golden boy of high aspirations and promising talents.

FOUR

THE MAKING OF A POLITICIAN

WHEN JOE AND Neilia Biden returned to Wilmington for a family visit over Easter break of 1968, the city was in turmoil. The assassination of Dr. Martin Luther King Jr. on the balcony of the Lorraine Hotel in Memphis, where he had gone to lead a strike of sanitation workers, had triggered violence and race riots across the country. Wilmington had not escaped the frightening and destructive phenomenon.

Observances for Dr. King began peacefully with weekend memorial services in Rodney Square, the downtown focal point. But on Monday, April 8, four days after the assassination, some African American students marched down Market Street trashing stores. They were dispersed by local black leaders who told them such action disrespected Dr. King's memory. That night, however, heavy violence broke out in the predominantly black section called "the Valley," leading city, county, and state offices to impose a night-long curfew. But vandalism, looting, shooting, and fire-bombing spread, leaving many residents homeless and resulting in fifty-four arrests.[1]

The next night, as local police sought to quell the outbursts in this and other city neighborhoods of growing African American population, Delaware Governor Charles Layman Terry Jr. mobilized all thirty-eight hundred members of the state's National Guard to pitch in. A week later, Terry ordered nightly patrols in those neighborhoods by the predominantly white army reservists who kept Wilmington under occupation, incredibly, until the following January—the longest such military presence in any American city since the Civil War.[2]

The decision aroused deep resentment in the black community, recalled James Baker, an African American social worker dealing with Wilmington youth groups at the time and much later mayor of the city. Terry was under the impression "there was this black army that was armed," Baker recalled. "These kids were armed only with what they had, which was rocks, basically, and bottles. There was some shooting, but the shooting was by older adults." The young people were bitter, Baker said, "because nobody felt there was need for them [the National Guard], and you had an army occupying your neighborhood, telling you what you can do and can't do. There were curfews and only so many people could be gathered on a corner at a particular time."

Terry believed, Baker said, "that there was this army; that if he pulled the troops out there would be a wholesale rebellion, violence with guns, a takeover by this black army as they called it. What had happened was they just caught two or three guys target-practicing out on [a nearby dump] and it just panicked everybody." The troublemakers were not, Baker said, outsiders from the black power movement who had brought violence to such other cities as Baltimore in the wake of the killing of Dr. King. "Joe felt it was an overreaction in having the National Guard there that long on the city streets," Baker said, but the governor was getting calls from business people downtown telling him they didn't feel safe.[3]

On Easter Sunday, quiet finally prevailed, but Governor Terry rejected the request of Mayor John Babiarz to remove the troops from Wilmington streets for the holiday. As intermittent violence continued in the city along with racial tensions and scuffles in its high schools, Terry, running for reelection, embraced the same law-and-order philosophy of repression that was then fueling the presidential candidacy of former vice president Richard M. Nixon. Even when the curfew was lifted in early May and the city ended its state of emergency, Terry refused to take the National Guard forces off the streets. He set out to counter local protests against the condition of black political powerlessness that had long existed in the city.[4]

Wilmington's Negro population, as it was then commonly called, had grown from 15.7 percent of the total in 1950 to 38.9 percent by 1967, as residents of the East Side moved gradually toward the center city, accompanied by the sort of white flight seen elsewhere in urban America. Construction of the new Interstate 95 linking New York and Washington spurred what was labeled "urban renewal" in the white community, while those obliged to relocate disparaged it as "black removal."

Kindling for racial conflagration came from high crime and poverty among blacks and Puerto Ricans in the precincts of the Valley, and conflict among residents of the nearby Little Italy section and the racially mixed population of the adjacent Hilltop area. Simultaneous riots in larger cities like Newark, Cleveland, and Detroit reinforced Nixon's argument that tough police action was justified to restore urban order.[5]

Also fueling a climate of protest everywhere was the spreading public disillusionment with the war in Vietnam, determinedly pursued by President Lyndon B. Johnson despite his recent decision to abandon an expected reelection campaign, challenged not only by Nixon but also by fiercely antiwar fellow Democrats Eugene J. McCarthy and Robert F. Kennedy. On April 27 in Wilmington, a protest against the war led to the arrest of twenty-three participants on grounds of violating the curfew prohibition against assemblies of more than ten people.[6]

Joe Biden, removed from all this turmoil in Wilmington as he was winding up his pursuit of a law degree in Syracuse, was taking no visible part in either the local racial unrest or the war protest. Bert DiClemente, a Delaware schoolmate, recalled that young Biden was against the unpopular war but didn't take to the streets against it. "A lot of young people were disgusted at where we were as a nation," DiClemente said, "and he thought he could help us extricate ourselves from this disaster. This needed to be brought to a halt, and the question was, how would we do it and do it intelligently? He was a coat-and-tie guy and would do everything correctly."[7]

Ted Kaufman, his longtime political adviser and later his replacement as U.S. senator from Delaware, agreed: "He was more of a system person rather than working outside the system. He wore a sport jacket, not a flak jacket."[8] Biden's sister agreed that her brother was never in the forefront of any street demonstration. "That's Joe," she said. "He wasn't the equivalent of a woman bra-burner."[9] Also, the deepening war in Vietnam was not touching him directly, as a result of routine student draft deferments, first at Delaware and then at Syracuse.

Biden himself said much later that coming from Claymont and Mayfield, heavily white communities, "in my neighborhood there were no African Americans. But I remember as a high school kid, even in my freshman and sophomore years, turning on the television and seeing Bull Connor and his dogs, a lot of guys in my generation, being confronted with what you never knew was going on. You didn't know how people were being treated." With Delaware segregated by law, he said, as a middle-class suburban boy he was somewhat insulated from the worst of that condition.[10] It was not until his exposure to black contemporaries as a lifeguard, he said, that he finally began to understand, and then he saw his best engagement as a lawyer, not a marcher.

That spring, Joe and Neilia were occupied with deciding whether to settle in Syracuse or in Wilmington. Joe Sr., managing the car dealership, had a business colleague in New Castle whose son, Bill Quillen, had been a special assistant to Governor Terry and was now a superior court judge, and he arranged for Joe Jr. to call on him. "He was debating in his own mind whether he should stay here or go to New York," Quillen recalled later of their first meeting. "I strongly suggested it would be a terrible mistake to go to New York when he ought to stay here, as he was interested in being a trial lawyer and there was great opportunity here. We had a nice chat and I can remember picking up the phone and calling Rod Ward, who was a roommate of mine at Harvard, and at the time was at Prickett, Ward, Burt & Sanders. They did a lot of trial and negligence work. Rod took him to lunch and he was hired."

Of this first meeting, Quillen said: "Joe's a sparkler. My impression was, this guy's good, he would be good before a jury. He was a little less talkative than usual, because he was trying to figure out something. Anyway, I argued for Delaware strongly. It's the greatest place to practice law in the country. It's small, we've got the best of all worlds, we've got the usual business and we've got the corporate business, and if you go to New York, you don't know what the hell you're going for."[11]

Indeed, Delawareans like to call their state the corporate capital of America, anchored by the DuPont Company, the chemical giant that once was, and still is to a lesser degree, a sort of feudal baron that keeps the state and its government under its broad wing. As such a corporate magnet, it has drawn the huge credit card business to Delaware, and politicians have been special protectors of its interest, a reality that later drew some rare criticism to Biden himself.

In any event, Joe wrote of his subsequent lunch with Quillen's Harvard colleague: "Ward took a long look at my resume, which didn't sparkle with academic achievement, and noted my photograph in the right-hand corner. 'Obviously, you're hoping to get a job based on your good looks,' he said. He was a wise guy but he was smiling."[12]

Joe did get the job after receiving what he later called "an insulting acceptance" that noted his modest performance at Syracuse and offered an equally modest starting salary incumbent on passing the bar in Delaware.

He and Neilia rented a little farmhouse in a middle-class neighborhood known as Brandywine Hundred in the Wilmington suburb of Mayfield, while looking for something better. He then bought a little house near the University of Delaware in Newark, just south of Wilmington for $11,000. "I got my brother Jim, who was in college, and he and I fixed it up," he reminisced, "and I sold it, made a $4,000 profit. So then I had five grand in my pocket."[13]

Thus began an odyssey of home-hunting around the Wilmington area as Joe, an admittedly frustrated architect, sought to find the perfect family base. In February 1969 their first son, Joseph Robinette Biden III, called Beau, was born, and Bobbie Greene from Syracuse became his godmother. When Neilia was pregnant again three months later, the young couple intensified their search for a larger place, and Joe found one on Woods Road in Wilmington.

With money short, Joe then somehow persuaded his father to buy the bigger house, and he would buy his parents' smaller house in Mayfield. The parents did move, but before taking over their house, Joe learned of another money-saving deal. A cottage on the grounds of a seventeen-acre country club could be had rent-free if he managed its swimming pool and served as lifeguard, so he rented out the house in Mayfield and took over the cottage. It was a bit of a squeeze, especially when Robert Hunter Biden, called Hunt, was born, but Joe built a small addition. Beau by this time had to sleep in a closet so the search for more space went on.

Joe finally found an eighty-five-acre farm in Elkton, Maryland, just over the state line, and with the profit from the first property, a loan of $5,000 from Neilia's father, and by selling off thirty acres, he was able to buy it for $55,000, with the thought that his sister,

Valerie, and brother Jimmy might join in, making it a Biden family compound. Meanwhile, Joe and Neilia continued to live at the rent-free cottage and tend to the swimming pool.[14]

Valerie said later: "Joe's life revolves around his family and his home. If politics hadn't worked, Joe and Jimmy could have gone into business as the Biden Brothers, Joe as a frustrated architect and landscaper, Jimmy as an interior decorator. Joe has a very symmetrical eye, and if he had a million dollars he wouldn't be traveling, he would be putting it into his house. And Jimmy is an amazing designer with a great sense of color."[15] Joe and Neilia now had three rental homes to help finance the renovation of the Elkton farm, and Joe and friends were soon getting involved in real estate on the side. "Joe had a very good sense of how to build equity," Bobbie Greene recalled.[16]

As if all this moving were not enough, Joe Biden decided before long that he could not stay at the Prickett firm. As a working-class Democrat he was not excited about defending big corporations, especially seeing daily how much of downtown Wilmington remained severely damaged by the riots. "The city was in turmoil," he remembered. By the time he got back the summer after graduating at Syracuse, he said, "the National Guard was on every corner with drawn bayonets, and the state police were in the city, all white and over six feet tall, and they'd march down the street, particularly on the East Side, three on a sidewalk and everybody had to get off. I mean they were just establishing they were in charge."[17] Wilmington was a constant reminder of the continuing local civil rights inequities.

One day in 1969 he accompanied William Prickett, the firm's senior partner, to a trial in federal court in which he was defending the Catalytic Construction Company being sued by a welder badly burned and crippled on the job. Prickett asked Biden to help write the brief and got a quick directed verdict for his client because the worker had failed to put on the required flame-resistant clothes and thus had been guilty of contributory negligence.

Young Biden didn't argue with the law, but he didn't care for the outcome. "I remember walking out of the courtroom," he recalled. "Mr. Prickett asked me to lunch, and I said I had to go to lunch with my dad. It's the only time I can ever remember lying. I know that sounds self-serving." He quit the job and walked across Rodney Square to the Office of the Public Defender and asked for a job. The administrator asked: "Don't you work for Prickett? Why the hell would you do this?" Biden said later: "I realized it was a combination of now wanting to do defense work, and all this stuff was going on around me in Wilmington and I was doing nothing."[18] He decided he needed to handle cases more suited to his temperament and objectives.

"So I became a public defender," he said, "which meant that 90 percent of my clients were East Side African Americans." On his first day he was sitting in a courtroom watching the judge assign cases. When another public defender told the judge he wasn't ready to handle a certain case, the judge ordered Biden to take it. "There's this guy sitting in the front row with hand shackles and ankle shackles on," he remembered. "I said, your honor, I'm not prepared. I haven't tried a case, I've just joined the office, I haven't seen the file." The judge replied: "Well, you draw the jury this afternoon. It will give you a day to look at the file."

As Biden recalled the scene, the judge turned to the defendant and asked: "Mr. Larkin, do you have any objection to Mr. Biden trying your case?" Larkin, Biden said, "just put his head down and two guards yanked him up to make him stand," and he shot back: "No, your honor, one motherfuckin' honky is just as bad as another."

Biden said he tried to prove mistaken identity in the robbery charge against a young sailor but failed. So he visited Larkin in jail, telling him he was going to appeal on grounds of insufficient time to build a proper case. Whereupon, the convicted man threw a table at him, saying, according to Biden: "God damn it, you dumb honky son of a bitch! I did it! I did it!"

A few years later, Biden was visiting another state prison where he was greeted with "Hey, Joe!" from a series of his former convicted clients. When an accompanying local reporter needled him as being a helluva lawyer, an arm reached out and grabbed the reporter, telling him: "I'll tell you one thing. Biden will stand up for a black motherfucker, unlike you!" The inmate, Biden said, was none other than his first client, Earl Larkin.[19]

Bert DiClemente recalled later Joe's early talents as a spellbinder and salesman who could liven up the most mundane legal task. "He was a struggling young lawyer, and I was a struggling young real estate guy, and our paths crossed several times," he remembered. "He did entertaining settlements, which is really hard to believe. After people bought property and signed documents, Joe had the ability to make it really entertaining for people. He was charming, would take his time going over the documents, which were very sterile, and make them come alive. He would tell a story about the document, what it really means, and means to them."[20]

The public defender job was only part-time, so Biden soon hooked up with another firm, Aerenson & Balick, which handled civil litigation more to his liking, and that of his conscience. His boss at the new law firm, Sid Balick, remembered that while his new hire efficiently handled all cases that came his way, "I sensed he was more interested in politics," he said. "He didn't mention it immediately, but at one point he came to me and said he wanted to talk to me. We went to a local restaurant and he said he wanted to talk about his future. I was pretty active in politics at the time, and he asked whether I was going to run in the next election. He suggested I run for governor. I had been a state representative and lost an election for attorney general. I sensed he really wanted to talk about himself." Young Joe, then only twenty-six years old, finally asked whether it would be okay to continue to practice law and run for public office. Balick said there was no objection if the office was only part-time and didn't conflict with business before the firm.[21]

At this time, Biden was still only thinking about getting seriously into politics and first registered as an independent. He did so, he later told Carl Leubsdorf of the *Dallas Morning News*, because "I was really upset with the Democratic Party in Delaware because of its perceived position on civil rights," referring to Democrat Governor Terry's lengthy National Guard occupation of Wilmington after Dr. King's assassination. In 1968 he voted for Republican Russell Peterson when he defeated Terry, "but I couldn't bring myself to register as a Republican because of Richard Nixon," he said.[22] While Biden was in law school in Syracuse, however, he did some entry-level volunteering for a Democratic congressional candidate, and finally changed his registration in 1969.

Back in Wilmington, he got involved in a citizens' fight against intrusive highway plans and was approached to run for the Delaware General Assembly, but he declined. Balick did, however, whet Joe's interest in the Democratic Party in the state by bringing him into a group called the Democratic Forum, which was seeking to reform the party, especially in the wake of Governor Terry's occupation of Wilmington. Young progressives in the party were increasingly concerned that it was dragging its feet in Delaware on such matters as school integration, open housing, and other civil rights issues. On Wednesday nights after work on Market Street in downtown Wilmington, Biden would walk to the nearby Pianni Grill and join the reformers in lively talk on what needed to be done to shake the party out of its discriminatory past.[23] Then he would rush home to his family.

Joe began building his own small law firm with his best friend at Archmere, David Walsh, who had gone to law school at Georgetown. "In 1969," Walsh recalled later, "he talked me into coming back [to Wilmington] and going into law practice with him, and we did."[24] Meanwhile, still missing his football days, Joe started up another touch football team called "the Orangemen" in Wilmington, patterned on the one he had organized in Syracuse, in which Mike Fay and other old teammates committed mayhem on each other on weekends.[25]

Around this time, another local Democrat named John Daniello, a member of the New Castle County Council with higher political ambitions on his mind, called on young Biden, then twenty-seven years old, with a proposal. "I was running for Congress at the time," Daniello recalled much later when he was the state party chairman, "although I still had a pretty heavy interest in the council, since I'd been on it for six years. A bunch of us thought we needed someone to run on our ticket from a certain district, and he was a young attorney starting out. One of the other attorneys on my campaign staff recommended him so I went to see him."

Daniello had not met Biden before, but his first impression, he remembered, was that he was "young, bright, a sharp attorney" with a very attractive wife. Work on the council, Daniello acknowledged, was not especially dramatic or exciting, dealing as it did with such mundane matters as issues of zoning, local construction regulations, and the like. And the seat would be in a strong Republican district. Biden at first was not interested.

"He was doubtful that he wanted to run or not," Daniello recalled. "He made it clear his interests were what they are today. He was interested in foreign affairs and history and really had no interest in 'local politics.' I think he used that terminology."

Daniello pressed him. "You've got to start somewhere if you're going to get known, even to other statewide officers," he said. "Joe talked about doing something at the federal level, and I thought he was talking about working in the federal government somewhere. I told him, 'If you want to get recognized, this was a hell of an opportunity for it.'"[26] When Joe told him he didn't have time, he was informed that the council met right across the street from his law office, and at night.

So Joe went home and told Neilia. Very politically aware herself, she joined in enthusiastically. Next he recruited his sister, Valerie, who without hesitation jumped in. Joe wrote in his memoir: "Valerie Biden did not go into any race to lose, Republican district or no.

. . . She got voter records going back several elections, had an index card for every block in every neighborhood, and started recruiting block captains." The Bidens, including brothers Jimmy and Frank, organized a volunteer army of young people who worked the strong Democratic precincts while Joe personally went door-to-door in Republican middle-class areas that resembled those he knew in Scranton and Wilmington. "I knew how to talk to them," he wrote. "I understood they valued good government and fiscal austerity, and most of all, the environment. I promised them to try to check the developers and fight to keep open space."[27]

Young Biden had two opponents for the county council seat, Lawrence T. Messick, a Republican, and Kenneth A. Horner, running on an American Party ticket. Joe also set himself up early as a crime fighter and defender of the police. In August 1970, when the New Castle County police released the latest statistics showing a 35 percent increase in major crime in the first six months of the year, candidate Biden accused the Republican-led county officials of "a deplorable lack of leadership," saying the public was "not receiving the police protection to which they are entitled" despite a 17 percent increase in the budget for that purpose. He proposed a four-point program that included an expanded police criminal division and the hiring of a full-time police director to achieve better cooperation between county and state police.[28]

No issue was too minor for the energetic, eager-beaver Biden to tackle. When he learned that the taxpayers in the towns he sought to serve were paying as much as two or three times as much for trash collection as those in other parts of the Fourth District, he proposed a reorganization of the county's Department of Public Works. He wanted the county to be rezoned and garbage collection let out to closed bids. He also pushed for a new landfill site "with the most modern compacting and recycling equipment."[29]

"He did a nice job," Daniello remembered. "His sister and two brothers really came together and worked the district, and he did all the right things."[30] Valerie Biden organized and ran the campaign from the basement of their parents' home. "We knocked on every door," she recalled later. "There was a Biden at every door."[31] In an election in which the Democratic Party was otherwise wiped out across the state, Joe Biden won the county council seat by two thousand votes.

Even before serving a day on the council, however, its newly elected member contacted Daniello, who had lost his race for Congress, and met him for lunch. "I thought we were going to talk about some information or transition, about me coming off the council and him going on," Daniello remembered. "I thought he was going to pump me on what some of the issues were. But he says he wanted to run for the Senate! I thought he meant the state senate, and I said, 'Why would you want to do that? You represent more people as a county councilman than you would do as a state senator.' He said, 'No, John, I'm talking about the United States Senate.' We were talking about running against the U.S. senator who was probably the most popular politician in Delaware at the time."[32]

Indeed, Republican James Caleb Boggs, born on a farm in Kent County, Delaware, in 1915, had by then compiled nearly thirty years in public office in the state and Congress—as a county judge, Delaware's only member of the U.S. House of Representatives for six years, governor of Delaware for eight years, and now in the U.S. Senate for his second six-year term. It was not too much to say, the astonished Daniello told Biden, that "Cale" Boggs had achieved the status of an icon in the state and the notion of a young upstart beating him was mind-boggling.

Daniello, though, didn't try to talk Biden out of the idea. "I told him to give it a lot of thought and do the best job he could on the council, start building a reputation around the state and decide later," Daniello said. "For the next six months to a year, he did what we all expected him to do. He did his job as a councilman, did it well, and it wasn't until the beginning of the following year, January of 1972, when he began to make it clear he wanted to run [against Boggs]."[33]

Biden said later he had been approached in 1972 to run for governor by a group that offered to open a campaign treasury of $50,000 for such a race. "I told them I wasn't interested," he advised *Wilmington News Journal* reporter Curtis Wilkie at the time. "There was only one office I'd rather have than any other. That was the Senate, and in terms of campaign and costs, I figured Boggs would be the easiest opponent." But he added: "The guy I would have been more interested in running against would be [Congressman Pierre S. "Pete"] du Pont, but I didn't think I could match money with him."[34]

At least one Delaware journalist saw twenty-seven-year-old Joe Biden's potential from the start. Jane Harriman, in the *Wilmington Evening Journal* just a week after his election to the county council, wrote that he might just be "a great statesman in his youth, Delaware's JFK, or would you believe somebody they can nominate in '72 [for the U.S. Senate]?" She went on: "The comments of the young [Biden] volunteers are often one squeak above the equal of a Beatlemaniac. And the loud effusive opinions of older Biden backers still sound like the interviews staged by Madison Avenue for the selling of the candidate. But the emotion behind them is undeniable."

Family man Biden also told the reporter that although his young wife had been instrumental in his election, he wanted her to stay home now to "mold my children." In the candor for which he soon would be famous, Biden blurted: "I'm not a 'Keep 'em barefoot and pregnant' man but I am all for keeping them pregnant until I have a little girl. . . . The only good thing in the world is kids."

He wasn't finished with that comment, either. Defying easy characterization on matters of racial bias and discrimination, he said: "I have some friends on the far left, and they can justify to me the murder of a white deaf mute for a nickel by five colored guys. But they can't justify some Alabama farmers tar-and-feathering an old colored woman. I suspect the ACLU would leap to defend the five black guys. But no one would go down to help the 'rednecks.' They are both products of an environment. The truth is somewhere between the two poles. And 'rednecks' are usually people with very real concerns, people who lack the education to express themselves quietly and articulately."

The free-wheeling remarks could have been taken as an early warning signal of self-inflicted controversies to come had they had wider circulation, and they certainly were an alert to Biden's willingness to say what was on his mind. He continued that his support of public housing had brought smears his way, but would not be deterring him. "The first time the phone rang and someone said, 'You nigger-lover, you want them living next to you?' I was shocked," he said. He said he replied, "If you're the alternative, I guess the answer is yes."

Biden also loudly proclaimed himself a teetotaler regarding smoking and drinking. "I don't use anything that could be a crutch," he said, then adding: "I use football as a crutch and motorcycle jumping and skiing—I ski like a madman. But those are crutches over which I have some control. I'm against chemical crutches." And he concluded with words that not long in the future would be emphatically confirmed by his actions: "The most important thing to me without question is to be a good father."[35]

One of New Castle County Councilman Biden's first announcements was the opening of a special constituency office in downtown Wilmington, to be paid for with his annual council salary of $7,000. The money, he said, would be used "to coordinate my research and volunteers and allow me to pay closer attention to constituency complaints." He acknowledged at the time that he probably would be accused by critics of having further political ambitions. "Well, I can't help what they think," he said then. "I just don't want to be put in a position of ever having it said that I've reneged on my campaign promises."[36]

Joe also set himself up early on the council as a foe of unmonitored development. When the Shell Oil Company stealthily bought up more than five thousand acres of shore property near the town of Delaware City, just south of New Castle, with plans to build a new refinery, he raised concerns about raw sewage and other pollution threats. While he favored job creation opportunities, he was in no mood to approve corporate free rides. "Let Shell prove to us they won't ruin our environment," he would say, and if the oil giant couldn't do so, "we'll rezone them right out of here."[37]

He introduced a rezoning ordinance aimed at blocking another refinery on farmland at Smyrna, farther south, but at the same time called for a hearing delay until Shell could mount its own case for the project.[38] Vince D'Anna, a council staff member who subsequently became a Biden political aide and adviser, recalled that "a lot of his opposition led to enactment of a state law prohibiting having industrial development including refineries," to protect Delaware's beaches, a major tourist attraction.[39] Biden also fought a superhighway construction project in the Wilmington area that threatened as many as seventy existing houses and many apartment buildings. "We've got to bring a screeching halt to these monstrosities," he said at one point.[40]

A particular target of Biden, D'Anna recalled, was Interstate 95, which split Wilmington and paved the way for high-rise apartments on Concord Pike. Among the tenants was builder and major Republican Party contributor John Rollins, who D'Anna said became a "lightning rod" there for a seven-story structure on which he had a helicopter pad that was seen by many as a local nuisance.

In all this, Biden was creating an image for himself as a workhorse for the people of New Castle County. "It gave him substance," D'Anna said later, "with which he was able sell himself elsewhere in the state where he was little known."[41]

In waging such battles as a freshman on the county council, Joe was getting minor recognition in the state Democratic Party, which was in pathetic shape. Until around 1968, there hadn't been much of a Democratic Party in Delaware. Banking, credit, and bankruptcy laws were catnip to the financial giants of the day, and the Democratic

Party, like many other things, was largely controlled by the du Pont family.

"There was a party, but it was not what you'd call a party," recalled real estate executive Henry Topel, the state chairman at the time. "It was a token. We organized a brand-new Democratic Party, and the first thing I did when I became chairman in 1970 was to establish a blue-ribbon commission to reform the party. The commission conducted hearings up and down the state involving new people and getting excitement for the new Democratic Party."

At the time, Topel remembered, New Castle Country Councilman Joe Biden was only twenty-eight years old when he dropped in on the state chairman with a notion that bowled him over. "He said, 'Henry, I would love to run for the United States Senate.' And I said, 'Well, Joe, I don't think you're old enough.' He just walked in cold, and I said, 'Joe, you have to be thirty years old.' And then I said, 'How in the damn hell are you possibly going to beat the most popular, most beloved man in the state of Delaware, Caleb Boggs?' He was a former congressman, a former governor, a former judge, and he knew everybody in a state with only 450,000 [people] at the time. And Joe says, 'Well, have you talked to your youngest son?' And I says, 'David is not even old enough to vote.' And Joe says, 'You need to talk to him.' "[42]

Young Topel, a friend of Frankie Biden, later wrote that his recruitment into the Biden Children's Crusade was spurred by the leadership of Joe's sister, Valerie, and because "Joe said he wanted the Vietnam War over. That was a biggie for my gang in 1971. Most of us were seventeen and eighteen and already had friends or knew of someone who had died in a war we simply couldn't understand.

"But I think it was the fact that the campaign was run by his family. . . . My specific job was to help contact each high school in the state so that we could get Joe into the classrooms. Whether it was speaking to a single civics class or to an entire auditorium of students, he began to excite and inspire young people by the thousands. No one had ever really asked young people to get involved before, not at this level. . . . It seemed more like a movement than a campaign, an opportunity to have our voices heard in a wartime atmosphere that had set the generations at odds."[43]

Young David Topel, his father recalled, "was one of Joe's soldiers. You know, you had Vietnam at the time. So Joe starts to explain to me how he's organizing the young people. Joe says, 'We've got an issue with Vietnam today, and the young people are concerned. I'm organizing all the young people in the state of Delaware and your son is on my committee.' "

To help the brash young man along, and to tap into his energy, Topel put Biden on the party reform commission, enabling him to travel the state and participate in hearings along with former governor Elbert Carvel, state senators, and, in Topel's words, "the best of the party." The Democratic elders, Topel said, thus had an opportunity to size up the young recruit from Wilmington. "He never asked for it. I put him on it," Topel said.

Tiny Delaware has only three counties—New Castle, where Biden was serving on the county council, and the two southern and more agricultural counties, Kent and Sussex, where he was hardly known at all. Biden often told Topel later: "Henry, it gave me the exposure I needed statewide" in that first race for the U.S. Senate. "You give Joe an opportunity," Topel recalled, "and he takes it all the way."[44]

Topel's teenage son David remembered later: "Despite the age gap between Joe and Frankie, Joe was around the house all the time, so we got to know him. I knew him to be a pretty unique guy actually. He didn't fit into any of the pigeonhole categories that we always used then back in the late sixties as teenagers. You were either a jock, an honor student, a hood, or maybe with a band.

"And although Joe was known as a really fine athlete—as I recall football and basketball the most—he was different than the other jocks. He voiced his opinions frequently, and they weren't the typical opinions you would hear. When we were young teens in junior high school and high school and he was the older brother, he was saying things that were surprising in our neighborhood. He was very vocal against things like segregation, and in suburban Wilmington you didn't have much exposure to integration back then. But I remember thinking, 'This is not the normal football jock.' What impressed me was Frankie telling me that Joe was going to an antisegregation protest somewhere in Wilmington. That stayed with me because we were in all-white schools in an all-white neighborhood. We didn't hear that before. It started us talking about what an antisegregation protest rally was."[45]

As the youngest member by nearly fifteen years on Henry Topel's party reform commission, Biden noted in his memoir, "I was the guy who kept the notes and turned off the lights. But I took it seriously because the other commissioners took it seriously. We knew we had to modernize our organization, our campaign techniques, and our substance." Over the winter and spring of 1971, Joe was able to meet every Democratic committee member in the state, and Neilia came up with the idea of hosting a series of blue-ribbon dinners at the senior Bidens' house, with Valerie and Joe's mother pitching in on the cooking and cleanup afterward.

In the course of this party renewal effort, one of Biden's Democratic friends, Bob Cunningham, and a few others called on him to help find a candidate to run for the U.S. Senate the following year, when the revered two-term Republican incumbent and former governor Cale Boggs would be up for reelection. Biden, though himself musing about a Senate run, apparently didn't let on about these thoughts at the time and agreed to join the quest. The group approached former governor Bert Carvel; Joe Tunnell, the chief justice of the Delaware Supreme Court; and some corporate leaders in the state, but none of them was interested. Nobody wanted to take on a Delaware icon like Boggs, who hadn't lost an election in the state for nearly three decades.

That summer, Biden was attending the state Democratic Party's off-year convention in Dover, the state capital, when there was a knock on the door of his room at the Hub Motel. Dressed in his shorts and in midshave, he opened the door and admitted Henry Topel and Bert Carvel, who had already rebuffed the efforts of Biden and the other young Democrats to get him to run against Boggs. They said they wanted *him* to be the candidate.

Joe wrote in his memoir later that "my initial reaction was: I don't think I'm old enough. I had to do the math," which found that he would reach the required age of thirty for U.S. Senate service thirteen days after the November 7, 1972, election, and several weeks before the swearing-in date in early January. But from what others said later, they had heard from him that he had thought about the prospect much earlier. In fact, he knew by this time that his service on the county council was limited as a result of a redistricting by the Republican majority that made his district even more heavily Republican than it had been when he was elected. "I heard them kid about it at the time," Joe wrote later. "They thought it was very funny. I had no place to go. It was up or out."[46]

Not that Biden was hankering for a long career on the county council. His friend and later Senate staff aide, John DiEleuterio, recalled him often saying, "I'd rather negotiate the toughest foreign policy treaty with any country than have to be involved with a county land-use issue."[47]

In the midst of all this, and all the real estate transactions already undertaken, the Bidens remained living in the country-club cottage and Joe continued to manage the swimming pool. And his search for the perfect house went on. Family outings often consisted of piling the kids into the car on Saturdays and touring country roads outside Wilmington, until they spied a 1723 stucco-over-stone Colonial in move-in condition on four acres in the village of North Star, near the Pennsylvania border, for $87,000. They quickly made an offer but first had to sell enough of the existing properties to make the down payment. And because the new house was not in Joe's county council district, the family could not move in until his term ended.

Once again Joe's ingenuity and chutzpah combined to solve the problem. On the following weekend, Joe, Neilia, and the two boys went to dinner at his parents' place, into which they had only recently been moved. It happened to be in Joe's district and, unabashedly, he promptly asked his father to switch houses with him! His mother was appalled, but in the end the switch was made. While Joe Sr. was at work a few days later, Jimmy Biden showed up with a truck and moved everything. Mom Biden phoned her husband and told him to report to the new address, where his bed was set up and waiting for him.[48]

The new North Star house was all part of Joe Jr.'s vision for a Biden homestead built around his growing immediate family, and the wider clan as well. It was soon enlarged again with the birth of a baby girl, Naomi, called Amy or Caspy for her likeness to television's Casper the Friendly Ghost. After the swimming pool was closed for the summer the family moved into the vacated house of his pliable parents. Now with all the musical chairs in place, Joe could address the impossible dream of challenging Delaware icon Cale Boggs for his U.S. Senate seat. He had been getting around the state and getting known, so he told Topel and Carvel he'd think about it, which of course he had already done.

Bob Cunningham encouraged him to run and so did Neilia, though she really preferred that he stay in the law. But if he was going to run, she told him he should forget about building the law firm and run full-time. At a family meeting, only Joe's mother expressed reservations, but he assuaged her by saying nobody expected him to win, so he had nothing to lose—though he really thought he could win.[49] By this time, D'Anna said later, with Carvel declining to take on Boggs, the attitude seemed to be, "Nobody wanted to do it, so let him [Biden] do it."[50]

The Biden clan, led by Neilia and Valerie, drew up a battle plan in which its members would be the generals organizing and directing a volunteer army, heavily dependent on students and other young folks, all to introduce little-known Joe Biden to the state. "I wanted to meet as many voters as I could," he wrote in his memoir later, "to let them see me and hear me, before I ever announced. All through 1971 and early 1972 it was coffee under the radar." The team leapfrogged around the state, doing as many as ten coffees a day, often with the small kids in tow; "we just carried them from house to house like footballs in wicker baskets," with Joe addressing more than three hundred people from early morning to late night.[51]

Valerie described the strategy this way: "The year before, we just snuck up on them." At a time when so many women stayed home, it was possible to generate thirty or more housewives to a single coffee, and by enlisting all the Bidens—the parents, Neilia, and Valerie—to show up on a rotating basis three or four days a week, "there'd be a Biden at every one." And afterward, Neilia and Valerie would team up to send a handwritten note to every single woman who came to hear Joe. "My motto or mantra as a campaign manager," Valerie said, "perhaps spoken like a true sister, was to know him was to love him. I was positive that if we could expose him and put him out there so people could see him and hear him, and he could talk and listen to them, that we could win. The key was to get people to want to listen. The power of the [television remote] clicker, you know, is you can turn it off in two seconds. And people wanted to listen to Joe, because when he started out he had just a couple of seconds to catch them. And he caught them."

Furthermore, she said, "there's a culture of civility here," and a small state like Delaware, where you could get around quickly, she knew, was ideal for this sort of retail politics. "The fact you can go from stem to stern in Delaware in two hours, and across the state in half an hour, made it a great level playing field for us where we didn't have the money for media. We're the fourth or fifth most expensive media market because we don't have our own TV. We have

Philadelphia [about twenty miles north of Wilmington], so 95 percent of our media is wasted [on non-Delaware voters] because it goes to people in the Philadelphia area."[52]

Meanwhile, Joe's brother Jimmy set out to raise start-up money for what was so widely regarded in the state as the impossible task of beating Boggs. Indeed, the Democratic nomination for the Senate seat was any reasonable party member's for the asking. The problem was the general election campaign, and the Bidens figured they would have to raise $150,000 to get Joe's name out, because the state party would be focusing on trying to take the governorship in 1972 from the Republican incumbent, Russell Peterson.

In late September 1971, Biden began to let the word out about his seemingly preposterous U.S. Senate ambitions. In a page-one story in the *Wilmington News*, he said that while he hadn't made up his mind, "it's more likely that I will run than I will not."[53]

One slight inducement was a rumor that incumbent Boggs after his long years of public service was ready at age sixty-two to retire. Even so, two other formidable Republicans, former Wilmington mayor Hal Haskell and Delaware's one congressman-at-large and family scion, Pierre S. "Pete" du Pont, were said to have their eyes on Boggs's seat if he were to step down. Du Pont said much later the party leaders feared a primary between them "would doom their Republican Party" in the state, "so they persuaded Boggs to run again."[54]

William F. Hildenbrand, then an aide to Boggs and later secretary of the U.S. Senate, said later in an oral history interview that Boggs indeed had decided not to seek reelection. Haskell and other leading Republicans in the state got together and said to Boggs, "Look, we can keep the governorship, you can get reelected, but we have to have you at the top of the ticket. . . . His heart wasn't in it. In July of the election year, he did not have one billboard up in that state."[55]

In any event, President Richard Nixon at one point flew into Delaware and met with Boggs at the home of state Republican bigwig John Rollins, urging him to seek reelection for the sake of the party.

Boggs, seeing an easy race ahead with no well-known Democratic challenger in sight, ultimately agreed to run one more time.

Edward "Pete" Peterson, the New Castle County Democratic chairman, told the local paper that Biden was "a good potential candidate. . . . He's young, but the party needs more young candidates. I believe in going with a winner."[56] To seasoned political ears, that comment had the ring of setting up a sacrificial lamb, a role that Joe Biden appeared ripe to fill.

Back in the old neighborhood in Scranton one day, sitting on Jimmy Kennedy's front steps, Biden told his old boyhood pal that he had plans to become the next U.S. senator from Delaware. The reason he gave for his confidence, Kennedy said much later, "was, basically, 'Well, I can talk now.' It's hard to say in a one-on-one conversation with a guy how good an orator is he? You can't tell that. But he certainly had an abundance of linguistic intelligence, linguistic talent. He could walk up with a lot of charm and say, 'Will you give me a chance?' " And that's what Joe Biden was now about to do.[57]

Another day, Biden was visiting with Sonia Sloan, a local Democratic Party activist involved in the effort to deal in a responsible and peaceful way with the desegregation of Wilmington schools. He bowled her over telling her he was thinking of challenging Cale Boggs for his presumably safe Senate seat. She told him, like others, that he was crazy. Sometime later she got a call from a friend associated with the Council for a Livable World, a peace-oriented group that supported Senate candidates in small and inexpensive states. He asked what she knew about a young fellow from Delaware named Joe Biden. Sloan praised him to the skies, and the next thing Biden knew, he got the first substantial contribution to his Senate campaign—a check for $25,000, which seemed at that stage of his long shot gamble like a million. He was on his way.[58]

Along with the financial help, the council sent a young fellow fresh out of Harvard named Patrick Caddell to do a baseline poll on Biden's chances against Boggs. Caddell was already making a name for himself as a polling wizard for the presidential campaign of Senator George McGovern of South Dakota, another long shot opponent of the Vietnam War. A wary Biden wondered whether it wasn't a bit presumptuous for the twenty-two-year-old Caddell to be polling for him. "What about you running for the Senate at twenty-nine?" Caddell snapped back.[59] Thus was born a political partnership and personal friendship that vacillated between smooth sailing and stormy weather over the next fifteen years and more, with the first challenge achieving the impossible against sure winner Cale Boggs.

The first Caddell poll in early 1972 predictably showed that the unknown young Biden would be shellacked. The handful of voters who said they would vote for him, Valerie acknowledged, "would have voted for Mickey Mouse if he was a Democrat."[60]

Joe fell into a panic. "Oh my God, I'm going to get killed!" he wailed. But Caddell explained that the low favorability numbers were to be expected, and there were grounds for optimism in other figures. There were indications, for instance, that voters might be ready for a younger senator than steady, likable old Boggs. Neilia, the coolest head in the family, told her husband to calm down and listen to the young pollster.[61] In any event, there was nowhere to go but up, and the Biden family proceeded to do what it could to make that happen.

FIVE

DAVID AND GOLIATH

ON THE MORNING of March 20, 1972, twenty-nine-year-old New Castle County Councilman Joe Biden packed the DuBarry Room of Wilmington's stately Hotel du Pont off Rodney Square, launching what would be the most improbable political campaign in the history of Delaware. "I am announcing today my candidacy for the United States Senate," he proclaimed to a crowd of wishful thinkers and some mocking disbelievers. The sentiments on both sides were wholly understandable, because young Biden was setting off to climb a mountain that seemed to veteran politicians of the First State to be about as easily conquerable as Mount Everest.

In challenging the two-term Senator Cale Boggs, Biden entered not only with a strong tailwind of personal confidence but also with a flair for the dramatic that played on his youth. At the Wilmington kickoff, he spoke for more than forty minutes in the manner that would become his trademark, as well as a magnet for much derision. Afterward, he boarded a small Piper Cub and, with young Beau cuddled in his lap in the copilot's seat, took off with two other planes following. The Biden campaign party made a fly-around about the state with the candidate preaching that Delaware needed fresh young blood in its politics and in the Senate in Washington.

Natty as always in a dark pinstripe suit and matching vest, white shirt, and brown wingtip shoes, he struck a professional picture belying his years. While arguing that Boggs should not be returned to the Senate, Biden took care not to speak ill of his venerable opponent. It was in keeping with the spirit of his tiny state in which everyone knew his neighbors and shared their insistence on civility, even in politics.

His platform was unabashedly liberal, even as he took pains to dissociate himself from the label that was in bad odor in the country at the time. "My issues were voting rights, civil rights, crime, clean water and clean air, pension protection, health care, and the war in Vietnam," he wrote later. "That day in 1972 I called for a comprehensive national health care program to protect families from the financial disasters of coping with catastrophic illness. [I said] we can no longer allow the wealthiest nation in the world to be a second-class citizen in providing health care for those who need it. Above all, I said that day, I still believed in the system, I wanted to make it work, and I could be trusted to try."[1]

Among the Democrats who campaigned with Biden in Wilmington was James Baker, who was then running for the city council there. "He was a great speaker," Baker recalled. "When he first started, you know people [later] said he was long-winded? He didn't have that then. I think he acquired it after a time, after being elected. Maybe he got it from other senators, I don't know. But he could raise a crowd up into this excitement about this guy Joe Biden and what he was saying. People liked him. He was very easy to like as a person. He could flash that smile on people and they just took him to heart, because he never had any pretentiousness about him. He was just like you, a normal person coming in; [saying] this is what I believe."

Biden, Baker said, "was also speaking to the times. People were very unhappy with what was happening in our country. There were a lot of things converging all at one time—the antiwar protest, the black power movement was getting so strong because it was pushing aside the integration movement, which had started with Martin Luther King, the NAACP, and all those groups. Now it became more of a rival. Younger people more or less came forward. They wanted change immediately and more rapidly. They did not want to wait this long period of time for change. So you had conflicting arguments going on in the black community between the integrationist movement and the black power movement. And within it there were just many groups, and they didn't necessarily believe the same thing.

"You had Pan-Africanists, you had separatists, you had nationalists; they were all coming into their philosophical beliefs. So people pretty much picked which side they wanted to be on. And shortly after that also began the women's rights movement, and they all came in. Out west you even had the Indian rights movement started. They were happening at the same time, not necessarily cohesively together. They were all just bubbling."

Laughing at Biden's balancing act, Baker said "he always seemed to be on the right side. He could deal with the radicals. It wasn't like he was afraid to go in and talk to people. During 1968, he'd sit down and talk to people who were radical on whatever they believed."[2] But Baker said he never saw him participate in any protest rallies or marches.

In his first week on the trail, Biden told a Young Democrats' Convention in Dover: "Truth happens to be in this year . . . one of the best ways to establish credibility, from a selfish standpoint, is to tell the people something they don't want to hear." He proceeded to tell the young group that he was against legalization of marijuana and, despite his own opposition to the war in Vietnam, against granting amnesty to anyone who fled the country to avoid the draft. And he flatly opposed school busing to combat segregation, calling it "a phony issue which allows the white liberals to sit in suburbia, confident that they are not going to have to live next to blacks."[3]

His observations on trust and truth-telling became a centerpiece of his campaign. At every stop along the way he told his listeners: "You may not agree with me, but at least you'll know where I stand."[4] The campaign made a commercial featuring "trust," which seemed to his Boston political consultant, John Marttila, a real gamble for a twenty-nine-year-old newcomer to make. "Can you imagine Joe Biden at twenty-nine?" Marttila asked with a grin years later. "He probably fired us two or three times and we probably quit two or three times."[5] But Biden persevered on the "trust" theme, hitting particularly hard in his opposition to President Nixon's conduct of the war in Vietnam.

That too was a gamble in light of the fact that Nixon was at the head of the Republican ticket, seeking reelection against a weak Democratic presidential nominee in George McGovern, himself an outspoken critic of the war. But Delaware was a state that was particularly sensitive to the impact of fighting in Vietnam. As Biden wrote in his memoir: "We were being told the war in Vietnam was winding down, but the casualties were arriving at Delaware's doorstep. Every week young American men were being shipped to the mortuary at Dover Air Force Base in body bags. How many mothers lay awake at night wondering how their own sons might return, and wondering what exactly they were risking their lives for?"

Boggs as a loyal Republican wasn't challenging Nixon on the war, and Biden took note of the fact without attacking him directly: "[M]aybe now we needed somebody to stand up to Nixon when he ordered the mining of North Vietnamese harbors, escalating American military operations even as he promised to withdraw troops. I didn't argue that the war in Vietnam was immoral; it was merely stupid and a horrendous waste of time, money, and lives based on a flawed premise. We were spending a billion dollars every two weeks, jeopardizing our entire international posture, and spending so much energy in Southeast Asia that we had left truly vital interests unattended. The president kept talking about how American honor was at stake, which I resented. If that was all we were now fighting for in Vietnam, I said, I couldn't see how that was worth anybody else giving up his life."[6]

Meanwhile, the mobilization of young Delawareans was in full swing. Henry Topel recalled how his son David would leave the house telling his mother: "I'm going to a board of directors meeting with Joe Biden." The elder Topel took it upon himself to check out what was going on. "I drove around to all the precincts and all the voting areas in the state," he said, "and I was never so surprised to see young people manning the phones—an army of youth. I shook my head in amazement."[7]

Typical of the recruits was Rich Heffron, a young Delawarean who had no political experience or involvement at the time but wanted to get his toe in. On hearing of Biden's long shot challenge to Boggs, he recalled later, "I thought, I don't think he's going to win, but that would be a good campaign to get involved with. I went and volunteered. I didn't know him at all. I just knew he was twenty-nine years old, was on the county council, and was running for Senate." Joining Biden on the trail, "we would campaign up and down the state. He was the Energizer Bunny. He'd never stop. If you went to a high school football game on Saturday morning, he was there. If you went to the Acme, he was there. If you went to the Delaware football game in the afternoon, he was there. He would go to those polka dances in the old Polish section of town. He'd shake hands. He had that smile, that grin; he was everybody's best friend. He would have the campaign staffers to his house. I guess we thought we were the Kennedys. We played touch football and swam in the pool. We were a bunch of kids. We were too stupid to think we couldn't win the race."[8]

Unopposed for the Democrats' Senate nomination, Biden served up fiery rhetoric at the party convention in Dover in late June, focusing on the war. "For God's sake, stop the Madison Avenue, sugar-coated garbage," he thundered. "Don't talk to us about a generation of peace when every day hundreds of planes cut through the skies of Indochina, and countless women and children and old men run from their liberators, their flesh burned with napalm—while the soul of America rises in torment and a generation of Americans believe that 'foreign policy' means only body counts and rubble in what were once peaceful hamlets."[9]

But the farthest he would go in criticizing Boggs on the war was to say the Republican incumbent "underrates" what the Senate could do to stop it. "After a bit of conventional hand-wringing about how terrible the war in Vietnam has been," Biden noted in a prepared statement later, "Mr. Boggs added only that he wished he could have done more toward ending the war. The fact is, Mr. Boggs has done everything he could do to keep the war going" by continuing to vote for the military appropriations financing it.[10]

A week after the sensational break-in of the Democratic National Committee headquarters at the Watergate complex, Biden also charged that the Nixon Justice Department "does not believe in the Bill of Rights and claims the right to tap the phones of law-abiding citizens, but fails to enforce the laws against . . . John Mitchell's Committee for the Reelection of the President."[11]

Eager to engage Boggs in a television debate, Biden offered to pay all costs for it, but Boggs remained aloof, as polls showed him to be an easy winner. By mid-September, Biden reported raising and spending more than $50,000 for his underdog race, the most of any Democratic candidate in Delaware. Meanwhile, the cruising, practically hibernating Boggs, with a larger campaign treasury, had spent only a shade more than $3,000.

Brother Jimmy Biden was in charge of fund-raising, and it was a tough task considering the candidate's youth, low profile, and inexperience. At first, Joe questioned whether his younger brother could handle the job, Jimmy recalled later, but Neilia told him: "Have Jimmy do it—and then leave him alone," and that was what happened.[12]

Valerie said one question Jimmy often encountered was "why take our money and put it in a black hole of Delaware when it could make a difference in California or someplace like that?"[13] One answer was that it took a lot less money to win a seat in little Delaware with the right candidate, and young brother Jimmy sold the argument hard that they had the right one in this state. By this time, with the Boggs campaign in unknowing slumber, Joe Biden continued creeping up,[14] and polls indicating that fact were being used with some success to lure contributors.

Complicating the task was Joe's stiff-necked attitude about being involved himself in asking for money, and a firm refusal to make commitments in return for contributions. When, sometime after Labor Day, Jimmy wrangled an offer of $5,000 from a leader of the machinists' union, Bill Holayter, the labor boss insisted that he meet the candidate first.[15] According to Biden later, the union leader asked him, had he been in the previous Congress, "how would you have voted on the SST [the supersonic transport plane]? And while you are at it, how would you have voted on bailing out Lockheed?"[16]

Jimmy Biden recalled later: "Joe looks at me like he's ready to kill me. He says to Holayter: 'Can I call you Bill? Bill, you can take that five thousand and shove it up your ass!' And he walks out of the room." Jimmy caught his brother at the elevator and urged him to come back. "Joe said, 'I want an apology,' " Jimmy recalled, and when they returned and Holayter tried to hand the check to the candidate, "Joe refused to take it. I grabbed the check and said, 'Thanks, Bill.' "[17]

Later in the campaign, Biden wrote, "when I began to show strength and it looked as though I might win, thirteen multimillionaires from my state invited me to cocktails. The spokesman for the group said, 'Well, Joe, let us get right to it. You are a young man and it looks as if you may win this damn thing, and it appears we underestimated you. Now, Joe, we would like to ask you a few questions. We know that everybody running for public office feels compelled to talk about tax reform, and we know that you have been talking tax reform, particularly capital gains and gains for millionaires by consequence of unearned income.' Then one man leaned over, patted me on the knee in a fatherly fashion and said—as if to say it was just among us—'Joe, you really don't mean what you say about capital gains, do you?' Again, I knew what was the right answer to that question [and it] was worth $20,000 in contributions. I did not give the correct answers . . . and accordingly I received no money."[18]

As for himself, Biden had no hesitation approaching the Democratic Senatorial Campaign Committee and other major Democratic contributors, though they had no idea who he was. Nordy Hoffmann, the head of the senatorial committee at the time, recalled Biden's visit later in an oral history interview:

> *He came in with his brother and he sat at my desk. He said, "I'm going to run for the Senate in Delaware." I had this poll. I knew a lot about this young man, but I wanted to find out if he had guts. So I really taunted him for the first fifteen minutes: "What makes you think you can win?" "Why should you be chosen?" "This costs money to do this." "We don't have all the money in the world, and I sure don't have it." I did this for about twelve minutes. Finally, he looked at me and said, "I don't have to take this crap from you." I said, "Senator [sic], we are going to go for you. That's what I wanted to know." He said, "What?" I said, "I just wanted to see if you had any guts. . . . That's exactly what I wanted to prove, and now here's why I said it," and I showed him this poll. He couldn't believe it. That was Joe Biden. He came through with flying colors. But he was going to tell me he wouldn't take my money and the hell with me, but I only wanted to know one thing: I wanted to know if he had the guts to turn me down, because I'm the guy with the money. He did, he told me he didn't have to take that.*[19]

Biden got the senatorial committee money.

Hoffmann and other Democratic moneymen provided Jimmy Biden with leads and on his own he hopped on planes to as far away as Alaska to talk to major Democratic contributors. Many gave, he said, "because I was there, and I asked. He was my brother. In sales, you have to believe in your product, and I was a believer. How were they going to say no?" He convinced many of these wealthy Democrats, he said, that in what it would cost to elect a U.S. senator from Delaware, they would be getting a bargain.[20]

Hoffmann recalled later: "When Joe Biden started, they didn't have a lot of money. . . . Delaware is not a big state. In order to do what they wanted to do, they would run mimeograph machines and they'd mimeograph all this stuff, and then they would take it by car to whatever areas they were going [to] around the state, and these people would hand-deliver it to every house. That saved mailing costs and everything else. This was all done by dedicated young workers. The only thing it cost was for the gas and money for paper. We would borrow or steal a mimeograph machine from somebody, and that's the way we prepared the information. That was probably one of the best operations I had ever seen, that Joe Biden had up in Delaware."[21]

Marttila had a young man named Mark Meyer in Boston who prepared small tabloid "newspapers" touting the little-known candidate, with text written by Marttila and Biden's old law school friend Roger Harrison, who moved to Wilmington to help. Valerie Biden's children's crusade then distributed them every Saturday morning from late September to election day to 350,000 doorsteps across the state.

The night before, Valerie remembered, "one of our guys would drive a rented U-Haul to Boston to get them and drive back in the middle of the night, and by six o'clock Saturday morning he would come into New Castle County [Wilmington]. We would have all the other vans here and they would take the hundreds of pieces they would need" to specific neighborhoods.[22] Parents would open their garages at daybreak to receive the bundles, sorted by prearrangement, and local kids, fortified with doughnuts and orange juice provided by the campaign, would walk or bike-ride their routes. Harrison called it "the Biden post office."

"It couldn't have been done in a large state," Marttila said much later. "But Delaware is one congressional district. It was a unique opportunity. We were all young and stupid. We were not intimidated by anybody. We were feeling our oats, living off the land, and flying under the radar the last three or four weeks." As the polling gap narrowed, he said, "We had the attitude we could do anything. The campaign had a certain rhythm. We weren't crazy, but we thought we could win."[23]

Meanwhile, Biden himself pounded the pavement without letup. Bobbie Greene, Neilia's roommate at Syracuse, came to Delaware to help in the campaign. She recalled later how Neilia would walk smiling behind Joe, prompting him about people coming toward them, so that he could make a personal reference to them about their families. "She would remember small things," Bobbie said, helping him provide the Biden touch in the event he might have forgotten the persons approaching.[24]

In late September the two candidates did hold a debate of sorts before a small local civic association, but few sparks flew. They even agreed in opposing a constitutional amendment that would bar busing in public schools to achieve racial balance. Biden said that while he agreed that busing was warranted to combat segregation by law as imposed in the Deep South, he was against busing to deal with de facto segregation based on residential patterns, as was the case in Wilmington. Boggs said he was opposed on grounds it was inappropriate to rush to an amendment as the solution to problems as they arose. Biden's position enabled him to cast himself outside the liberal mode and hence be more acceptable to a broader constituency.[25]

In early October, as Biden continued to narrow the gap between himself and Boggs, the Republicans began to show signs of concern.

The GOP chairman in Biden's home county of New Castle accused the Biden campaign of showing a "phony" union bug, or identifier, on a piece of its campaign literature in "a very crude attempt" to indicate falsely that it was using union labor. The accusation was obviously designed to erode Biden's organized labor support and to paint him as dishonest. It turned out that some of the flyers were printed out of state with a different union bug, somewhat smudged. Biden shot back, in reference to the Watergate scandal: "In view of their extensive experience with 'bugging,' I'm surprised that the Republicans can't tell a real union bug when they see one."[26]

As Biden steadily crept up on Boggs in the polls, a few local political observers began to muse about the supposedly impossible. One headline on a column in the *Wilmington Journal* proclaimed: BIDEN ACTS AS IF HE CAN BEAT BOGGS: "SACRIFICIAL LAMB" MAY REWRITE SCRIPT.[27]

The Republicans also labored hard to tie Biden to McGovern in light of their shared opposition to the Vietnam War. But Biden pointed out significant differences, including his opposition to blanket amnesty for Americans who fled to Canada to avoid the draft. And he made a point of saying the youth volunteers showing up for him all over the state were his own recruits, separate from the young McGovernites, who were encountering resistance in conservative parts of Delaware.

The Biden campaign got a late helping hand when Biden called his old Syracuse Law School friend Jack Owens, who had moved to Pittsburgh and become a close aide to Pennsylvania Governor Milton Shapp. Joe asked Jack whether he could get Shapp to come into Delaware and campaign for him in the large Jewish community that didn't know Biden well. Jack arranged it, and Shapp spoke at a large reception at the Hotel du Pont in Wilmington. During the course of the night, Owens remembered later, he or Joe said to Shapp, "It's unfortunate but there's a big Jewish wedding over at the Brandywine Country Club tonight," meaning they might otherwise have drawn a bigger crowd. Thereupon Shapp said, "Let's go," and he, Joe, and Jack drove right over to the club. "They were thrilled to have the first Jewish governor of Pennsylvania there," Jack recalled. Joe would go on to win a large share of the Jewish vote.[28]

"With about ten days to go it was about a dead heat," Marttila said, citing polls by Caddell.[29] As election day approached, the Biden and Boggs campaigns continued to joust over radio or television debates. Through it all, Biden continued to speak well of his opponent's personal qualities. "I don't think [Ralph] Nader or anybody else can find anything unethical about Senator Boggs," he said at one point. "He's just a very ethical guy." While "I don't think the senator has any intention of misleading the public," he said, he wasn't so sure about Boggs's advertising agency and might need to question their ads. Boggs returned the observation, telling the Wilmington Kiwanis Club shortly before election day: "I like [Biden] very much, but I can't say the same for some of his advertising."[30]

But Biden's ads in mentioning Boggs at all were for the most part very easy on him. One Biden flyer late in the campaign said:

> *What keeps Senator Boggs from fighting unfair taxes? No doubt Senator Boggs would be pushing for fairer taxes if he truly believed we needed them. But Senator Boggs is not pushing. And he has a perfectly honest reason not to. Senator Boggs believes our taxes are fair. When there was talk of closing one of the big capital gains loopholes, Senator Boggs resisted. When big business asked for $8 billion more in new tax breaks last year, Senator Boggs went along with it. Senator Boggs doesn't see what we're all complaining about. He believes we should be satisfied. One of his campaign leaflets says, "Senator Boggs's legislative proposals to stem inflation and expand employment has led to much of our record economic gains." This sounds like a man who believes we've already solved our problems. He can't help us.*[31]

The flyer was part of the massive literature drop distributed by the young Biden volunteers. "It was like setting up a newspaper route every Saturday in the entire state," said Ted Kaufman, a key organizer of the effort with Valerie Biden. "You had to go out and organize by towns. It was a county council campaign scaled up."[32]

Meanwhile, the parrying about a television debate continued. In the Sussex County seat of Georgetown, Biden said that he was very angry "for the first time" at Boggs over refusing to debate—and then added that if he were the incumbent senator "maybe I'd do it too."[33] The Delaware "culture of civility" of which his sister spoke remained alive and well. Biden took to wearing a Boggs campaign button that said I LOVE CALE pinned inside his jacket lapel and playfully turning it outward when he met a Boggs supporter.

Finally, in a debate before the Jewish women's organization Hadassah, Boggs showed up late and neither candidate really engaged the other. According to Biden's memoir, each was asked for his position on the Genocide Treaty against mass annihilation of peoples, approved by the United Nations in 1948 and at the time not yet ratified by the United States. Biden wrote that Boggs apologized that he wasn't familiar with the specifics and would have to get back with an answer. Whereupon Biden, according to his own account, "knew the answer cold" but elected not to show up Boggs because "it would have been like clubbing the family's favorite uncle."[34]

Joe Biden's stealth campaign was now beginning to register on local Democratic leaders who had not previously paid it much mind. About two weeks before the election, he got a phone call from a conservative state senator, Curtis Steen of Dagsboro, just west of the seaside resort of Bethany Beach, inviting him to campaign with him there. At a rural Sussex County rendezvous, Steen arrived in a big Lincoln Continental to find Biden in a beat-up, tinny Chevette. Steen was troubled, especially by a report that the young candidate was, of all things, a Catholic. Biden readily confessed, but added that his wife was a Presbyterian. Local author Celia Cohen in her book *Only in*

Delaware reported Steen's reply: "We'll just have to go with that."[35]

Five days before the election, the *Wilmington Morning News* ran a story on page one quoting a young former Biden campaign volunteer saying the candidate had tried to get him to write an intentionally misleading position paper on Israel to court Jewish contributors. Biden irately denied the allegation and said his accuser was "either stupid, emotionally mixed up, or a downright liar." Victor E. Livingston, a student at the Columbia Graduate School of Journalism and a summer intern at the *Wilmington News Journal*, said he refused to write the paper "because he felt he would have been 'helping to sucker Jews out of their money.' " In the discussion also involving other campaign aides, Livingston said, Biden agued that the only solution in the Mideast was for a negotiated settlement between the United States and the Soviet Union and the internationalization of Jerusalem, both of which Israel opposed. Biden flatly denied he had advocated either position and in fact opposed both, as did Boggs.[36] It was a one-day story that didn't seem to arouse Delaware voters.

Throughout the campaign, the Biden team realized that its candidate's tender age of twenty-nine gave the Republicans one obvious target. They emphasized the heavy experience of the sixty-two-year-old incumbent and former governor Boggs. One day when Valerie accompanied her brother to the Senate for a meeting with Senate Majority Leader Mike Mansfield, she spied a portrait on the wall of the reception room. "Henry Clay!" she called out, a lightbulb going on in her head. She remembered that Clay had also been under thirty when he was first elected to the Senate, a fact Joe could use thereafter to counter his age as a barrier.[37]

Money was such a problem down the homestretch that Biden had to take out a $20,000 bank loan on the family's new house in order to keep radio ads on the air. Then, at the eleventh hour of the campaign, a serendipitous circumstance occurred. In the final days, Valerie remembered, the Boggs campaign had suddenly "awakened to the fact that Biden was real. They were going to run a full-page newspaper insert the Sunday before the Tuesday election and we didn't have any money to counter whatever they had. The Republicans were like the German army; they knew how to do things. They would roll through like a Nixon landslide."[38]

But just then, because of a sudden newspaper strike over the final campaign weekend, the insert on the virtues of Cale Boggs never got to the voters, and Joe Biden pulled off the greatest upset in Delaware political history with a victory margin of 2,986 votes. The final tally was Biden 115,528, Boggs 112,542, positioning the twenty-nine-year-old winner to become the youngest member of the U.S. Senate.

Beyond Boggs's long campaign hibernation, Biden was helped by a couple of other factors. One was that the 1972 election was the first permitting eighteen-year-olds to vote. The second was Delaware's abandonment of straight-party voting by virtue of pulling down a single lever in the voting booth. Its elimination avoided Biden's being buried in the overwhelming loss of Democratic presidential nominee George McGovern in Delaware.[39] (Many years later, he would joke to political audiences that he had been elected to the Senate for the first time "on McGovern's coattails.")[40]

In his victory remarks at the Hotel du Pont that night, he exhibited humility and the same respect for the man he had beaten that he had shown throughout the campaign. He paid tribute to Boggs's "twenty-six years of faithful service in public office" and called him "a real gentleman," saying "I'm proud to have campaigned against a man like him."[41]

When Boggs in good grace called and conceded by telling Biden, "You ran a good race," the winner expressed sorrow that the loser had to be his distinguished elder.[42] There is a Delaware tradition called Return Day two days after the election in which victors and vanquished ride together in a horse and buggy in through the center of Georgetown, the Sussex County seat. Joe tried to beg off in deference to Boggs, claiming he had caught a bad cold. But the retiring senator insisted, telling Biden: "Joe, I went every time I won, and I'd be proud to ride with you in this loss."[43] And so they did, a visual evidence of the spirit of goodwill that has long reigned in the First State.

The lead story in the *Wilmington Journal* the morning after the election said Biden had naturally "drawn a strong youth vote. But that alone could not have elected him. In order to swing it, the young Democrat had to pull from older voters of both parties," including inroads into Boggs's customary stronghold in the rural downstate.

Biden's campaign manager, sister Valerie, had her own take on the outcome. "I attributed it," she said years later, "to the parents who said to us, 'Anybody who could get my child out of bed at six o'clock on Saturday morning [to deliver the tabloids] before their football game at ten o'clock in school; to do this, I've got to take another look at this person, because there must be something good about him.'" At the same time, she said, she and Neilia Biden, both teaching school at the time in Republican areas of Wilmington, spread the word in their own networks where, Valerie said, no parent in that school had ever voted for a Democrat.[44]

Still another element in their success, Kaufman said, was that the Biden family operation under Neilia and Valerie was so efficient that the candidate himself had full confidence to leave the nuts and bolts of the campaign to them and spend every waking hour meeting voters. As a candidate, Kaufman said, "Joe never came to the headquarters. He spent all his time on the hustings."[45] On the road, Neilia often drove the car and Joe sat beside her, jumping out at every red light and running back to shake the hands of Delawareans in the car behind, then racing back and climbing in next to Neilia before the light changed.[46]

When Boggs finally had agreed to run, word had soon gone out, Hildenbrand later wrote, "that he was only running to keep the seat. Then he was going to resign and the governor was going to appoint [Wilmington Mayor Hal] Haskell. That's all they [the Democrats] needed to hear. That, plus Cale's reluctance to do any campaigning, since his heart wasn't in it. He didn't want to come back down here [to Washington]. Those things, plus Joe, who was young, had a beautiful wife and kids, was articulate, was a fresh face on the horizon. Cale had been before the electorate for what, seven times statewide. He was old hat. All of those factors played into the ultimate defeat of Cale. And also Nixon never did one thing to help him."

Boggs and at least three other Republicans seeking reelection repeatedly asked the president "to go and help them," Hildenbrand said. "He would not do it. He flew over Delaware on his way from Rhode Island, where he was helping [John] Chafee, to North Carolina where he was going down to help Jesse [Helms], and wouldn't even sit down for an airport stop for Caleb Boggs. . . . And that didn't help, because Cale didn't lose by that many votes. . . . [T]hat was Nixon's style. Nixon didn't believe he needed anybody really but himself. He certainly didn't think he needed anybody in the Congress. That was just [his] attitude."[47]

(According to Hildenbrand, Boggs earlier had gotten better treatment from a Democrat, Robert F. Kennedy, in his previous reelection race. "Cale was on the [Senate] floor one day, sitting in his seat, and Bobby Kennedy came over and sat down next to him," Boggs's former aide recalled. "They talked for a while and after it was over, Cale came back and said that it seemed that the Democrats in Delaware were trying to get Bobby Kennedy to come up to Delaware to do a fund-raiser for Cale's opponent. Cale said that Bobby came over to him and said: 'Cale, I'm under a lot of pressure to go to Delaware and raise some money for your opponent.' He said, 'I don't know what I should do. What do you think I ought to do?' Cale said: 'Well, Bobby, it's really up to you. If you go there, I hope you won't say too many bad things about me. But you have to do whatever you think you need to do.' He said that Bobby thought for a minute and then said, 'Oh, hell, Cale, I'm too busy. I don't think I have time to go to Delaware.' And he did not go. At that time, in those days, the biggest draw that the Democrats had was Bobby Kennedy, but he did not go in. He made some excuse and they never were able to get him into Delaware.")[48]

Election night at Wilmington's showpiece Hotel du Pont was particularly sweet for Valerie. The previous June, she later recalled with a Cheshire cat grin, "I went to the Hotel du Pont just to rattle things up" and reserved the ballroom for a highly unanticipated Biden victory party in November. When the Republicans dropped by routinely in September to arrange for the traditional GOP election night celebration, she said, "and they found it was booked to Biden, they went nuts!"[49]

Biden's uncommon zeal and ambition apparently were recognized from the start. On election night, Henry Topel recalled, "we were all at the Hotel du Pont and I get a telephone call from Vance Hartke of Indiana, in those days the chairman of the Senate campaign committee. And he says, 'Henry, I understand your man just won. What kind of a guy is he?' I paused for a moment, and Hartke says: 'You mean he wants to be president?' I busted out laughing."[50]

About a year later, Topel recalled, he was at a party dinner where former governor Elbert Carvel told one of the U.S. Senate elders, John Pastore of Rhode Island, "Senator, we just sent you a nice young man from Delaware." To which Pastore replied, "Yes, he's a nice young man, but he cannot be president yet!" Biden by then had reached the ripe old age of thirty-one, four years short of the constitutional requirement for the Oval Office. Was Biden telling people already that he had his eyes on the presidency? "Possibly," Topel said with a broad grin many years later. "Joe's very honest, and he's a kid at heart."[51]

As for himself, Biden in a sense declared himself independent of ideological rigidity and party discipline. "My obligation is not to the Democratic Party, not just to the people who voted for me," he said. "My obligation goes far beyond that. It goes to the people who got involved, and maybe believed a little too much in me. I hope I don't let you all down. I may go down and be the lousiest senator in the world. I may be the best."

In any event, he told a local reporter on election night, "I won't be toeing any party line or listening to the majority leader when I don't agree. I'm in a unique position. I'm thirty years old, the youngest senator down there. I might be able to sit there for another forty years if I'm a good boy and play my cards right, and level with my constituency, and I'm not going to jeopardize that for party wishes."

He closed by saying "I've reached my ultimate goal."[52] But anyone who had seen Joe Biden's swift and meteoric rise was unlikely to believe that on this historic night. For the new junior U.S. senator from Delaware, and for his beautiful and politically astute young wife, Neilia, and their three small children, the future looked limitless as they approached a new life in the nation's seat of power.

SIX

THE DREAM SHATTERED

ACTUALLY JOE BIDEN had to await his thirtieth birthday thirteen days after his election to reach the constitutional age of eligibility, and a few weeks after that to be sworn in as a member of the first session of the 93rd Congress. But that birthday on November 20 was a joyous milestone for the senator-to-be and his picture-book young family. A big party was held at the Pianni Grill in downtown Wilmington, where he had acquired his first strong taste of politics at the weekly meetings of the Democratic Forum. He and Neilia cut the cake before a barrage of news cameras, amid much talk about Joe's impending swearing-in and service as the Senate's youngest member.

Buoyant and confident, Biden wasted no time familiarizing himself with the lay of the land, going down to Washington in advance and introducing himself—even to an elevator operator in the Old Senate Office Building. Turning to an accompanying Delaware reporter, Curtis Wilkie of the *Wilmington News,* Biden observed: "At first they're going to think I'm a page." He stuck out his hand, grinned at the operator, and said: "Hi, I'm Joe Biden, the new senator from Delaware." Wilkie subsequently wrote: "Biden's Irish forebears might call it 'cheek,' but there is a Yiddish word that best describes it: 'chutzpah.'"[1]

There was much more to do before the opening of the session on January 6—finding a place to live in Washington (while keeping the new house in Wilmington), choosing schools for the boys, staffing the Senate office, and, of course, getting ready for a big family Christmas in Delaware. In the interim, Biden spent long hours in Washington interviewing prospective staff members from a rush of fifteen hundred applicants who wanted to work for this new young sensation. He screened many of them personally, asking why they wanted to work for him and warning them he was not among the party's flaming liberals. And he pointedly declared his independence, noting the state party had given his campaign nothing and that he had gotten the nomination because nobody else wanted it.

Often, Neilia went with him to Washington to look for their new home away from home. Neilia's father had offered to provide the down payment, and the young couple found a small colonial house on Chevy Chase Circle in the District of Columbia's fashionable Northwest section, near an appropriate school. On December 15 their purchase offer was accepted, with closing set for the middle of the following week. Then Joe and Neilia returned to Wilmington for a quiet weekend.

On Monday morning, Joe boarded the commuter train at the Wilmington station for the eighty-minute ride to the nation's capital, to do more staff interviewing. Riding with him was a Republican friend, Mike Harkins, who was going to Washington to help his new boss, incoming congressman Bill Cohen of Maine, select his own staff. They talked about having dinner that night but decided to wait until after the first of the year because Joe wanted to look at another house in Georgetown before closing on the one he was about to buy. Neilia had planned to accompany Joe that day but decided to remain in Wilmington with the children, getting ready for Christmas there. Later in the morning, after having breakfast with Jimmy Biden, she set out with the three kids in her Chevrolet station wagon on a shopping trip that would include buying the family's Christmas tree.

Joe as a senator-elect had not yet been assigned his own office. So, joined by Valerie, he was sitting in one provided by Senator Robert Byrd in midafternoon when a phone rang. It was Jimmy calling. He asked for Val, and she picked up the receiver. She listened a moment and then hung up. Joe Biden in his memoir wrote: "She looked white. 'There's been a slight accident,' she said. 'Nothing to be worried about. But we ought to go home.' "

He described his reaction: "Was it something in the way Val's voice caught? Something in the way she set her mouth? What I felt was something jarring, something stronger than a premonition. It was a physical sensation, like a little pinprick at the center of my chest. I could already feel Neilia's absence. 'She's dead,' I said, 'isn't she?' "[2]

The stark details were these: Neilia Biden had piled sons Beau, four, and Hunt, three, and daughter Naomi, thirteen months, into the vehicle and had driven off to Tim's Corner, an intersection of Valley and Limestone roads near the town of Hockessin in suburban Wilmington. She was pulling out when a tractor-trailer loaded with corncobs came rolling down Limestone on her left and plowed into the side of the station wagon, sending it spinning into a ditch and demolishing it. The truck skidded and flipped on its right side. The driver pulled himself out of the cab and stumbled to the car. Rescuers extricated the four Bidens and rushed them to the Delaware Division of the Wilmington Medical Center, where Neilia and the baby were pronounced dead on arrival. The boys, Beau with a broken leg and other injuries, Hunter with head injuries, were treated and admitted in fair condition.

Joe and Val Biden were flown to Wilmington. "All the way up," Joe wrote in his memoir, "I kept telling myself that everything was going to be okay, that I was letting my imagination run away with me, but the minute I got to the hospital and saw Jimmy's face, I knew the worst had happened."[3]

The wrecked car left sorrowful evidence of the victorious Senate campaign that now receded in importance. Campaign literature was strewn across the road, and index cards bearing names and telephone numbers of voters meticulously gathered by Neilia cluttered the inside of the crushed vehicle. Ironically, she had first planned to accompany her husband to Washington that day to be with him for a dinner of the American Newspaper Women's Club at which he was to be honored. But she had decided to stay home and make the trip the next day for the closing of the contract on their house on Chevy Chase Circle.[4]

When the accident occurred, the driver of the tractor-trailer, Curtis C. Dunn, forty-three, was headed home to Kaolin, Pennsylvania. Two days later, Delaware Deputy Attorney General Jerome O. Herlihy issued a statement absolving the driver of wrongdoing, saying there was no evidence he had been speeding, drinking, or driving a truck with faulty brakes.[5]

A shattered and numb Joe Biden began a vigil at the bedside of the surviving boys. He was given sedation, later describing in his memoir the next few days: "I felt trapped in a constant twilight of vertigo, like in the dream when you're suddenly falling . . . only I was *constantly* falling. In moments of fitful sleep I was aware of the dim possibility that I would wake up, truly wake up, and this would not have happened. But then I'd open my eyes to the sight of my sons in their hospital beds—Beau in a full body cast—and it was back. And as consciousness gathered again, I could always feel at least one other physical presence in the room—and there would be Val, or my mom, or Jimmy. They never left my side. I have no memory of ever being physically alone."

Neilia's parents from upstate New York and their two sons also arrived to join the Biden family, as did Bobbie Greene from Syracuse. "Joe was totally devastated," she remembered. "He was a wreck, but he pulled himself together as best he could." She immediately pitched in with Valerie to attend to the necessary details accompanying the tragedy.[6]

Joe reflected on "how despair led people to just cash it in; how suicide wasn't just an option but a *rational* option. But I'd look at Beau and Hunter asleep and wonder what new terrors their own dreams held, and wonder who would explain to my sons my being gone, too. And I knew I had no choice but to fight to stay alive."[7]

The bereaved husband and father slept on a couch in the boys' room and left the hospital only to attend the private family burial of Neilia and Naomi, and two days later a memorial service at Saint Mary Magdalen Catholic Church. To the surprise of the nearly seven hundred attendees, he spoke about the unimaginable loss he had suffered. "The night before [Neilia] died," he said according to the account in the *Wilmington Journal*, "she was writing Christmas cards. We were both in the living room in front of the fire and I was sitting in my lounge chair, a pompous young senator thinking about the big things I was going to do in Washington." His wife, interrupting his pensive mood, said, "'What's going to happen, Joey? . . . Things are too good.' We had known something was going to happen. We had decided not to have a fourth child because of a fear that something would happen to it. . . . We had three beautiful children. Now I have two."

The account had him telling his hushed listeners: "Please don't be sad. I'm pretty proud of her. . . . She had a principle—she treated everyone the same, and that worked both ways. Those who were poor, black, minority, affluent or socially esteemed, she made no distinction among them. . . . I was probably one of those phony liberals . . . the kind that go out of their way to be nice to a minority, and she made me realize I was making a distinction. But in dealing with minorities, she made no subtle condescending gestures . . . she made no distinctions. I'm going to be that way. . . . I'm going to try to follow her example." He closed by quoting John Milton: "I waked, she fled and day brought back my night."[8] And with that he hurriedly left the altar. Then he stood in the rear of the church and greeted those who filed by.

It was not at all certain that Biden, in light of the tragedy, would accept the Senate seat he had won in November. At first, he informed the Senate majority leader, Mike Mansfield of Montana, that he would not be taking it. He continued to stay at the hospital with the boys. In the next trying days over the Christmas holidays, Mansfield and Senator Hubert Humphrey of Minnesota repeatedly called the senator-elect, Mansfield urging him to reconsider his decision and Humphrey inquiring about him. "He just wanted to know how I was doing," Biden wrote in his memoir, "but usually he never got past Jimmy, who was screening my calls and knew I didn't much feel like talking to anybody outside the family. Jimmy was also talking to Delaware's governor-elect, who would have to appoint a new senator."[9] Among the calls that did get through at the hospital, according to Roger Harrison, was President Richard M. Nixon, who offered his condolences.[10]

Although the doctors soon reported that both boys would make full recoveries, their father's mental condition as he continued his hospital vigil remained in doubt. "I began to feel my anger," he wrote. "When the boys were asleep or when Val or Mom was taking a turn at their bedside, I'd bust out of the hospital and go walking the nearby streets. Jimmy would go with me, and I'd steer him wordlessly down into the darkest and seediest neighborhoods I could find. I liked to go at night when I thought there was a better chance of finding a fight. I was always looking for a fight. I had not known I was capable of such rage. I knew I had been cheated of a future, but I felt I'd been cheated of a past, too." Even his strong Catholic faith was shaken. "I felt God had played a horrible trick on me, and I was angry," he wrote.[11]

Pat Caddell, who had forged close personal ties with both Joe and Neilia as a near contemporary, recalled later the combination of grief and anger that gripped and paralyzed the young widower. He joined the counsel that advised him to put off the critical decision to give up his Senate seat until he could pull himself together.[12]

His recent law partner, David Walsh, said later that it would have been against Biden's whole being to just give up when his sons needed him, especially with "the fact that he had this great family that just kind of wrapped themselves around him during that period of time. I know there were a couple of situations where he was very, very angry, not just emotionally upset. He had this perfect wife and perfect family, and all of a sudden . . . these things happened to other people, they didn't happen to him. But he's got great instincts and judgment and knows what to do at the right time. When it happened, he just dealt with it. His family is the most important thing in the world, and that's where he went to get through it all."[13]

In any event, Mansfield would not give up on bringing him to the Senate. He called the hospital every day to convince Biden he needed the new senator and his vote on key committee assignments. The majority leader put him on the Senate Democratic Steering Committee, an unheard-of appointment for a freshman that smacked of Mansfield's cunning as well as his compassion.

Sister Val also recognized the catharsis that going into the Senate could be for her brother. "We were on the top of the world on November 9 and then on December 18 the world ended," she recalled, and for Joe "immediately it was like, 'Let me out of here. I don't want anything; I'm taking care of my boys.' He wasn't going to go. But he had to get up and keep moving because of Beau and Hunt.

"I moved in with him, just exactly what my brothers would have done for me; nothing heroic. But he had to do things. I mean, he was talked into it, but it wasn't that they twisted his arm; it made sense. He had good counsel. He had friends who cared about him and said, 'Look, you and Neilia worked too hard for this. Give it time. Just try it, and if it doesn't work we understand your priorities.' Once he realized that his priority was in being a good father and then a good senator, he could achieve that. He could achieve one without doing harm to the other, then he was okay with it. His whole focus was on the kids. But then he threw himself in, when he did get down there, and was able to live a little bit. The boys were okay and he put one foot in front of the other."[14]

On the following Friday, January 5, 1973, Joseph Robinette Biden Jr. was administered the oath of office of the U.S. Senate by Frank Valeo, secretary of the Senate, in the small chapel of the medical center in Wilmington. Beau was still bedridden with his fractured left leg in traction, but the bed and boy were wheeled into the room. His brother, Hunter, released from the hospital after treatment for a slight skull fracture, was brought into the room by his grandfather, Neilia's father, and sat on Beau's bed to watch the ceremony.

Dorothye Scott, an administrative assistant to Valeo, accompanied him and recalled later: "We went into the room where the little boy was and his leg was elevated. Senator Biden's wife's handbag was down there on the floor, and oh, my goodness, it was such a scene. They opened up a double reception room with doors that slid open, and Senator Biden's mother and father were there, and [Neilia's] mother and father, and it was very touching when Frank swore him in."[15]

The room was crowded with witnesses and television cameramen as Biden took the oath, twice. Because his back was to the cameras of the three national networks and the Philadelphia station, a prominent NBC correspondent, Peter Hackes, asked that Biden and Valeo switch places and repeat the swearing-in, which they did. Then Biden spoke.

"I hope that I can be a good senator for you all," he said. "I make this one promise: If in six months or so there's a conflict between my being a good father and being a good senator, which I hope will not occur . . . I promise you that I will contact Governor-elect [Sherman W.] Tribbitt as I had earlier [who would by law fill the vacancy] and tell him we can always get another senator, but they [his sons] can't get another father."[16]

Biden recalled much later that one of his thoughts in the depth of his despair and depression about the loss of Neilia and Naomi was to reconsider the idea of becoming a priest. He went to the local Catholic bishop in Wilmington, he remembered, "about getting a dispensation. From the Catholic Church you could be married and have lost your spouse and have children, and you can get a dispensation to go to the priesthood. I didn't ask him to get it, I asked if he could, would he, et cetera, and he said, 'Look, Joe, why don't you take a year to think about this? I don't think this is the right thing for you, but if you still want to do that, I will initiate the procedure.' I think I was just, I don't know, it was a lousy time, and I never followed up on it. . . . It just went away. It was the only other thing I ever thought about, but it was obvious I didn't have the vocation or I would have done it." Then, laughing, he added, "Girls got in the way."[17]

So began what soon would be a familiar daily sight to train passengers on the run between New York and Washington, in both directions—the Amtrak life of Senator Joe Biden of Delaware.

"Commuting was not a burden," his sister said. "Taking care of the boys, they needed each other. It was a touchstone. He'd come home to the boys, he'd get up in the morning, get them awake, take them to school and drop them off. And Aunt Val would pick them up and take them home; he'd come home and if you weren't there to tuck them in, he'd be there to kiss them when he walked in the door and put them to bed. That was his core. It was much easier than if we had been in California. He was able to be here, to be part of their world, to see their games and go to their plays and do the parent-teacher meetings. Delaware afforded him a great opportunity to do that, and Delaware embraced him and the boys."[18]

The earliest memories of Biden's elder son, Beau, was of his father "jumping in bed with us, being held and kissed" no matter how late at night he arrived home, and in the mornings being dropped off at school by him on his way to the train to Washington. He and younger brother Hunter "were aware that he had an important job," Beau said much later, "a meaningful job he cared about and that had impact. . . . I remember his entire focus was us and everything else was second. Never in my life did I doubt it for one second."[19]

On the first occasion Biden stepped into the Senate chamber as a duly sworn member, looking his tender age, he was stopped by a Capitol policeman. "Senator Biden, do you remember me?" he asked. Biden confessed he didn't. Whereupon the officer reminded him of a morning in 1963 when as a student at Delaware Biden had for the first time entered the empty chamber. He was an unescorted solitary sightseer and was taken into custody for a short time. "I'm the fellow who stopped you ten years ago," the Capitol cop said, grinning. "I'm retiring tomorrow. But, Senator, welcome. I'm happy you're back."[20]

The schedule that Joe Biden now faced would have been a heavy load for a senator of whatever age. For a young man of only thirty, with no experience at all in the ways of Washington nor with the national agenda, not yet remotely recovered from personal tragedy and laden with uncommon personal family obligations, it was another mountain to climb.

SEVEN

A FRESH START

THE U.S. SENATE had already been in session about a week when Joe Biden finally arrived to take his place on the bottom rung in terms of seniority and experience at the national level. He was greeted warmly by his new colleagues, but also with a certain awkwardness from some who felt ill at ease in how to deal with his well-publicized personal tragedy. Mike Mansfield, who had pressured him hard to take up his seat in the Senate, kept a protective eye on him, to the point of asking Biden to drop by every week to discuss how he was doing. Or, as Biden put it, "to take my temperature."[1]

Democratic giants like Hubert Humphrey and Ted Kennedy took him under wing, with Humphrey including him early on a Senate delegation to a conference at Oxford, and Kennedy dragging him to the Senate gym for more casual conversations with some of the other leading members of both parties. On his own, Biden also took the opportunity to introduce himself to fellow senators, always properly deferential as a thirty-year-old novice in a body of experienced gray hairs. Among them was John Stennis of Mississippi, chairman of the Armed Services Committee and an unyielding defender of segregation. As Biden related the encounter in his memoir, when Stennis asked him why he had run for the Senate, he blurted out: "Civil rights, sir." Whereupon Stennis simply smiled and replied: "Civil rights? Good . . . . Good. Good. Glad to have you here."[2]

For some reason, Biden got along well with many of the southern Democrats, who still asserted considerable power in the Senate through many of the most important committee chairmanships in the days when seniority was the sole determinant. After he made his maiden speech on the Senate floor, Stennis sent him a note. "I watched you today as you took the floor," it said. "You stood tall—like a stone wall. Like Stonewall Jackson."[3] But the other Mississippi senator, James Eastland, chairman of the Judiciary Committee, did not first express any such sentiment. When Mansfield a bit later assigned Biden to discuss before the Senate Caucus a tough campaign finance proposal he was advancing along with freshman senator Dick Clark of Iowa, Eastland was not pleased.

The bill sought public funding for all congressional elections, with challengers receiving ten percent more than incumbents to compensate for their obvious advantage. The proposal met with silence in the caucus until Eastland spoke up. "They tell me you're the youngest man in the history of America ever elected to the U.S. Senate," he said, still chomping on his cigar, as remembered by Biden in his memoir. "Actually, I was the second youngest ever elected," Biden wrote, "but it didn't seem a good time to correct him." Eastland continued: "Y'all keep making speeches like you made today and you gonna be the youngest one-term senator in the history of America."[4]

In time, however, Biden smoothed things over with Eastland to the point that he would often sit in the old lion's lair after a committee meeting and probe his senatorial wisdom. On one occasion, Biden asked him what was the most significant thing that had occurred in all the time he had been in the Senate. Eastland looked at him and said: "Air conditionin'." As Biden recalled the reply, the old man went on: "You know, Joe, before we had air conditionin' all that recessed lighting all used to be great big pieces of glass like in showers. Come around May . . . that darn sun would beat down on that dome, hit that glass, act like a magnifying glass and heat up the Chamber, and we would all go home in May and June for the year. Then we put in air conditionin', stayed year round, and ruined America."[5]

During the time Mansfield was "taking his temperature," Biden encountered another Dixie demon of the Senate, Jesse Helms of North Carolina, as he walked across the Senate floor en route to Mansfield's office. As Biden related the incident years later, Helms "was standing in the back excoriating [Kansas Senator] Bob Dole for [sponsoring] the Americans With Disabilities Act." Steaming, Biden strode into Mansfield's office. "Joe, looks like something is bothering you," Mansfield said. "Mr. Leader, I can't believe what I just heard on the floor of the Senate," Biden told him. "I can't believe that anyone could be so heartless and care so little about people with disabilities. I tell you, it makes me angry, Mr. Leader."

Whereupon, Biden recalled, Mansfield told him: "Joe, what would you say if I told you that four years ago, maybe five, Dot Helms and Jesse Helms were reading, I think the *Charlotte Observer*, the local newspaper, and they saw . . . a piece in the paper about a young man in braces who was handicapped at an orphanage. He was in his early teens. All the caption said was the young man wanted nothing more for Christmas than to be part of a family. What would you say if I told you Dot Helms and Jesse Helms adopted that young man as their own child?" I said, "I would feel like a fool, an absolute fool." What he learned from that incident, Biden told his fellow senators later, was never to question anyone's motives, because you can always find good in "those who are willing to look for the good in the other guy, the other woman. . . . This approach allowed me to develop friendships I would never have expected would have occurred."[6]

Hoping, unrealistically for a freshman, to land a seat on the

Foreign Relations Committee, Biden in his first Senate days went to see J. William Fulbright, the committee chairman. Fulbright suggested he talk to John McClellan, his gruff fellow Arkansan who headed Appropriations, where he could have a say on foreign policy. Encountering him in a Senate dining room, Biden introduced himself and got a chilly reception. McClellan, in reference to Biden's recent losses, matter-of-factly told him of losing his own wife to spinal meningitis in his first term in the House, a son in the same way eight years later, and then two other sons. He told him to bury himself in work as the only recourse.[7]

But the advice was easier said than done. Biden persevered but found himself constantly thinking of his losses and of his own boys. They had obvious apprehensions about his daily departures for Washington and about his return no matter how much he reassured them, or how diligent he was being home every night to put them to bed. Their father had mobile phones installed in his car so they could reach him anytime and told them they could go to work with him anytime they chose. Together or separately the boys would frequently ride the train to and from Washington with him, attending hearings, hanging out in his Senate office, or playing in the Senate gym. The staff, Hunter Biden recalled, became a sort of extended family for them.[8] When the boys stayed in Wilmington, senators and their wives repeatedly invited Joe to dinner; he went sometimes not to offend, but more often hustled home by car or train to be with Beau and Hunter.

Beau long after remembered his father telling him of one occasion when a major oil executive came to call as the boy was spending the day with him. "I was sitting in his office with him and being a good boy, drawing or coloring on a book, and they were having small talk. Then it's clear the executive wants to get on with the meeting and says, 'Well, can we begin?' [My father replied] 'Yeah, let's go.' " But the visitor gestured toward Beau and said, "Well, when your son leaves, I guess." Whereupon the new senator from Delaware told him: "He's being a good boy, he's not doing anything. . . . He won't cause a problem." As Beau told it, "Well, they went back and forth and to make a long story short, the meeting never happened, and I was the one who stayed and the executive left." The episode, he said, illustrated how the boys were "his total focus. Now, again, we weren't allowed to sit there and be precocious kids who interjected in a meeting or wanted to cry. If that happened, he wouldn't allow for that. I don't want to pat myself on the back, but we were pretty good kids. We knew the drill. . . . You were hanging with Dad, and you let Dad do his work."[9]

Through this most difficult time, Biden's younger brother Jimmy also provided a close continuing family presence, often accompanying him to Washington for the day. Joe's old Syracuse Law pal Jack Owens by this time had moved to Wilmington to join Joe's law firm, and he spent long weekend nights with him, going to a movie and then sitting up late talking into the early hours. Some weeks after Joe entered the Senate, Jimmy and Jack took the fledgling young senator Joe on a skiing trip in Vermont, and Joe expressed an interest in moving there with the boys to make a fresh start. But the Wilmington ties apparently remained too strong, and its closeness to Washington made the daily commute feasible.[10]

Among those who commiserated with Biden was fellow Democratic Senator Birch Bayh of Indiana, whose wife, Marvella, had been diagnosed with breast cancer in 1971. Her illness led him to abandon his quest of the presidency in 1972, as Joe and Neilia Biden were busily campaigning for the Senate. Marvella Bayh subsequently died of the disease at age forty-six, leaving her husband to raise their young son, Evan, much later to become governor of Indiana and then a U.S. senator.

"I was one of those who talked to Joe about the tragedy that had befallen him," Bayh recalled long afterward. "I said, 'Nobody can fully appreciate what you're going through, but I know she [Neilia] would want you to continue. She's no longer here. You can't do anything but grieve, but it will be easier being a senator than not doing anything.' " It was a question, Bayh said he told Biden, of whether he was "willing to find a way to get this tragedy behind him." But Biden, he recalled, "was angry. He couldn't understand it. Why him?" Bayh told him: "I remember losing my mother at age twelve. My father couldn't understand it. She was forty." Recovery for those left behind, he said, "was a slow process."[11]

William Bader, chief of staff of the Senate Foreign Relations Committee at the time, recalled that in Biden's first years in the Senate, "Joe was distracted. He was not genuinely comfortable with what he had decided to do. The getting home every day; all the conflicts and internal things in leaving the Senate at a time his presence could have been important. Some senators thought, really, he should get over it and get on with it, for his sake and the Senate's sake, though God knows they couldn't have been more sympathetic. There was more than one senator who lived with that kind of pain, being away too long or whatever, and Joe turned this into something that was almost ritualistic about it. The Senate was a place where you had a strong responsibility to your colleagues to be there for major issues, and the Amtrak regime, if you want to call it that, [produced] a sort of mixture of understanding and not understanding. In those first days, he was not considered by many senators as 'reliable.' "[12]

At the same time, Democratic Senator Patrick Leahy of Vermont, who was elected two years after Biden's arrival in the Senate, recalled that there was widespread sentiment for his young colleague's trial. Leahy said he and other young senators also had small children. "His desire to get home to the boys was very, very real," he said. In the Senate chamber, Leahy remembered, "you could see him pacing the floor when it was late in the day, looking at his watch. I don't know what time the train left, but say it was leaving at 6:10, and he could make it if he voted by 6:01." Rather than resenting Biden's attitude, Leahy recalled, "it was a question of admiring him for doing that."[13]

Through this trying period, Jimmy Biden continued to be at his tortured brother's side with the assistance of concerned senators like Hubert Humphrey, even when Joe was persuaded to join trips on essential committee business abroad. "You've got to get him out of here," Jimmy recalled Humphrey urging him. "I'll cover for him." On the trip to the annual British conference at Ditchley, Humphrey arranged for Jimmy Biden to hook up with his brother and then go off on their own to various European destinations, through Belgium, France, Portugal, Spain, and Italy. "We were just killing time," Jimmy recalled, as Joe wrestled with his loss and contemplated the future. Other times, Jimmy would tag along on his own in an unofficial capacity. "I became basically a Senate wife for the first year," he jested later, traveling and rooming with his brother while Valerie tended to the boys' needs at home.[14]

"It was so Joe didn't have to go to an empty room every night," Valerie explained. "Jimmy was his confidant. He was a morale booster, a sounding board for Joe. And he will always deliver the message, if Joe wants to hear it or not. And he can't get rid of him, and Joe values that. He can't fire him. He's his brother."[15]

Jimmy taking Joe in hand was another case of the Biden sense of family responsibility, demonstrated earlier when Joe, not yet in politics, helped get young Jimmy, the smart but somewhat rambunctious and unfocused sibling out of Mt. Pleasant High School, into the University of Delaware. Joe one day simply took him to see the dean of the college of agriculture. Though he had no interest whatever in farming, Jimmy recalled later, "it was the easiest one to get into," and Joe got him enrolled. Thereafter he moved into the college of arts and sciences and thrived.[16] That was the Biden way, and for Joe especially, it extended beyond the family tie. Later, as he regained his bearings, Joe resumed his practice of regularly putting in appearances at baptisms, weddings, bar mitzvahs, and funerals as if he were the whole of Delaware's parish priest or rabbi.

As the youngest member of the Senate, Biden was given seats on two fairly unglamorous committees: Public Works; and Banking,

Housing, and Urban Affairs. With proper deference, he knuckled down to yeoman tasks on each. In his first week, he was back in Wilmington on the doorstep of 517 Vandever Avenue, a house in a deteriorating neighborhood being homesteaded, with a boyhood friend, Mayor Tom Maloney, who had started the nation's first urban homesteading program. Biden announced plans to introduce his first bill, authorizing the Federal Housing Administration to turn over abandoned houses to cities for similar rehabilitation.[17]

But it didn't take long for him to plunge into the national discussion. He was one of only three senators to vote in committee against Richard Nixon's second-term nomination of James T. Lynn as his secretary of housing and urban development. And he called on his colleagues to slash the president's budget for the White House in half in retaliation for Nixon's cuts in housing spending. "If he wants to play the game, we'll play the game," he told a reporter for the *Wilmington Journal*,[18] demonstrating the same feistiness that marked his oratory in the senatorial campaign.

In those first weeks and months in Washington, however, Biden was never able to shake the grim personal prelude to his new political responsibilities. Early each morning, he faithfully boarded the Amtrak train at the Wilmington station, and each night he raced to Union Station at the foot of Capitol Hill to head back to his boys. (One morning years later, when he arrived late at a Judiciary Committee hearing, he observed to the audience that "I keep telling my colleagues—if they fully funded Amtrak, I would not be late. . . . And some suggest that's why they don't fully fund Amtrak!")[19] Religiously, he attended to his political obligations as well at local meetings and dinners around his small state.

Nor did he in his grief forget his friends in Scranton. A few days after his election, and days before the accident, Jimmy Kennedy had called him to ask a favor about the approaching Saint Patrick's Day. "Joe," he told him, "I don't belong to the Friendly Sons of Ireland or whatever the hell it is, but they need a speaker. You're an Irish kid with an upset victory, you're the guy." Without hesitation, Biden replied: "I'll be there." Then came the family tragedy, Kennedy recalled, and he heard that the new senator was saying he wasn't going to take the job. The organizers of the dinner were fearful they would be without a speaker, until Biden confirmed to Kennedy: "Yeah, I'll be there."

Kennedy remembered: "So that was announced, and of course because of all the emotion and tragedy attached to him, the place was a sellout—there's no women, an all-male deal. I remember he's walking in. It looks like *Gunfight at the O.K. Corral.* His dad's next to him and three uncles. They're all walking in together and sit down at his table. And he was electrifying. There was an amazing amount of curiosity and sentiment for him because he had just lost his wife and his daughter. So he spoke, a long speech with three standing ovations." He never mentioned what had happened to them, Kennedy said. Of Biden's demeanor that night, he said: "His old man would say as a kid and throughout his life: 'If you get knocked down, get up.' " That indeed was Joe Biden Sr.'s repeated message to his eldest son, prior to the tragedy and when he would suffer subsequent personal blows.[20]

(After Joe Jr.'s election to the Senate, his sister Valerie later told the *New York Times*, their father in a demonstration of empathy "gave up car sales and went into real estate. He didn't want a United States senator to have a used-car salesman for a dad."[21] But it was that particular occupation, so often the brunt of jokes, that had enabled his son to commute so often and so cheaply to Syracuse in his days of courting Neilia.)

Back in Delaware in Biden's first months in the Senate, he also went to a covered-dish supper by the Sussex County Democratic Committee in Georgetown, the rural county seat. He eschewed the stage with the governor and other high officials for a seat on the floor, and listened to them as they addressed the crowd of more than four hundred of the party faithful. The master of ceremonies suddenly introduced a "special, special speaker," and Biden went to the stage amid thunderous applause.

Eyes moist, he fought back emotion. "I very seldom have trouble speaking, as you know," he began, "but this is where we started." He was referring to the first stop on his Delaware fly-around nearly a year earlier upon his declaration of Senate candidacy. As tears streamed down his face, he halted and reached for his handkerchief, as many women in the crowd joined him in crying. "My wife had a great deal of respect for you all," he began again, but broke off, head down, and ducked behind the stage curtain.

It was Biden's first breakdown in public since Neilia and Naomi had died. As he composed himself out of sight, Beau and Hunt Biden played nearby in the church hall nursery. After a few minutes, their father reappeared and finished his speech, lauding his late wife for her role in his election.[22] A short time later, after a brief trip to Syracuse, where he and Neilia had their first home, he wrote to law school classmate John Covino: "I don't particularly like coming back 'home' because I just think too much, and when I think, it doesn't feel good."[23]

Joe tried to put Neilia's death behind him, but he increasingly talked about it now, giving rise to some criticism that he was trying to capitalize politically on what had happened. In an interview with Wilmington reporter Norm Lockman months later, he sought to explain. "I can't help it," he said. "I can try not to talk about it, but if it comes up, then I can't be talking about Neilia without *talking* about Neilia. I can't help but be personal. I mean, I just can't treat her memory like some campaign worker who is gone and it's all over with. I'm trying to work it out."

The effort, he suggested, probably accounted for occasional outbursts of anger. "All I know is I've been looking for a fight. It's been like, 'Go ahead, smack me, give me a reason to let go.' " He recalled about how as a new senator he was in New Orleans for a speech at Tulane University, accompanied by brother Jimmy, when he narrowly averted a street fight.

"We were walking the street late at night talking about Neilia and looking for a place to eat," he said. "These four guys were coming toward us, taking up the whole street, looking for trouble, and for a split second it flashed through my mind, 'Take 'em on.' We banged into each other. Nothing had happened, nothing had been said, but, you know, we were ready. I'm no fighter, and I'd probably have gotten the stuffing kicked out of me.

"At that moment, a New Orleans policeman walked around the corner, and as soon as I saw him, it clicked: What the hell am I doing? I'm a United States senator letting my emotions get to the point that I'm willing to take on four toughs on a side street in New Orleans just to let the frustration out."[24]

Biden's attitude did not go unnoticed by some in the Capitol. Frank Valeo, who had administered the oath of office to him in that Wilmington hospital room, recalled later:

> *You can't put yourself in the shoes of somebody who's been through a tragedy like that. . . . But whatever the reasons, he came in with a chip on his shoulder, and Mansfield did everything to assuage him. He showed him all kinds of special attention and privilege. Again, I think, [he was] trying to neutralize the effect of the tragedy in some small way. But it seemed to have no effect. Biden would continue to go his own way and be skeptical of everything that came up in connection with the Senate. He was very critical of most of the things that happened in the early caucuses and in some of the luncheons that Mansfield had for the younger members. Then I began to realize that probably, it had nothing to do with the tragedy. This was his characteristic. This was his nature.*

Having said that, Valeo said of Biden: "He's bright, he's able." And he recognized compassion in him. Compared to some other senators, he said, "I think Biden knows a little bit more about poverty and economic anxieties, partly because he comes from Wilmington and that's an industrial city where space is limited and wages aren't that high, and there's unemployment."[25]

As Biden settled into the routine of the Senate, his competitive spirit seemed to revive. At a testimonial dinner for Boggs, the man he defeated, the new senator praised him profusely as a gentleman and worthy opponent. Then he turned to Republican congressman Pete du Pont, challenging him to try to take the Senate seat away from him, though it would not be up for another five years. Biden had promised Mansfield to try the Senate for six months before deciding whether to stay, but already he was giving indications that the place was growing on him.

"I'm probably going to stay the whole term," he told a Delaware reporter in early June. "My boys are doing well. Things are better than can be expected at home," he said. "I don't want to say anything definitely. It was probably a mistake to set a time on it. Some people think I have no right to impose my personal grief on the public. . . . They're probably right."[26]

One of the major reasons things were better at home was that the Biden family by now had been enlarged by one member through marriage. And that one was Jack Owens, whom Neilia and Joe had once tried to fix up with Valerie on a blind date in Syracuse. It had set off the wrong kind of sparks at the time, but eventually they turned into the right kind, and she and Jack got married. When Jack first proposed, Valerie said she couldn't marry him because she could not leave Beau and Hunter, who were in her care. Joe solved the problem by—as was his fashion—adding a wing to his house and having Val and Jack move in.[27]

As the youngest member of the Senate and one of the youngest of all time, Biden soon became the recipient of a host of speaking engagements, which older colleagues, including Hubert Humphrey, urged him to accept. One was to join other senators at a major Cook County Democratic fund-raiser in Chicago at which Humphrey was billed as the main speaker. When the legendary Mayor Richard J.

Daley introduced the head table, he referred to Humphrey "and the rest of the senators who came from Washington with him." The brash young Biden, rising to say a few words, turned to Daley and wisecracked: "Actually, Mr. Mayor, Hubert Humphrey came out here with me. I didn't come out here with Hubert Humphrey."

That crack fell flat, especially with Daley. So Biden decided he might as well go for broke. He told the gathered Chicago pols how lucky they were to have him, and Daley especially. The stone-faced mayor finally laughed, and then his obedient fiefdom with him, whereupon Biden made his own little speech. Humphrey, impressed, remarked with admiration: "Ah, to be young again, and to be able to speak like that!"[28]

In the Senate, meanwhile, Joe also was beginning to get his bearings. Without challenging Jesse Helms's motives, he didn't hesitate to square off against him over the issue of a Senate pay raise—a subject not commonly engaged in by a freshman. Biden said he didn't disagree in terms of economic conditions that it might not be the best time, but he took issue with what he saw as Helms's argument that the senators didn't deserve more than their $42,000 annual salary. Biden himself pointedly had no stock investments, as most others did, and he empathized with colleagues who had to maintain two homes, one in their state and one in Washington, although he himself had no residence in the capital.

In floor debate, Biden argued, "It seems to me that we should flat-out tell the American people we are worth our salt. The American people would understand because they are a lot smarter than we give them credit for. . . . I do not think many of the visitors sitting up there in the public gallery or outside the Capitol feel that they want people in the U.S. Senate who are not worthy of a high salary."[29] At the same time, he called for eliminating all outside income to members of Congress.

Especially from freshman senators, who tradition held should be seen and not heard in their first term, Biden's remarks were surefire headline-grabbers. The *Evening Journal* in Wilmington blared on its front page: "I'M WORTH MORE MONEY," BIDEN TELLS COLLEAGUES; $42,000 NOT ENOUGH.[30]

The story got coverage far beyond Delaware. In New Hampshire the archconservative publisher of the *Manchester Union Leader*, William Loeb, ran an editorial on its front page: "The voters of Delaware who elected this stupid, conceited jackass to the Senate should kick him in the rear to knock some sense into him, and then kick themselves [for] voting [for] such an idiot." Biden had it framed and hung in his Senate office.[31] He wasn't quite a household name yet, but he was being noticed.

Much of the new senator's first year in Washington was dominated by the investigation of the break-in the year before of the Democratic National Committee headquarters at the Watergate complex by a group of Cuban exiles hired by the committee to reelect President Nixon. Biden had little to say about it. But when Nixon, in a desperate effort to save his presidency, forced the resignation of his attorney general, Richard Kleindienst, Biden had one opportunity to weigh in. Nixon's defense secretary, Elliot Richardson, was drafted by Nixon to take over the embattled Justice Department, and Biden was one of only three Senate Democrats who voted against his nomination. Explaining his vote, he offered that "the impartiality of the attorney general is essential, and I'm afraid Mr. Richardson's extremely close ties to the Nixon Administration unfortunately have tainted the possibility of his establishing the impartiality and gaining the confidence of the American people."[32]

The next night, Biden warned attendees at the annual dinner of the New Democratic Coalition of Delaware not to make the Watergate case a partisan matter. "Let's not hang him before we have a trial," he said. "If we [prematurely] hang Watergate around the Republicans and the people buy it, the system goes under."[33] And he questioned the televising of the Watergate hearings. As a former defense lawyer, he said at a Delaware town meeting, he feared anyone indicted in the case might be able to claim legitimately that he had been denied a fair trial, and if any are guilty "I want to see them nailed."[34]

As for Richardson, six months later Biden had reason to reflect on his judgment when the new attorney general, in what came to be known as the "Saturday Night Massacre," resigned rather than carry out Nixon's order to fire his nemesis, Watergate special prosecutor Archibald Cox.

On domestic issues, Biden demonstrated surer footing. Near year's end, he was approached by a group of truckers complaining about increasing fuel and freight rates varying from truck stops and states over their long-distance hauls. So on a Sunday night he doffed his usual natty attire, put on old clothes, climbed into the cab of a rig, and made a fourteen-hour run from Washington to a small town outside Cincinnati, Ohio. En route, he got stuck in a truck stop blockade and had a captive audience for a speech sympathetic to their woes.[35]

The year concluded with an announcement from the New Castle County Department of Parks and Recreation that construction would begin the following spring or summer on a new neighborhood area of ten and a half acres near Prices Corner, in Wilmington. At a cost of $126,000 appropriated by the council on which Joe had once served, it would feature a football field, a Little League baseball field, two basketball courts, bicycle racks, and playground apparatus for little kids, and would be named Neilia Hunter Biden Park.

EIGHT

FINDING HIMSELF, AND JILL

AFTER ONLY A year in the U.S. Senate, Joe Biden was already catching the attention of appraisers of the Washington scene. In late January 1974 he traveled to Mobile, Alabama, where he was honored at the annual awards dinner of the Junior Chamber of Commerce as one of America's Ten Outstanding Young Men.[1] Obviously it was his tender age, and perhaps his widely reported family tragedy, that brought him the recognition rather than anything notable he had achieved as the Senate's youngest member.

Around the same time, he was listed in the *San Francisco Chronicle* among the ten best-dressed men in the Senate. It was a distinction less surprising to his constituents in Delaware, who had already taken note of his crisp sartorial style, if only in comparison to many of his older and rumpled colleagues.

Not surprisingly, such credentials put Biden as a widower in the social columns on the list of the Senate's most eligible bachelors, though he seemed to show no interest, still grieving the loss of his wife and infant daughter. What had started as his commitment to Senate Majority Leader Mansfield to give the Senate a six-month try had by now become open-ended. His near-daily routine of commuting between Wilmington and Washington when the Senate was in session continued, with his dedication to his young sons as resolute as ever.

In early 1974, however, United Press International and the Wilmington newspapers reported that Biden had started dating Francie Barnard, a strikingly attractive twenty-seven-year-old Washington reporter for the *Fort Worth Star-Telegram*, published by her father, igniting rumors of an impending marriage.[2] Biden and the young woman quickly denied the rumors, though the story flared anew a few months later in a long article in *Washingtonian* magazine. In it, he was highly complimentary of her, but he was quoted as telling the writer, Kitty Kelley, "I can look you straight in the eye and say that I have no present or future plans of getting married." He went on: "Besides, why would someone like Francie marry a guy like me who is still in love with his wife, who has a political constituency and a readymade family? She deserves better than that." She was quoted in the article saying that she was "very fond" of Biden but wasn't interested in getting married, either.[3]

In any event, as the new year unfolded, he found himself essentially in the Senate's shadows as the presidency of Richard Nixon faced increasing likelihood of impeachment. Elliot Richardson, the attorney whose impartiality had been questioned by Biden, had overseen a Justice Department investigation into bribery charges that led to the resignation of Vice President Spiro T. Agnew. Richardson engineered a plea bargain with Agnew, whereby he gave up his office and his position as next in line to succeed the beleaguered Nixon in return for escaping conviction and imprisonment.

Then came Richardson's own resignation after his refusal to fire special Watergate prosecutor Archibald Cox. The deputy attorney general, William Ruckelshaus, also refused and quit. So the unpleasant deed was then done by the department's solicitor general, Robert Bork, a figure who later would cross Biden's path in circumstances critical to the futures of both men. But the Watergate investigation went on, under a new special prosecutor, Leon Jaworski, who proved to be just as tenacious as Cox had been.

As for Biden, he continued to caution his fellow Democrats, and the press, to proceed in a way that would assure Nixon a fair hearing that would not in any way jeopardize the ultimate administration of justice in the matter. From the Senate floor in mid-April, he said, "the American people have a right to expect members of the judicial and legislative branches of the federal government to conduct themselves circumspectly under whatever duress they may labor. The press, too, should be expected to perform with restraint at this critical period in our history. Impeachment is too important a matter to be left to the press, legitimate as its objectives are."[4]

At the same time, Biden indicated a certain skepticism toward Nixon's offer to release edited transcripts of the tapes of White House conversations covering the period after the Watergate break-in, demanded by the Senate investigating committee and the special prosecutor. "I don't understand," he said in late April, "since he has abandoned [the claim of] executive privilege, why he just doesn't give up the tapes."[5]

The controversy dragged on, paralyzing the government, until the tape bearing the famous "smoking gun" made clear that Nixon had been aware and privy to payoffs to the Watergate burglars for their silence. Biden joined the chorus of demands that the president resign or be impeached, stating his preference that the constitutional route of impeachment be followed.[6] When Nixon finally chose resignation, Biden's statement declared the outcome was not political, as the departing president's staunchest defenders would continue to claim.

"For the sake of history," he said, "the issue must not be confused. The issue is not how well President Nixon conducted foreign policy over the past five and a half years, but whether the President of the

United States violated high public trust of that office to such a degree as to warrant his forcible or voluntary removal from that office. For the sake of our Constitution and for our children's sake the record must be clear that his resignation was not the consequence of political pressures, but solely as a consequence of a violation of that high public trust. There should be no mistake about that."[7]

Biden expressed his best wishes to incoming President Gerald Ford and pledged his cooperation. But when, only a month after taking office, Ford issued a blanket pardon to Nixon for his Watergate crimes, Biden joined the chorus of outrage. The nation's first unelected president, he said, "is playing blind man's bluff" with the judicial system.[8]

Though obviously interested in this turmoil on the national stage, Biden was determined as a freshman senator to focus on Delaware affairs. Asked in an interview how he appraised his performance so far, he emphasized his service to Delaware and how his daily commuting was enabling him to stay in touch with the needs of its citizens. But he was hardly unaware of the attention that continued to come to him as the Senate's youngest member with an independent streak and flair as a public speaker.

"All I can do is continue to do what I think I should be doing and keep as much contact at home as possible," he said. However, he went on, "one thing that worries me, quite frankly, is that down here [in Washington] there is a tendency to try to make me a national figure—to make me a would-be presidential candidate for 1980, that sort of thing. I want to make sure the people of Delaware realize that my first priority is Delaware. I feel the important thing is for me not to change from the way I was before I was elected. . . . If you hang around Washington, it's easy to start thinking you're important, so it is a blessing in disguise that I commute every day and get out of this city. I do my work and get involved, but I steer clear of the social circuit. I prefer being home with my kids, and that way I'm home with my constituents too."[9]

The situation was a politically serendipitous one for Biden. The commuting underscored his commitment not only to his family but also to the voters of Delaware. It came at the price of being regarded by most of his fellow senators only as a colleague, not a bosom buddy the way others who were anchored in Washington and its after-hour obligations and social engagements were regarded. While other senators spent many nights eating and drinking together in their homes away from home or in various Capitol Hill and downtown watering holes, Biden was regularly absent, taking that train back to Wilmington. But his generally upbeat, avuncular nature kept him on good terms with most of the others. In addition to the father figure of Mansfield, he developed warm ties with several of his seniors like Hubert Humphrey, Ted Kennedy, Fritz Hollings of South Carolina, Birch Bayh of Indiana, and Dan Inouye of Hawaii.

In the same interview, after expressing worry about the "tendency to make me into a national figure" and attributing it to his extreme youth, he went on revealingly: "I have no desire to run for those offices, but I'd be a damn liar if I said that I wouldn't be interested in five, ten, or twenty years if the opportunity were offered. I think it is totally unrealistic that it should be offered. If it were, well, anyone who runs for public office has a desire to affect what happens, and there is no place you can have greater effect than as president. So you're being phony to say you're not interested in being president if you really want to change things. But I'm certainly not qualified at this point. I don't have the experience or background."

Having said the obvious, Biden confessed: "I do want to become a national political figure in the sense that I want to be someone who can affect things in the U.S. Senate. If I return for a second term, I would not want to be 'just another senator'—I would want to be a power in the U.S. Senate. Otherwise, why be here?" As for coming from tiny Delaware, he said size was not so important in the electronic era, and old-time party bosses and machines didn't matter that much, either.[10]

Biden didn't have to worry about becoming just another senator. If nothing else, his tender age, along with a growing reputation as a crowd-pleaser, put him much in demand on the party fund-raising circuit in the first off-year congressional elections of his Senate tenure. The Democratic Congressional Campaign Committee drafted him to speak for House candidates in more than two dozen cities, from Vermont to Hawaii. And as he shuttled across the country and back through the long campaign season, with regular detours to the Senate and to home and family in Wilmington, he gradually coped with the lingering grief that had nearly aborted a Senate career.

As for speculation about a grand future, Biden wrote much later in his memoir that during his first term in the Senate he was talking to a Catholic elementary school class "and one of the children asked me if I wanted to be president. When I started to tell the class I was perfectly happy being a senator and had no plans to run for the White House, I could see a nun at the back of the room stand up. 'You know that's not true, Joey Biden,' she said as she pulled from the folds of her habit a paper I'd authored in grade school. I'd written I wanted to be president when I grew up, she said. So I was caught red-handed, guilty of schoolboy musings. The thing was, I didn't think that sort of paper separated me from other kids. Don't a lot of twelve-year-olds write the same thing?"[11]

Meanwhile, Majority Leader Mansfield, still taking Biden's pulse, had given him the plum he had hoped for upon entering the Senate—a seat on the Foreign Relations Committee. The chairman, William Fulbright, had been defeated in the Democratic primary in Arkansas, and Edmund S. Muskie of Maine had given up his seat to take the chairmanship of the new Senate Budget Committee. Biden had often said foreign policy was his priority interest; the appointment was considered a special prize for someone so young, and a showcase for one of the party's most outspoken rising stars.

In the new arena, Biden proceeded cautiously at first, allowing himself only mild criticism of the Ford administration. One of his early laments was his sense that under Secretary of State Henry Kissinger, President Ford appeared to be conducting "a one-man foreign policy" for a task beyond the capabilities of any one man.

His first encounter with Kissinger, shortly after joining the Foreign Relations Committee, was a memorable one. Biden received an embossed invitation to a briefing for committee members only, with no staff permitted. Thinking the briefing would take place in the committee hearing room, he went there but found no one. He called his office and was told the meeting was in the Capitol itself, in a room with which Biden was not familiar. He arrived late and as he headed for the door, he later wrote, "an armed Capitol policeman grabbed me by the shoulder, spun me around and pinned me against a wall. 'Where d'ya think you're goin', buddy?'"

Biden had to pull out his Senate ID to get by. He raced in, noisily slamming the heavy door behind him, then ran into the back of Kissinger's chair and managed to take a seat on a dais as the witness was winding up. When Mansfield, as acting chairman, opened the floor to questions, Biden was recognized, but as he began to speak, Kissinger cut him off.

"Mr. Chairman," the witness appealed to Mansfield, "I thought no staff was allowed." As Biden recounted the moment, "I could see one of Kissinger's deputies frantically scribbling on a piece of paper, 'Biden, D-Del.'

"'Oh,' Kissinger said, looking at the note. 'I apologize, Senator Bid-den.' His pronunciation wasn't even close. 'No problem,' I said, 'Secretary Dulles.'"[12] Even under those circumstances, he was Joe Biden, smart-ass.

In March 1975 the hectic but lonely life of the young and still grieving senator from Delaware took a sudden, sharp turn. He had just returned to Wilmington from Washington by plane and was walking through the airport when he saw a wall poster for the New Castle County Park. It bore snapshots of various park scenes adorned by a beautiful young blonde. He went home, where the family was waiting for him to go out. Brother Frank gave him the phone number of a "girl" he knew and could call to join them, saying, "You'll like her, Joe. She doesn't like politics." The weary senator begged off.

The next afternoon, a Saturday, he decided to call the number and ask the "girl" to go out that night. She said she had a date. He pressed her. He told her he was a U.S. senator and only in town for the night. Could she break her date? She agreed to try and did. When he called for her that night, he saw she was the young woman in the airport poster. They had dinner in Philadelphia, a short drive from Wilmington, and went to a movie, and that's how it began with Jill Tracy Jacobs, who became the second love of his life.

At twenty-four, she was eight years his junior. "That was sort of refreshing," Biden wrote later in his memoir. "There weren't too very many places I went where people wondered if I was too old. At dinner that night, Jill showed no interest in politics. She didn't ask a single thing about my career, about Washington, about the famous people I'd met. I didn't want to talk about that stuff anyway. So we talked about family and mutual friends in Delaware. We talked about books and real life. That night, for the first time since Neilia, I felt something like absolute attraction—something like joy. And we just kept talking."

They went out again the next two nights but she made it clear to him that she was not looking for a serious relationship. "She'd married young, was separated, and in the process of getting divorced," he wrote. "Jill liked her life as a single woman; she was looking forward to starting her first teaching job in the fall; and most of all she did not want to be involved with somebody in politics, let alone a United States senator. We should just think of this as fun."

That was agreeable to Biden at first, but as he came to realize how strongly he felt about her, he was compelled to push his luck. He phoned her the next day or two and asked her "not to go out with anybody else." As he recalled the conversation, she said, "Okay, I'm willing to try it. But I have a date to the Philadelphia Flower Show next weekend that I can't break." She kept the date, but as he put it, "that was a start."[13]

Jill, a native of Willow Grove, Pennsylvania, and a junior majoring in English at the University of Delaware at the time, had heard around the campus of the young senator from the same school. She had cast her first vote for him but had little interest in politics. On the night in 1972 he was elected, she remembered later, she and her dinner date had stopped by his celebration at the Hotel du Pont, where she shook hands with Joe and Neilia.

Of his request that she not see anybody else, Jill said later, "I thought it was a little bold, but he seemed to have some logical reasons for it—that he didn't want, I guess, to be perceived as someone who was one of many people I was dating. He was pretty hard-hitting. We went out on the date, [and] he was, like, what are you doing tomorrow night? What are you doing the next night? What are you doing the next night? I mean, he was fast. I could tell he was definitely interested in me. So because he was so aggressive, maybe, it didn't surprise me."[14]

Joe Biden and Jill Jacobs continued to see each other often, but with her satisfaction with her single life and aversion to politics, and the fact that he had two young sons, he did not think of marriage at first. But when he finally introduced her to the boys, "they hit it off," he wrote, "and she was happy to include them in some of our dates." As for Jill, she said later, "I felt like I dated three guys. . . . As time went on, I dated Joe for two years. I ate dinner there almost every night. The boys were truly part of my life."[15]

Eventually Jill took Joe—and the boys—to meet her family, including her grandparents. By Christmastime, he wrote, "Jill was already integrated into our lives,"[16] which now included sister Valerie and her husband, Jack Owens, ensconced in a large new Biden family home they called the Station. In a never-ending quest for the ideal homestead, Joe had found and remodeled an old estate mansion that had fallen into disuse in suburban Wilmington.

Son Beau recalled much later: "My dad is a frustrated architect, not only of houses but landscape architecture. One of my early memories of my dad was, either coming home on the train, with a piece of scrap paper designing houses, or on a trip. He'd always have a drawing of a house he'd envision." But for most of his early adolescent years, Beau said, the family occupied the Station, and for twenty years his father never stopped fixing up and improving it.

"Painting and planting trees, that's what we would do a lot of weekends," Beau remembered. "We spent a lot of time with my dad going between the hardware store and the house, either painting or putting a fence around it or planting hemlock tree after hemlock tree. But those are some of the best memories of my life: hopping in the station wagon or a pickup truck my Dad borrowed, or getting a sub and eating it in a parking lot and heading back to the house. . . . Driving around, looking at houses: it was part of dreaming, and home was the central place."[17] The Biden clan, like many Irish families, maintained the old-world tradition of family solidarity and heritage. Joe's parents had housed blood relatives from both sides, the Bidens and the Finnegans, under their roof in Scranton, and Joe in his turn continued it in Wilmington.

One early morning in 1976, the senator was shaving when the boys nervously accosted him. Beau, now seven, turned to Hunter, now six. "You tell him, Hunt," Beau said, as their father later recounted the conversation. "No, you tell him," Hunter said, finally blurting out: "Beau thinks we should get married." Beau took over: "We think we should marry Jill. What do you think, Dad? . . . D'ya think she'll do it?" So their father popped the question, only to be put off. Jill was not sure she was ready to take on parenthood, and she still had that aversion to becoming a public person as wife of a U.S. senator. But he kept asking—at least four more times, she reckoned later—and she kept dodging.[18]

One reason, she said later, was her concern for the boys. "They were so easy to love. They were such great kids and so warm. So that's why when Joe kept saying, 'Will you marry me? Will you marry me?' I really had to make sure it was going to work, because I could not break their hearts if it didn't work."[19]

In the meantime, with the 1976 presidential election approaching, the freshman senator, with typical brashness, began to express concern over what he saw as his party's hesitancy to challenge Republican President Ford, weakened by the Watergate scandal and his pardon of Richard Nixon. Convinced that Senator Ted Kennedy would not run, Biden deplored the paucity of Democratic candidates. In the summer of 1975 he had surprisingly endorsed Pennsylvania's colorless Governor Milton Shapp, who at the urging of Jack Owens had so generously campaigned for him for the Senate three years earlier. Shapp had earned modest acclaim in helping to settle a nationwide truckers' strike, and Biden equally generously called him "one of the most qualified men to be president."[20] Biden later also called on the party to "look at some of the other unknowns" such as former one-term governor Jimmy Carter of Georgia and other southern leaders.[21]

Carter was indeed largely unknown at the time despite spending a remarkable 260 days in 1975 running like a determined tortoise in Iowa, where the first precinct caucuses of 1976 would kick off the quest for Democratic presidential nomination delegates. On the night of January 19, 1976, he jolted the party's political community by running first among the contesting candidates with 27.6 percent of the caucus vote, twice the tally for runner-up Birch Bayh. An uncommitted slate actually finished first with 37 percent, but Carter's showing was more than enough to grab national headlines for the first time.[22]

About three weeks later, Biden disclosed at a meeting of the Milford Chamber of Commerce in Delaware that he had met with Carter in Atlanta, had been asked "to take a major part in his campaign," and was likely to accept.[23] By now Shapp had dropped out of the race and Carter had annexed another victory in the New Hampshire primary. Biden in late March became the first senator to endorse him and agreed to chair his national campaign steering committee.

As the Carter campaign rolled along, Biden took on surrogate candidate chores for him around the country. He was an attractive stand-in, more so as a result of submitting to hair-transplant treatment to cope with his thinning mane. It was the cause of much derision in some quarters, as well as speculation about future political aspirations of his own. As Carter closed in on the Democratic nomination, Biden on occasion would remind crowds that he had been invited to speak only because he was the only member of the Senate who, at thirty-three, wasn't eligible to run for either president or vice president—yet. Carter's election generated some early talk about a cabinet post, but Biden had long before declared that he intended to stay in the Senate and seek a second term in 1978.

From the start, Biden's relations with the Carter administration were often contentious. When Carter nominated former JFK speechwriter and aide Ted Sorensen to be director of the Central Intelligence Agency, Biden, now also member of the Senate Intelligence Committee, had a run-in with the nominee. Sorensen had given an affidavit in the trial of Daniel Ellsberg, who leaked the Pentagon Papers chronicling the American involvement in the Vietnam War, criticizing overclassification of documents. As Sorensen later put it in his memoir, "in the early 1960s, it was common practice in Washington for government officials to take classified documents home to review, that it was not uncommon for some officials to leak classified information to the press, and that documents of far greater importance to national security than those in the Pentagon Papers had been leaked to the press without criminal prosecution."[24] Sorensen elsewhere acknowledged that he had taken White House documents home in preparing his bestselling book, *Kennedy*.

As a consequence of these and other complaints against Sorensen, and behind-the-scenes pressures from Carter, the old JFK speechwriter agreed to have his nomination withdrawn, while declaring he had acted properly. An aide to Biden said the senator had no intention of detouring the nomination, but Sorensen in his memoir later wrote: "The prize for political hypocrisy in a town noted for political hypocrisy went to Joe Biden. On my first courtesy call to his office, he could not have been more enthusiastic, supportive, and gracious, calling me 'the best appointment Carter has made!' At the opening of the hearing, he changed both his tune and his tone, stating: 'Quite honestly, I'm not sure whether or not Mr. Sorensen could be indicted or convicted under the espionage statutes . . . whether Mr. Sorensen intentionally took advantage of ambiguities in the law or carelessly ignored the law.' After listening to my statement of defense and withdrawal, he said: 'Ted, you are one of the classiest men I have ever run across in my whole life.' "[25]

In February 1977 Biden attained another ambition with assignment to the Senate Judiciary Committee, where he became a longtime and staunch defender of civil rights and justice, and foe of constitutional abuse and crime. In another argument with the Carter White House, he sharply questioned the nomination to be solicitor general of a black appellate judge, Wade McCree, who had played a role in a landmark desegregation ruling in support of public school busing into the suburbs. The order later was overturned by the Supreme Court, and Biden said he would oppose the choice unless he got a firm commitment from McCree that he would oppose busing in the Wilmington school system.[26] McCree was confirmed anyway.

(A Biden ally in the fight to resist school busing in northern Delaware was previous solicitor general Robert Bork, the man who had agreed to fire Archibald Cox, in the infamous Saturday Night Massacre of the Watergate scandal. Bork filed the Justice Department brief with the Supreme Court arguing that a U.S. District Court had gone too far in requiring racial balance in the Delaware schools. Eventually, the paths of Biden and Bork would again cross in a momentous Judiciary Committee collision.)

Even in this committee post, which he had aggressively sought out, Biden could be heard at times grumbling at his lot. To a *Wilmington Journal* columnist he complained of getting unsatisfactory assignments from Chairman Eastland and being trapped as an independent between committee factions. "They have conservatives, they have liberals, and they have Biden," he lamented.[27] This kind of talk inevitably nurtured some speculation that he might have had enough by now of the daily commuting and being an outsider in an insider game, and he might decide after all not to run for a second term in 1978.

On the Intelligence Committee, Biden was occupied with cosponsoring and helping to write the Foreign Intelligence Surveillance Act. FISA, as it was known, established procedures to obtain warrants for certain critical domestic wiretapping and other intrusions of privacy that had run rampant during the Nixon years. Joe Biden had a full plate for a first-term senator and widower with small children to whom he rushed home in Delaware nearly every night.

About a year had now passed since he first proposed to Jill Jacobs. Departing on a ten-day trip to South Africa, he finally told her he had to have an answer when he got back. While he was away, he decided that if Jill would agree, he would indeed not run for reelection, though his commitment to the work in the Senate and his rise in the ranks and in public celebrity augured an even more promising political future. And so, upon his return from South Africa, he told her of his decision. "He gave me an ultimatum," she recalled later. "He said, 'This is it. I'm not going to ask you again. I have two boys here. I love you, but my responsibility is to them. This [courtship has been] going on; I want a family,' and it was obvious the boys wanted a family. At that time it was five years after the accident . . . and it was time."[28]

When she indicated she didn't think he meant it, he reached for the phone and told her he was going to inform the chief political reporter of the *Wilmington Journal* that he was not going to run. As he related the story in his memoir, he could hear the phone ringing, "and then I heard a dial tone. Jill had her finger on the phone cradle. She'd cut off the call. . . . She told me later why she did it: 'If I denied you your dream,' she said, 'I would not be marrying the man I fell in love with.' "[29]

Hunter Biden had a different recollection. "My brother and I desperately wanted my dad to ask my mom [as the boys called her almost from the first] to get married," he said later. "The way I remember it, my brother and myself, with my dad standing there, [asked] my mom to marry us, and she said yes. She may well have been asked many times before, and I don't know whether we were the ones who sealed the deal or not, but I remember she couldn't refuse Beau or me."[30]

By now Jill realized she would also be marrying a family with a very lofty goal. Joe's brothers, Jimmy and Frank, took her to dinner. "They told me," she recalled later, "it was a dream of this family that Joe would be president, and did I have any problem with that? I guess I thought it would go away."[31]

Joe Biden and Jill Jacobs were married on June 17, 1977, at the United Nations Chapel in New York by a Catholic priest, with sons Beau and Hunter standing at the altar with them. Only their families and close friends were there—nearly forty of them, and after a luncheon reception the newlyweds were off on their honeymoon in Manhattan—Joe, Jill, Beau, and Hunter. The foursome went to see *Annie* on Broadway, then repaired to Blimpie's for cheeseburgers and back to their hotel and bed. "We took the honeymoon suite, part of it, Hunt and I, and left them the other bedroom," Beau recalled.[32]

Years later, Beau said: "I've been lucky in many respects but no more lucky than in having my mom coming into our life. What I say is that I've been lucky in having two moms. When my mom [Neilia was and still is "Mommy" to her sons] came into our lives, she came not only into my dad's life but she came into my brother's and mine, too. We were there at each of the meaningful moments on the way to reforming our family. I don't think my dad needed any prodding, but we prodded him in that direction. Led by my mom as much as my dad, we rebuilt our family."

From that first night of the marriage of Joe and Jill, Beau said, "we were all together and we've been together ever since." The foursome began a new tradition by going to Nantucket for Thanksgiving weekend. Three years later the Bidens became a quintet with the birth of daughter Ashley, who, Beau said, "lit up my dad's life."[33] For now, the Joe Biden family was squared away in plenty of time for a family reelection campaign just ahead.

NINE

CIVIL RIGHTS, JIMMY CARTER, AND REELECTION

FROM THE START of his political career, Joe Biden had taken special pride in his support of civil rights, and he was rewarded with solid support from the African American community in Delaware. From his days as a student at Archmere Academy, when he walked out of the off-campus hangout, the Charcoal Pit, because a black fellow football player would not be seated there, he had been a foe of racial discrimination. But as noted, he avoided street protest or anything else that smacked of civil disobedience, eschewing its tactics and garb in preference for making his concerns known through established legal avenues.

Ever since the Supreme Court in *Brown v. Board of Education* had ordered desegregation of public schools in 1954, Wilmington's racially segregated school system had been living on borrowed time. Early efforts to thwart the decision got nowhere, and for more than twenty years various plans to implement desegregation had stalled or been rejected by lower federal courts.

For all the uproar in the country over Watergate and the Nixon pardon, the issue was now demanding attention in Delaware, a northern state replete with southern traditions and attitudes, especially in the two more rural counties, Kent and Sussex. As a New Castle County councilman in 1971–72, Biden had expressed early concerns about school busing, which since 1968 had been authorized for the De La Warr District by order of the federal Health, Education and Welfare Department (HEW).

Now as a U.S. senator, his votes on busing were watched diligently by Delaware civic groups. In May 1974 members of the local Neighborhood School Association met with him in his Wilmington office. Some came away critical of a vote he had just cast against an antibusing amendment in the Senate, despite his declarations against the practice. It had failed by a single vote. The coordinator of the group, John Trager, who had not attended the meeting, nevertheless labeled Biden "the number-one phony in Delaware."[1]

Biden, defending the vote, said the amendment in question "would have allowed anyone affected by a civil rights decision from 1954 onward [the year of *Brown v. Board of Education*] to reopen their court case. Ninety percent of those cases had nothing to do with busing. It would have created havoc in our court system. . . . It was one of those political flag-waving things to show the folks, nothing more. I've gotten to the point where I think the only recourse to eliminate busing may be a constitutional amendment."[2]

Biden did in fact vote for another amendment that eliminated the provision for reopening a court case ordering desegregation, while reaffirming the right of federal courts to enforce means of countering racial discrimination in the schools. This amendment passed by one vote, and Biden subsequently acknowledged that "the courts had to be able to stop government-enforced segregation," while he continued to oppose busing simply to achieve desired racial balance.

But in 1975 New Castle County, which included Wilmington, was ordered to create a plan to integrate all middle schools and high schools by September 1977 and elementary schools a year later. The black population was concentrated in Wilmington, and students there would have to be bused to schools in the predominantly white surrounding suburbs, with many of the white students in turn bused to Wilmington schools.

Memories of the riots in the town of Milford in 1954, after the *Brown v. Board* decision, and in Wilmington in 1968, following the assassination of Martin Luther King Jr.—and the long National Guard "occupation" of the city that followed—at first stirred fears of more racial violence accompanying actual school desegregation. Suburban white parents were overwhelmingly opposed to having their children shipped daily for miles into the city, and some black families weren't pleased either at having their kids transported, even to superior schools.

A citizens' antibusing group calling itself the Positive Action Committee began asserting political influence in local elections. But this group largely avoided street protest and in fact by its demeanor helped defuse prospects of violence. At the same time, an appointed Delaware Committee on the School Decision pointedly labored to find a peaceful and harmonious resolution.

Biden, to the chagrin of Democratic liberals and many members of the black community, came down on the side of opposition to the busing of students to achieve integration. In 1975 he introduced and won Senate approval of two antibusing amendments. One prohibited the Department of Health, Education and Welfare from requiring school systems to assign teachers or students to schools or classes on the basis of race. The other barred any federal funds for student busing unless specifically ordered by a court. But both amendments were dropped by House-Senate conferees.[3]

"It was a big vote that became a gigantic vote soon after," Biden wrote in his memoir, "when the federal courts ordered school officials in New Castle County, Delaware, to come up with a plan to desegregate the schools—and then ordered them to make a vigorous cross-district busing scheme the centerpiece of the plan."

Busing, he wrote, "was a liberal train wreck, and it was tearing people apart. The quality of the schools in and around Wilmington was already suffering, and they would never be the same. Teachers were going to be transferred without consultation to new school districts. In some instances they would have to take a pay cut." He noted the county had about two-thirds of the school-age population in Delaware and all of those students were going to be bused "to a new school on the basis of racial balance" when the new term began in September 1978. "Nobody was happy," he wrote. "I kept introducing legislation to try to keep busing as a last resort, to be used only when school districts had worked actively to segregate children by race—de jure segregation."[4]

In an interview at the time with a local paper, Biden was categorical: "I oppose busing. It's an asinine concept, the utility of which has never been proven to me. I took that position, along with Howard Brown, a black candidate for mayor, long before the '72 election [for the Senate]; we were the only Democrats on record as opposing busing. Many people forget that, conveniently, now for their own political reasons."[5]

Biden's position by now had hardened. He made a distinction between de jure integration, required by a court order to end segregation, to which he acquiesced, and de facto integration, motivated by the desire to alter the racial composition of a school absent a court order, which HEW espoused and he strenuously opposed.

At the same time, Biden was aware of the political risk he faced in the civil rights community in which he had a growing reputation as a champion. "The unsavory part about this," he acknowledged, "is when I come out against busing, as I have all along, I don't want to be mixed up with a George Wallace. I don't want anybody to give me credit for sharing any point of view George Wallace has."[6]

The danger, he also wrote, was that frustrated white parents would blame white, liberal civil rights advocates and neighborhood blacks, who in turn would fear that rejection of busing would lead to regression in combating discrimination in other areas such as housing, job opportunity, and higher education.

Biden had joined Delaware's other senator, Republican William Roth, in Roth's proposal for a national commission to study busing that would also require a halt to compulsory busing until the study was completed. In the meantime, Biden said, if busing was ordered at all, it should be the court doing the ordering, not a federal agency. He advocated legislation prohibiting the busing of children beyond their own school district. And in defense of de jure integration only, he came down hard on courts that "have gone overboard in their interpretation of what is required to remedy unlawful segregation."

He said he did not "buy the concept, popular in the sixties, which said, 'We have suppressed the black man for three hundred years, and the white man is now far ahead in the race for everything our society offers. In order to even the score, we must now give the black man a head start or even hold the white man back to even the race.' I don't buy that. I don't feel responsible for the sins of my father and grandfather. I feel responsible for what the situation is today, for the sins of my own generation. And I'll be damned if I feel responsible for what happened three hundred years ago."[7]

In a speech on the Senate floor near year's end, Biden expressed further disillusionment with busing. He called it "a dubious triumph of technique over substance. By and large our children's education suffers," he said, "and our energies are diverted from finding formulas and ways of achieving the goal of fair and open and equal opportunities for all our schools." He said he would be offering antibusing legislation of his own in the next Senate session.[8]

Although Biden had been the first senator to support Jimmy Carter's presidential campaign, when Carter was elected he reminded reporters that "I don't want to be anybody's man" and that he had so told the new president. "When I think he's right, I'm gonna be there and he can ask me to carry his water," he said. "There are going to be places where we disagree, and I would have no hesitation in taking a different point of view."[9] That was the case with the busing issue.

As already noted, long before Carter entered the White House, Biden had been in the thick of the fight on busing, particularly as it so directly affected Wilmington schools. In 1977 he again joined Bill Roth, his senior Delaware Republican colleague, in sponsoring a bill to bar any federal court from ordering busing unless intentional racial discrimination existed to maintain segregation. Specifically, they hoped to head off the start of busing that September in New Castle County. The Senate Judiciary Committee was then in the hands of probusing liberals, and Biden testified for the bill, almost immediately running into criticism from civil rights leaders. Veteran NAACP leader Clarence Mitchell accused Biden and Roth of being motivated by "racism . . . deep in the state of Delaware."[10]

The allegation had Biden seething. "The first time I disagree with the civil rights people on an issue of substance," he told Mitchell, "the question of racism is raised. I thought my record, albeit brief, was enough to get [this discussion] beyond insinuations of racism." The bill also stipulated that no busing order could be executed until all appeals had been exhausted, which—if passed and signed by Carter—would have delayed busing in Delaware in the fall term until the Supreme Court ruled. Mitchell said the bill was unconstitutional, and he was joined by Senator Edward W. Brooke of Massachusetts, the Senate's only black member.[11] Sociologists from Harvard testified that busing was inspiring white flight from inner-city school districts, making them more segregated than before any busing occurred.

Biden and Roth, along with the third member of the tiny Delaware congressional delegation, Republican Representative-at-Large Thomas Evans, took their case directly to President Carter but got nowhere. He concurred with the opinion of his attorney general,

Griffin Bell, that the bill was arguably unconstitutional and unnecessary, because the Supreme Court had already ruled on the imperative of desegregation, and without further delay.[12] Biden held more hearings and the Judiciary Committee finally reported out a bill in September, but the threat of a filibuster by liberals kept it off the Senate floor for the remainder of the session. Busing continued in New Castle County.

This issue and others soured Biden on the president for whom he had gone out on a limb. At a dinner meeting of the Delaware State Chamber of Commerce, he said sarcastically of Carter's first nine months in office: "Nixon had his enemies list and President Carter has his friends list. I guess I'm on his friends list, and I don't know which is worse." He added about the Carter crowd: "They got to Washington and didn't know how Washington worked. . . . The president is learning, but not fast enough."[13] Biden was still a freshman senator, but that wasn't stopping him from running off at the mouth to emphasize his independence. It was, as veteran Delawareans often said, Joe just being Joe.

As Biden's 1978 bid for a second Senate term began, Carter hinted at his coolness. He made a quick and short incursion into Wilmington by helicopter for two fund-raising receptions that together lasted less than two hours. At the first, about sixty couples paid a thousand dollars each to snack, drink, and have their pictures taken shaking hands with the president at the Hotel du Pont. It took thirty-six minutes of Carter's time. Then he was rushed to a nearby hall, where he spent another twenty-eight minutes meeting and speaking to a crowd of about 750 of the faithful paying $35 a head. The two events yielded $60,000 for Biden's reelection campaign and about $20,000 for the Delaware Democratic Party.

Carter praised Biden for his early support, called him what he called everybody—"a friend"—and said he wanted Delawareans to keep him in the Senate. But he also took the occasion to remark that Biden was "independent almost to a fault."[14] In other words, although

Carter admitted to the larger crowd that Biden had campaigned for him in thirty states, he was also Jimmy just being Jimmy.

The busing issue in Delaware was in the forefront as Biden pursued his reelection campaign. In conjunction with Democratic Senator Thomas Eagleton of Missouri, he won approval of an amendment to a Senate appropriations bill that would bar the Department of Health, Education and Welfare from ordering school busing in Delaware schools. But the latest Biden-Roth antibusing bill fell short in the Senate, and with Clarence Mitchell's criticism stinging him, Biden wrote in his memoir, "the truth is, I was shaken . . . I started to get the feeling that busing might cost me my seat in the coming election. Basically, I was being buffeted by every side, and with the busing plan [in Delaware] due to take effect two months before the election, it was hard to talk about anything else."[15]

On one of his daily commutes from Wilmington to Washington with Ted Kaufman, Biden wrote, "we put together a strategy for a Republican nominee for my Senate seat. How would a challenger, we asked, take down Joe Biden? It wasn't complicated. Point out Joe Biden's strong commitment to civil rights and civil liberties. Point out his statements that he'd gotten into politics largely because of civil rights. Joe Biden's a card-carrying liberal, a challenger could tell people, and his campaign to minimize court-ordered busing is just a Trojan horse. Once Joe Biden was safely elected for another six years, he'll pull the rug out from under the antibusing lobby. By the time we got off the train, Ted and I had convinced ourselves I could be beaten."[16]

Jeffrey Raffel, a University of Delaware professor specializing in civil rights matters and a member of the Delaware Committee on the School Decision, remembers asking Ted Kaufman, Biden's chief aide then: "Do you spend much time worrying about or talking about busing?" Kaufman replied: "Do we talk about it? That's all we talk about."

Raffel continued, "This was their most difficult political issue, because Biden certainly was for civil rights, there's no doubt about that. But like a lot of other liberals, he was caught. . . . He was against forced busing, [but] you've got to put Biden in the context of the general tenor of the times and the attitude of other civil rights advocates and liberals. The position of our committee was, we have to do what the court order says. You can be against it, you can be upset about it, you can favor it—whatever your position, you don't want to tear this community apart."

Beyond the convenient rationale that this view gave Biden and other politicians in defending their position with strong liberals and civil rights advocates, Raffel said, "Biden is a guy of very strong convictions. He will tell you where he's at," and at the same time he could see others' point of view. Also, Raffel said, Delaware having had a long history of segregation, there was no strong voting bloc actively arguing for busing likely to chastise Biden for his position, generally accepted as reasonable in the state.[17] So if Biden had any real political concern about the position he took, it wasn't really warranted. In any event, Biden's constant presence in the state over the previous six years of commuting insulated him from allegations of ignoring his constituents. Years later, Biden said he could have lost his reelection bid in 1978 "because I would not support stripping the court of jurisdiction over school busing. I thought busing was a mistake and I still think it was a mistake. I think it was overkill, but the court had a right under the Constitution [to seek a remedy to segregation]."[18]

In the primary fight among Republicans to challenge Biden, Delaware's most outspoken foe of school busing, James A. Venema, a leader of the Positive Action Committee, by comparison made Biden look moderate and reasonable on the issue. Eventually, however, Venema was defeated for the right to face Biden by a Sussex County poultry farmer, James H. Baxter, on the strength of downstate votes.

In any event, the general election from the outset never was about Baxter; it was about Biden, whose peripatetic campaigning continued. As Baxter charged the incumbent was too liberal for Delaware,

Biden's only challenge was justifying to liberals why as a strong advocate of civil rights he was adamantly opposed to busing. It was, after all, the simplest means of giving black children their right to attend schools as good as those available to the white kids. His repeated lectures about the difference between de jure and de facto integration were hardly the stuff designed to placate the emotions and anger of black parents—or of white parents who were fearful, or resentful, or just plain angry about busing black kids into "their" better-equipped, often better-staffed schools.

Absent congressional action to halt the busing in Delaware, it appeared the big yellow-bus cavalcade would start rolling in early September, when the New Castle County schools were to open. Legal challenges, however, and then a teachers' strike in protest of the imminent busing, kept schools closed into October. But near the end of the month, and only days before the election, a Supreme Court ruling set the buses in motion and the kids into the integrated schools.

Baxter sought to attack Biden's attendance record in the Senate, running an ad in the Wilmington newspapers noting he had missed 14 percent of the votes cast since he was elected. But a check of the *Congressional Record* showed that that was the average attendance record for senators of both parties, and besides, he was spending all that time back among the voters in Delaware.

Faced with Biden's growing reputation as Delaware's most visible and popular politician, Baxter near election day told the Wilmington Kiwanis Club that Biden was "a desperate man" who was "nowhere as good for Delaware has he says he is." Biden responded: "Sounds like we're nearing Desperation Junction."[19]

He was right. For all the concerns about busing and his stand on it, Biden defeated the unimpressive Baxter by 58 percent to 41—an emphatic victory, especially when compared to his razor-thin election over the venerable Cale Boggs six years earlier. Once again, the family-oriented Biden campaign prevailed, with energetic and crafty

Valerie in charge of the same young army of more than four thousand door-knocking volunteers across the little state who had delivered the upset over Boggs. Even Baxter's home county, Sussex—which Biden had lost in 1972—gave the incumbent nearly 53 percent of its vote.

However, with an increasingly unpopular Democrat in the White House bedeviled by economic woes at home, the off-year elections were bad news for Biden's party, and for a number of prominent Senate colleagues who lost their seats: Dick Clark, in Iowa; Tom McIntyre, in New Hampshire; Bill Hathaway, in Maine; Floyd Haskell, in Colorado; and Wendell Anderson, in Minnesota.

Some years later, looking back at his first Senate term, Biden said he had been "treading water" as he sought to cope with the tragic event that had followed so quickly on the heels of that first election. "The first two years I didn't know where the hell I was and what I was doing," he told *Congressional Quarterly*. "The second two years I was trying to make up for the stupid things I had done, and the third two years I was trying to get reelected." All that time, he said, "I paid more attention to getting my family and me together than I did my job." He acknowledged also that "I came here with a chip on my shoulder," and that because of "this black Irish mood I was in . . . I had done myself some damage in terms of relationships, in terms of what my image was."[20]

Now, however, with his family life restored and one full Senate term under his belt, Joe Biden was more at peace with himself and with his circumstances. Entering his second term, he could spread his wings a bit, and he intended to do so.

TEN

MOVING ON TO THE NATIONAL STAGE

FROM BIDEN'S EARLIEST days in the Senate, as a lawyer he had had his eye on a seat on the Judiciary Committee. His political godfather, Mike Mansfield, had put him originally on Banking, of particular interest to Delaware in light of the heavy corporate, investment, and credit card interests drawn there by favorable state laws, and on Public Works. In 1975 Mansfield had moved him to the Foreign Relations Committee, an unusual plum for a freshman, but he continued to yearn for Judiciary. Early on, as already noted, he struck up a strange-bedfellows friendship with James Eastland, chairman of that committee, and in 1978 Mansfield negotiated Biden's seat on the then heavily conservative body. He took with him his staff counsel on the Intelligence Committee, Mark Gitenstein, who later would play a prime role in some of the most critical fights over Supreme Court nominations.

Delaware's junior senator was drawn to Judiciary by his concern over civil rights. His new committee assignment gave him a front-row seat in debates on the school-busing issue so heated in his state, but his opposition to it put him in conflict with customary civil rights allies. Nevertheless he helped shape an omnibus drug bill that boosted penalties for violations and more money for education, and he proved to be a pragmatic negotiator within the committee. Unusual circumstances subsequently led to a rapid Biden rise in the committee's ranks. In 1979, upon Eastland's retirement, Ted Kennedy became chairman for two years, with Biden joining him in the majority.

Meanwhile, the presidency of Jimmy Carter was on the rocks as a result of severe economic woes—soaring gas prices, unemployment, and double-digit inflation—and a seemingly endless crisis in Iran in which American employees of the embassy in Tehran were held hostage. Ted Kennedy was already challenging Carter for the 1980 Democratic nomination, and a group of young Democratic political consultants called on Biden to join the fray.

John Marttila, who had helped Valerie Biden run her brother's upset victory against Caleb Boggs in 1972, and media expert Bob Squier were among those arguing a Carter-Kennedy fight would split the party and open the way for Biden as a compromise. They talked about how he might fare in the Iowa caucuses and the New Hampshire primary, but Marttila cautioned him that unless he knew why he was running for president and what he would do if elected, he should not run. Biden concluded he didn't yet have the answers.[1] So in 1980, as he had done in 1976, Joe Biden campaigned for Carter, though by then their relationship had cooled. He was seldom asked to speak and focused on Delaware and on Pennsylvania, where he boasted being called its third senator.

In April Biden and other members of the Senate Foreign Relations Committee traveled to the Persian Gulf region to assess the hostage situation. In the party was a naval escort officer named John McCain, who proved to be the life of the party and Biden's warm friend. Also along was Bill Bader, the committee's chief of staff. At a stop in Oman, Bader, an old navy officer himself, asked Biden one day whether he would care to join him on a visit to the aircraft carrier *Nimitz* anchored offshore. Along with McCain, they set off for the carrier, and once aboard, as Biden and McCain watched landing drills on the deck, Bader split off on his own to explore the ship. He went down to a below-decks hangar, which, to his surprise, he found jammed with eight very large Sea Stallion helicopters.

Bader hadn't noticed that the carrier's executive officer had followed him down to the hangar. "What are you doing here?" the ship's number two man demanded to know. As Bader recalled the one-sided conversation later, he was told in no uncertain terms: "Dr. Bader, I want to tell you, and this is an order: you are not to mention what you have seen, ever, to the senators." Bader said later: "He sensed I might have an idea what the helicopters were for, and I did, because we were off the coast there of Oman and this was just the time we were shaping up to try to rescue the diplomats who had been taken by the Iranians."

Without hesitation, Bader found Biden. "I had come to trust Joe Biden in every sense of the word," he said later. "When it was important, he was discreet and not loquacious, and didn't blow things. And I told him, and only him; I said, 'I'd hope you'd keep this to yourself because I'd like your advice.' I told him what I'd seen down there and what I suspected. I said, 'I think this is the carrier [that's going to send helicopters in to rescue the hostages],' which they were. What else were they doing out there?"

Bader said he asked Biden whether he should tell Frank Church, the Senate Foreign Relations Committee chairman, when they got back. "He was already pressing the administration on whether they had any designs on Iran," Bader said. "'Joe, what should I do about this? Should I tell Church?' I was a little concerned before I started the conversation that he would take it in a kind of amused and frivolous way, but he didn't. He took it enormously seriously and said, 'Bill, it's your responsibility and your duty for the country's sake.' Right then I had the sense, this guy is going to be good, taking his responsibilities in the Senate in a serious way. It was a defining moment for me."

When the committee party returned to Washington, Bader told Church, who "turned on the heat. He went on television and used it, without disclosing what he was told to the press, on whether we had any such plans." Church introduced a resolution saying Congress had to be informed of any such military action, but to no avail. On the night of the botched raid, Secretary of Defense Cyrus Vance called Bader at his home in suburban Virginia looking for Church, who was away at his private cabin. Vance asked him to inform the chairman that "unbeknownst to me and against my best judgment, Jimmy Carter has sent in this array of special forces to go rescue the hostages Israeli-style, and the mission has already failed."

Vance told Bader he "wanted Church to know he was going to resign, even before he told the president."[2]

Afterward, there was speculation as to why Vance hadn't gone public with his opposition and possibly could have averted the failure, in which a helicopter and a huge C-130 transport collided on the ground, another helicopter got lost in a sandstorm, and the mission was aborted. But Vance held his tongue until afterward rather than risk jeopardizing the military operation and the lives of the men involved.

Three months later, when the Democratic National Convention in New York renominated Carter without enthusiasm, Biden was a minor player. Some Delaware delegates hoisted a BIDEN IN '84 banner as he addressed the hall, and more than anything substantive he had to say, Biden was heard at one point declaring on the floor: "I'm one of the most important people in America." Apparently, it was a typical Biden wisecrack to stir up attention for his speech. Reporters for the Wilmington papers said later the senator had complained to their editor that the flip remark was reported as serious. In any event, the reporter who quoted the remark acknowledged later he knew Biden had meant it as a joke.[3]

Enough local interest was stirred up in Biden at the convention, however, that he was moved to report back home that some folks had been encouraging him to run for president in 1984, and that he wasn't interested—yet. He said the matter had come up at a weekend visit at the Martha's Vineyard home of Carter's young pollster, Patrick Caddell, who had become a very close friend.

A Kennedy delegate at the convention named George Mitrovich distributed buttons proposing BIDEN-SCHROEDER, 1984, linking the Delawarean with Colorado Congresswoman Patricia Schroeder. She laughed it off, and the Wilmington papers quoted him as saying, "I don't want to be president in 1984. I've got a lot to learn. . . . I have no intention of running for president or vice president in 1984." He added, according to the paper, "If I wanted it, I think I could take it," and Delaware Democratic Chairman Henry Topel proclaimed he would be with Joe if he tried.

It was hardly, however, the kickoff of another Biden campaign. Biden said he intended to seek a third term in the Senate in 1984 "and my kids will be fourteen and fifteen in 1984. If I were serious [about a presidential race] I'd have to devote two full years to running flat out. . . . It's difficult to be the kind of father I want to be and go flat out for two years."[4]

Such talk, however, inevitably fanned speculation at least in Delaware of a Biden presidential bid sometime down the road. In the fall, Carter had little chance once Reagan demonstrated on the stump and in debate that he was more than a faded movie actor. The Senate again went Republican, with Ted Kennedy losing the Judiciary Committee chairmanship. The Reagan landslide also wiped out a dozen Democratic senators, propitiously enabling Biden to move up the seniority ladder. Kennedy thereupon elected to step aside as ranking minority member on Judiciary for the same position on the Labor and Human Resources Committee. At age thirty-nine, Biden found himself Judiciary's ranking Democrat.

Many civil rights leaders were not enthused about Kennedy giving up that position for the similar one on Labor, and being succeeded by

Biden. At any rate, under Senate tradition Kennedy would have the option of reclaiming the Judiciary chairmanship if the Democrats later regained control of the Senate. However, in light of new Senate rules stating no senator could serve on more than two major committees, he was about to be eased off Judiciary. Biden, a member of the Democratic Steering Committee in the Senate, to the surprise of some colleagues intervened and won approval to keep Kennedy on. "I started thinking to myself," he said later, "that's not right. For all the things people can say and do about Ted Kennedy, he to Hispanics and to blacks is in fact the embodiment of hope. It somehow smacked of being a refutation of the Democratic Party to say Ted Kennedy could no longer be on that committee."[5]

Kennedy also was a member of the Steering Committee, and with him present, Biden successfully made the case for an exception. He said later fellow Democrats ribbed him for an action that might cost him the chairmanship later, though that never happened.

With Republican Strom Thurmond as chairman, Biden would have an ally in his fight against school busing. But on other issues of civil rights, the Delawarean could look forward to some tough fights in committee. Thurmond threatened to reverse parts of the Civil Rights Act of 1964 barring discrimination in public accommodations and employment; the Voting Rights Act of 1965; and the Civil Rights Act of 1968, guaranteeing open housing. "If Strom Thurmond is serious about eliminating the Voting Rights Act," Biden said, "I'm going to fight it. I'll be visible in that fight."[6]

When he first replaced Ted Kennedy as leader of the minority on Judiciary in 1981, there had been grumbling among civil rights activists that Biden had neither the gravitas nor the aggressiveness required on a committee now chaired by Thurmond. But he managed to achieve a cooperative relationship with South Carolina's rock-ribbed segregationist on completion of an eleven-year rewriting of the country's criminal code. Biden also successfully defended for a time the composition of the U.S. Civil Rights Commission against Reagan administration efforts to water it down. His greatest challenges, and public notoriety, would come however when the committee addressed critical confirmation hearings of high Justice Department officials and federal judges, including nominees to the Supreme Court.

Now that Carter was out of the White House after a single, controversial term, speculation immediately began over the leadership of the Democratic Party, and the 1984 presidential nomination. Kennedy and former vice president Walter Mondale were regarded as early front-runners, and although Biden had said he would not run, his name continued to pop up, pushed by his pollster friend Patrick Caddell and some others.

Even when Kennedy declared in December 1982 that he would not be a candidate, Biden stuck to his guns, reiterating he would seek Senate reelection in 1984. "I'd much rather be a senator than a vice president," he said then. Sherman-like, he declared: "If asked, I will refuse. If put on the ticket, I will refuse. And if I'm elected with the president, I'll refuse to serve."[7] At only forty years of age, he had plenty of time to reach higher, as many in Delaware expected he would.

Nevertheless, Caddell kept coaxing him to think about a presidential bid in 1984, with an elaborate theory on how both the Democratic Party and the country were ready for a distinct generational change in leadership. He argued that Mondale could not beat incumbent Reagan, and he built a profile of a "Mr. Smith," a fresh young Democrat with strong support from the baby boomers who could win in 1984. It snugly fit Biden, and Caddell beseeched him to run. It wasn't that Biden lacked the nerve to take on high odds; doing so was, after all, what had gotten him into the Senate. But there was always the consideration of his family.

For the time being, however, Biden pressed on with the message of generational change. In September 1983, in a speech at the New Jersey Democratic Convention in Atlantic City, he wowed the crowd with a ringing call for that change, and an implied rejection of Mondale as a figure of the past. "America is in serious trouble, and so is the Democratic Party," he said. "As every hour passes, our ability to shape the American future is slipping from our grasp. . . . Instead of thinking of ourselves as Americans first, Democrats second, and members of special-interest groups third, we have begun to think in terms of special interests first and the greater interest second."

The Democrats, who once had been "the engine of the national interest," he said, had become "perceived as little more than the broker of narrow special interests." And they "went astray by failing to remember what got us this far and how we got here—moral indignation, decent instincts, a sense of shared sacrifice and mutual responsibility, and a set of national priorities that emphasized what we had in common. The party whose vision produced the greatest economic prosperity and the broadest commitment to social justice the world has ever seen became complacent and unwilling to adjust its programs to accommodate a rapidly changing world. . . . Suddenly the party that had been the party of ideas, change, and experimentation became perceived as the party of the status quo."[8]

These remarks from the conspicuously independent-minded senator from Delaware were hard allegations for many special-interest Democrats to swallow, particularly activists in civil rights and abortion rights groups. The sting was softened, however, by typical Biden optimism about recovering from adversity, and by his talent for strumming on the heartstrings of young Democratic yearnings for the inspirational 1960s days of John and Robert Kennedy and Martin Luther King Jr.

"The cynics believe that my generation, having reached the conservative age of mortgage payments, pediatrician bills, and concern about our children's welfare, are ripe for Republican picking," he said. "These experts believe that, like the Democratic Party itself, the less-than-forty-year-old voters are prepared to sell their souls for some security, real or illusory. They have misjudged us. Just because our political heroes were murdered does not mean that the dream does not still live, buried deep in our broken hearts."

Biden ended by saying of RFK, "His words still echo, reminding us that only with our idealism, commitment, and energy can we ever hope to achieve the destiny history has laid before us. Not as blacks or as whites; not as workers or professionals; not as rich or poor; not as men or women; not even as Democrats or Republicans. But as people of God in the service of the American dream."[9]

The electrifying speech sent sparks of a 1984 presidential candidacy through insider Democratic ranks, and Caddell saw the reaction as validation of his reading of what the party needed and wanted. He kept pushing Biden. "Caddell would pull from his briefcase the dot-matrix printouts that showed the latest polling data," Biden wrote later, "and then he'd make a hard case as to why I could win this in '84, and why this was the time. Pat was relentless."

Shortly before the filing deadline for the New Hampshire primary, Biden recalled, "I kept saying, 'Damn it, Pat, I don't want to do this thing.' But he kept at me. 'Just in case,' he said, 'sign the filing papers.' . . . So I signed them, almost as a lark, but I told everyone in the room that the papers were to remain in Val's possession. Nobody else could touch them. If I decided to run, Val could fly north to file the papers. Then Jill and I got on a plane to take a short vacation."[10]

At the time, Caddell anticipated that Gary Hart, who as campaign manager for presidential candidate George McGovern in 1972 had given him his start in national politics, was going to ask him to join his own campaign for the 1984 Democratic presidential nomination. It was just before New Year's Day, he said later, so he tried one more time to convince Biden to run, contacting him at the airport in New York just before he and Jill were to take off. Biden, torn over the decision, finally agreed to run, Caddell said much later, but later that night he called and changed his mind again, apparently for family reasons.[11]

In the end, Biden judged that neither he nor his young family was yet ready to take on a national campaign. "Jill and I had a serious talk on the flight down to the islands," he wrote later. "I wasn't worried so much about getting beat. Nobody would expect me to win in the first place. The chances of my winning were miniscule. But what if, unlikely as it was, I did win? I still had not answered the big questions: Why run? To do what? I simply could not visualize myself running the bureaucracy of the federal government. . . . By the time our plane touched down in the islands, I knew what I had to do. I called Val. 'Don't file that thing,' I told her. 'I'm not running.' "[12]

On New Year's Eve, John Marttila got a phone call from Biden telling him of his decision. Although there was what he called "a monster opening" for a candidate such as his friend from Delaware, Marttila said later, "Joe made the right decision. He wasn't ready [for a presidential run]."[13]

Contrary to a general impression that Biden was a loose cannon ready to plunge over a cliff out of unfettered ambition, he mused openly about the chances of his generational message playing against other prospective Democratic candidates in the field, and in the end decided prudence was his best course. That and a sense that his family might not yet be up to handling the rigors of a national campaign. "There was a lot of push from guys around me," he said much later, "but I just didn't feel right about it at all. I listened, but I didn't come close at all. I didn't think it was the right time, and I wasn't interested."[14]

So Caddell moved on to Gary Hart, who also fit his "Mr. Smith" profile, and Hart became the surprise of the 1984 Democratic primary campaign. He threw a major scare into Mondale before finally falling short, but he positioned himself well for another try in 1988, should Mondale lose against Reagan, which predictably happened. But when Hart had defeated Mondale in the New Hampshire primary, Caddell recalled later, "Joe was totally happy. He was going around the Senate saying to others [who also had declined to run, such as Arkansas Senator Dale Bumpers], 'You could have been president.' He just felt others were better positioned. It wasn't a logical thing for him at the time."[15] For himself, however, Caddell said much later, he was convinced that had Biden run he would have been the Democratic nominee, and even if he had lost, would have had a good chance of being nominated again in 1988.[16]

Biden therefore settled in in 1984 for a reelection campaign for his third Senate term, confidently holding off until June to declare his candidacy. As in 1972 and 1978, he kicked off his quest by announcing in each of Delaware's three counties, in part playing again on the theme of the Kennedy era interrupted. "It's time to start to mend the broken hearts of my generation," he said. "It's time for a new dawn to break over America." Ridiculing the Reagan line of 1980 that had helped sink Carter, Biden said he was running "to challenge the idea that the ultimate standard for America is whether we feel better off today than we did four years ago, and to establish once again the ideal that we ask not what our country will do for us, but what we will do for our country."[17]

At the Democratic National Convention in San Francisco that nominated Mondale, Biden was in much greater demand for speeches to state delegations and other party events than he had been four years earlier. Still voicing no real enthusiasm for Mondale, he continued to urge the party to expand its base beyond the old New Deal horizons and to court the younger Democrats who had rallied around Gary Hart. To more questions at the convention about future presidential plans of his own, he focused again on his Senate reelection challenge. But one delegate named Keron Kerr, a schoolteacher from Bath, Maine, cast her presidential vote for him.[18]

Biden's Republican opponent in the fall was a jovial former Delaware House majority leader, John M. Burris. In effect, he was drafted to run against Biden by the state's GOP governor, Pete du Pont. At least since his second term, Biden had been baiting du Pont to run against him for the Senate, but du Pont had his eye on nothing less than the presidency. So, according to Burris, du Pont asked him to take on Biden, enabling him to avoid a possible detour from his White House ambition.

"I didn't decide to run against him," Burris said later. "In 1980, Pete had approached me to run with him as lieutenant governor and then governor [when du Pont would have left that office]. I said no for family reasons and coaching [kids' sports]. People thought I was crazy because it was mine for the taking." Burris said he then "term-limited myself" and dropped out of politics, only to hear from du Pont a few years later.

"I get a phone call from Pete," he recalled, "to meet him and his wife, Elise. He said he had a favor to ask of me." Over olive sandwiches, Burris said, the governor told him: "I have a dilemma, John. I'm planning to run for president in 1988. Howard Baker [then President Reagan's chief of staff] won't consider the idea unless I run against Biden in 1984 or find some other credible Republican [to do so]."

Burris said he expressed astonishment at the proposal, believing he never could beat Biden. But as a good party man, he relented and agreed to be, he said, "a sacrificial lamb. It freed Pete up to run for president. Pete was always good to me, and I thought he would be a good president."

Du Pont apparently wanted no part of a race against hotshot Biden.

"Du Pont–Biden would have been a classic race, and I was caught up in high-stakes poker," Burris said. "It wasn't the smartest thing to do, but I enjoyed it." In debate with Biden, he said, "we had mutual respect. I didn't look at him as an evil guy. Our whole strategy was to get Joe mad because he had such a short temper." In one debate, he later recalled, he succeeded by asking Biden how he was against the U.S. invasion of Grenada in the morning but against it in the afternoon as it appeared to be going well. "Biden came up out of his chair and began screaming at me," Burris said. "When his trigger goes, and his staff is not there to protect him, that is his weakness."

Burris attacked Biden's Senate attendance record and thought he might possibly ride on the coattails of Reagan's easy reelection. But the tactic didn't work. Biden's team ran a radio ad on the need for a campaign watchdog. It had a dog barking at each Burris allegation. The Republican challenger responded with a barking-dog ad of his own. But in the end, Biden's constant presence in Delaware, his army of volunteers, and this time a healthy campaign treasury all were too much for Burris to overcome. Biden won by 20 percentage points. Of Biden's firm ties to Delaware by now, Burris said, simply, "I was John Burris; he was Joe."[19] Later, in the Delaware nature of politics, Democrat Biden and Republican Burris became golf-playing pals.

Much later, du Pont professed to having no recollection of urging Burris to run against Biden so that he wouldn't have to. In an interview, he acknowledged there had been a lot of talk in Delaware about the prospect of a Biden–du Pont contest. But, he said, "once you've been in the Congress as I was for three terms [in the House of Representatives] and once been governor [for eight years], governor is a much better job. The thought of going back and being a junior member of the legislature didn't really appeal to me very much. It had nothing to do with Joe Biden, but I don't think you would have ever gotten me to run for the Senate."

It was true, he said, "that everybody in the party wanted me to run," but he said he'd never heard that Biden had needled him to enter the race against him. "I think I used to tell them [the party leaders] I'd be happy to run," he said, laughing, "if they promised me [that] if I won, I wouldn't have to serve."[20]

Joe Farley, a former Democratic Party state chairman in Dover, said that in fact "Joe egged Pete on. Joe's notion is to take on all comers."[21] But Vince D'Anna contended that for all of Biden's reputation as a risk-taker, when it came to getting elected and reelected "he knew how to manage risk." In 1983 in the midst of speculation that du Pont might indeed challenge him, D'Anna said, "we were doing everything we could to discourage Pete from running." The Biden operation organized a huge picnic outside Newark that drew an estimated twenty thousand people, D'Anna said. It caused a tremendous traffic jam that could not have escaped du Pont's attention.[22]

Burris turned out to be a very good candidate and gave Biden his toughest challenge as an incumbent, D'Anna said, but still fell far short. Biden's election to a third six-year term at age forty-one still left him as a junior figure in the Democratic Party as it absorbed its second presidential defeat in a row.

Gary Hart's unexpectedly strong showing in the 1984 presidential primaries did appear to give him an early edge for the 1988 Democratic nomination, but with continued speculation that Ted Kennedy would try again, or New York's Governor Mario Cuomo, whose oratorical skills had won much higher acclaim than that yet garnered by Biden. But the junior senator from Delaware had by now alerted his party that he had an appealing message and the style and fervor with which to deliver it. And twelve years earlier he had emphatically demonstrated that patience in reaching for a higher rung was not always his strongest suit.

In the meantime, Biden would be kept busy in the Judiciary Committee. Early in 1984 Reagan had nominated his old California friend and White House counsel, Edwin Meese, to be attorney general. The nomination was stalled, however, when a committee investigation surfaced allegations of ethical violations by Meese, and the matter dragged on through the year as an independent counsel was appointed to resolve it. The confirmation would be one of the committee's first orders of business in the new year.

In Biden's twelve years in the Senate, his reputation as more of a glib talker than a man of intellectual substance had clung to him, a perception of personal frustration. The Meese confirmation, and subsequent judicial nominations coming up, would provide opportunities for lawyer Biden to demonstrate his cross-examination skills and his own command of the legal terrain in what figured to be contentious Senate Judiciary hearings.

ELEVEN

JOE BIDEN FOR PRESIDENT

AS 1985 BEGAN, Joe Biden's daily commuting from Wilmington encouraged an impression of limited engagement in the demanding business of the Senate. But he had proved himself by now to be a legislative workhorse, and as he continued to fulfill his responsibilities to family at home, he remained fully committed on Capitol Hill, as well as undertaking foreign fact-finding.

For all his diligence, however, some colleagues acknowledged that his rigid schedule came at a price in Senate collegiality. Though his gregarious nature minimized that cost, the fact was he wasn't much in the social mix there. "It's a club here," one of his closest friends and advocates, Senator Fritz Hollings of South Carolina, remarked at the time. "The atmosphere of club is missing with Joe. You never really get to know the others unless you go on a trip or to parties with them."[1]

As a member of the Foreign Relations Committee, Biden made such trips, but he wasn't a party-goer and once bragged that he had attended only one embassy reception in his Senate career. To this observation, Biden readily acknowledged that his family "has always been the beginning, the middle and the end" with him.[2]

Bill Bader, who had reservations about Biden's early commitment to the work of the Senate, said later a gradual change came about as his family situation changed. "Joe grew up in the Senate without having the discipline and strong sense of public service and its requirements that the Kennedy family had," he said. "It took about three years, and his remarriage had a great deal to do with it. Jill traveled a lot with him," reinforcing his strong sense of family involvement as he met his Senate duties.[3]

Among Biden's first tests as ranking Democrat on Judiciary was grilling Ed Meese in his confirmation hearings to be Ronald Reagan's attorney general. An independent counsel had explored reports of a questionable intervention by Meese while he was Reagan's White House counsel. He concluded there was insufficient evidence to bring charges of a Foreign Corrupt Practices Act violation against Meese. But Biden was not about to let the nominee off without a parting shot.

While concluding the nominee had "done no criminal wrong, and I don't believe you're unethical," he told Meese he considered him to be "beneath the office" of attorney general. Of his opposition, Biden said, "It relates less to you than it does to the office of attorney general. I think some would say . . . that I have maybe an idealistic and unrealistic view of the office of attorney general. But I think it should be occupied by a person of extraordinary stature and character."[4]

Biden's observation drew uncommon wrath in a *Wall Street Journal* editorial, calling it "one of the smarmiest, most reprehensible and self-indulgent speeches delivered in a Senate hearing room in at least thirty years."[5] Biden in the end failed to detour Meese's confirmation, which proved instrumental in Reagan's push for his congressional agenda of social issues.

Within the civil rights community, there were reservations about Biden's interest in, or willingness to, challenge the Reagan administration in this area. Later in 1985, however, under prodding by the civil rights community,[6] Biden was more successful in derailing the promotion of the controversial head of the Justice Department's Civil Rights Division, William Bradford Reynolds, to the job of associate attorney general.

Reynolds dismissed his civil rights foes as "the probusing, pro-quota lobby," but Biden sidestepped those issues and instead focused on a congressional redistricting case in New Orleans, in which white local officials had drawn lines that prevented formation of a majority-black House seat. Reynolds declined to challenge the plan; it was later rejected in court.[7]

Meanwhile, for all of Biden's disavowals about running for president in 1988, he began to explore the prospects. In two speeches over the summer of 1985 to black audiences, one in Atlanta to PUSH, Jesse Jackson's organization, and another before the NAACP national convention in Baltimore, he warned of racial division in the civil rights movement. In the first, he cited "misplaced" priorities pushing employment quotas and school busing, and in the second he said "voices in the movement who tell blacks to go it alone . . . and that only blacks should represent blacks" were destructive of racial harmony.[8]

Such observations did not go unnoticed by the combative Jackson. He went to Wilmington and Washington a few days later and took off on the centrist Democratic Leadership Council, which Biden had joined. He said its members were "riding with the Kennedy credentials on the coattails of Reaganite reaction." He said they were "combing their hair to the left like Kennedy and moving their policies to the right like Reagan." He called the DLC "Democrats for the Leisure Class . . . who didn't march in the sixties and won't stand up in the eighties." Asked by a Wilmington reporter whether he was making a "thinly veiled" swipe at Biden, Jackson said there was nothing "thinly veiled" about it. He added: "I did not go up there and attack him, but I did make a clear distinction on what constitutes the progressive direction the party should take."[9]

In November 1985 Biden accepted the invitation of the Iowa

Democratic Party to be keynote speaker at its annual Jefferson-Jackson dinner in Des Moines. It was the same event that in 1975 had lifted Jimmy Carter from national obscurity by running first in a straw poll deftly orchestrated by a few early campaign operatives. Biden chose this major forum, assured of drawing a large contingent of national political writers, as an opportunity to dust off and polish the generational theme that had stirred crowds for him in 1983.

Casting himself as the next young and rising John F. Kennedy, Irish Catholic Biden told the cheering Iowa Democratic faithful: "It's time we heard the sound of the country singing and soaring in the dawn of a new day. It's time to restore America's soul. It's time to be on the march again. It's time to get America on the move again. Our time has come." And later: "I think 1988 is going to be about 1960, about our country and not causes, about idealism and not ideology, about the future and not the status quo. . . . It's almost as simple as 'Let's get America moving again' [JFK's precise 1960 campaign cry]. My generation is really ready, and I want to be a part of it."[10]

It was at least Biden's fourth visit to the state that kicks off the national convention delegate-selection process with caucus meetings in towns across the Hawkeye State on a single Monday night. He sandwiched the Des Moines event between side trips to Detroit to touch base with Democratic Mayor Coleman Young, to Lansing to say hello to Democratic Governor James Blanchard, and on to Chicago. On another swing he went south, where he could temper his liberal reputation with accounts of his energetic opposition to school busing.

Biden continued to insist that he was merely testing the water temperature for some future presidential swim. In December, for example, he said this: "I'm not going to run in 1988. Someday I may want to do it, and I don't think you should shoot blanks. I'm forty-three years old and there's lots of time." He continued also to say members of his immediate family had to be aboard a presidential run, and he demonstrated it, insisting that his now famous daily commuting between Washington and Wilmington since the death of his first wife and infant daughter was no opportunistic political stunt.

"One of the things people think about me is that my attitude toward my family was born in the crucible of tragedy," he said. "That's not so. I've always had that kind of commitment to them." Noting that his children were traveling with him often as he explored his chances for a presidential bid, Biden said, "I'm taking my family out and letting them get a taste of what it would be like. If I decide to go for it, we're going to have to handle it as a family."[11]

After another round of explorations on the road with wife Jill and sons Beau and Hunter often in tow, Biden said: "They've got to answer the question: How comfortable will they be out there? So far the answer is that they're pretty comfortable." But with four-year-old Ashley now in the mix, he added, "Can we do this without Jill becoming a single parent? That question is not answered yet."[12]

But Biden's relationship with his kids still applied. Ashley recalled much later: "I just always remember Dad being present. He was present every morning, I talked to him two times a day by phone, he was always home at night, most of the nights, to catch dinner and to tuck us into bed. He was around on the weekends. I think Dad was around more than some of my friends' fathers who lived two blocks away. You could always count on Dad to be there, if not physically, to be there by a phone call. He had a rule that no matter where he was, no matter what he was doing, if us kids called, that was it. You got him out of the meeting."[13]

Over the next few months, Biden continued at state party dinners to strum the strings of 1960s nostalgia with repeated references to the Kennedys and Dr. King, insisting that the generation of protest politics could be inspired anew. "The cynics believe that my generation has forgotten all of this," he told a party dinner in Richmond. "They believe that having reached the conservative age of mortgage payments, pediatricians' bills, and saving for our children's education, that we are ripe for Republican picking. But they've misjudged us." And again he segued into the line denying that the murder of their heroes had erased the middle-class dream of a more just society.[14]

At the same time, Biden's position as ranking Democrat on Judiciary kept him in the spotlight through a series of high-profile confirmation fights with the Reagan administration. But many in the civil rights world continued to criticize him for not taking Reagan on soon enough. He fell one vote short of blocking the nomination of a conservative nominee to the federal appeals court, Daniel Manion. After that, he brought a lot of energy to the Democratic argument against Reagan's elevation of Justice William Rehnquist to chief justice. Rehnquist had been accused early in his career as a private lawyer with intimidating black voters in Arizona, when he was a foot soldier in the Barry Goldwater ranks. But the charge had not prevented his confirmation as an associate justice nor did it now bar his confirmation to lead the court.

Biden in his opening statement made the point that while many Supreme Court nominees historically had been examined in terms of their character and personal lives, others had also been questioned legitimately about their "judicial philosophy and vision of the Constitution." It was a practice that ran counter to the view that there should be no ideological "litmus test" applied.

Probing on a question of ideology, Biden pressed Rehnquist to say whether he believed the Court was correct thirty years earlier, when he was a law clerk for Justice Robert H. Jackson and the Court struck down the sixty-nine-year-old ruling in *Plessy v. Ferguson* upholding racially segregated public facilities. In the same regard, he asked Rehnquist whether he thought that the *Brown v. Board of Education* decision, requiring school desegregation, had been wrongly decided.

A verbal merry-go-round ensued between the two men, after which Rehnquist said "I thought *Plessy* had been wrongly decided at the time, that it was not a good interpretation of the Equal Protection Clause [of the Fourteenth Amendment] to say that when you segregate people by race, there is no denial of equal protection."

Biden pressed him: "[L]et us try to finish that thought. If you got that far, then it seems your conclusion must have been that it was the Congress's business, not the Court's, to change *Plessy*?" Rehnquist replied: "Senator, I do not think I reached a conclusion. Law clerks do not have to vote." Biden shot back sarcastically: "No, but they surely think."[15]

The Republican-controlled committee voted on Rehnquist favorably by a 13–5 vote, with Biden and four other Democrats opposing. In their minority views, they wrote that while "there is no doubt about the legal ability and competence of the nominee . . . the hearings raised a number of serious questions about the appropriateness of Justice Rehnquist for the office of Chief Justice. In particular, questions were raised about his credibility, his commitment to equality and racial justice, his decision to participate in the resolution of certain litigation before the Supreme Court, and his commitment to fundamental constitutional values."[16]

In separate individual views, Biden wrote that "while technically Justice Rehnquist's decisions do not indicate that he has rejected the basic fabric of contemporary constitutional law, he made no effort in the confirmation hearings to allay legitimate concerns that he might wish to move the Court away from these fundamental and respected constitutional doctrines. . . . Pertaining to the sensitive issue of race and gender discrimination, his answers on the record are, at best, incomplete." Biden specifically cited as troublesome to him Rehnquist's answers on *Brown v. Board of Education* as law clerk to Justice Jackson.[17]

Nominated to take Rehnquist's seat on the Court as an associate justice was the acerbic conservative Antonin Scalia. In that confirmation, interestingly in light of later Supreme Court developments, Biden said he could not object to him because he did not find him "significantly more conservative" than the man he was replacing. "Therefore I do not have undue concern about the impact of this appointment on the balance of the court."[18] It was an argument that eleven years later would serve Biden in good stead. He invoked it in reverse in his firm opposition to the nomination of Judge Robert Bork, the darling of the conservatives at that time, to replace retiring Justice Lewis Powell, who had a more moderate voting record.

On the Foreign Relations Committee, Biden unleashed an uncommonly sharp tongue in July 1986, accusing mild-mannered Secretary of State George Shultz of softening economic sanctions against South Africa for its policy of apartheid. "We ask them to put up a timetable [for remedial action]," he thundered, waving a fist. "What is our timetable? Where do we stand morally? . . . I hate to hear an administration and a secretary of state refusing to act on a morally abhorrent point. . . . I'm ashamed of this country that puts out a policy like this that says nothing, nothing. . . . I'm ashamed of the lack of moral backbone to this policy."[19]

The diplomatic Shultz at one point countered rather incongruously: "Just because I'm secretary of state, you can't kick me around. I'm a taxpayer." He insisted that the administration's policy had "tremendous moral backbone. . . ." "What we want is a society that they all can live in together. So I don't turn my back on the whites and I would hope that you wouldn't." Biden countered grandly: "I speak for the oppressed, whatever they may be."[20]

To many, Biden was discourteously browbeating the secretary of state and tooting his own horn in the process. Republican Senator Nancy Kassebaum of Kansas said she thought he had gone "too far" with Shultz. "I think it stems from a passion that he can feel toward the subject, but he simply doesn't know how to control it."[21] The exchange garnered wide newspaper and television coverage, adding a dimension of abrasive intensity to Biden's reputation as the Senate's most loquacious member. As he was simultaneously stepping up his travel to key 1988 presidential primary and caucus states, skeptics saw political motivation in the committee performance.

Biden told his hometown paper afterward that his political advisers had cautioned that he had come off as "too hot" in the exchange, to which he replied: "There are certain things worth getting mad about." And a vice chairman of the Democratic National Committee, Lynn Cutler of Iowa, was quoted in the same paper: "A senator raising his voice and his fist at the secretary of state and articulating an important point is not something you forget right away."[22]

But Joe Biden was not a man reluctant to duck an argument, in his fashion. Appearing before the annual NAACP convention in Baltimore, he seemed to be taking on Jesse Jackson without mentioning his name. He preached to the moderate African American choir that "you must reject those voices in the movement who tell black Americans to go it alone, who tell you that coalitions don't work anymore . . . that only blacks should represent blacks." In a news conference afterward, Biden said Jackson "does not know what he's talking about on economic policy," but then tempered his criticism by adding he "has done some really phenomenal things."[23]

It was Joe at his balancing-act best. Through the fall, speculation rose about a Biden declaration of presidential candidacy, especially as he indicated confidence that after a family discussion over the Thanksgiving holiday it would have the blessing of his wife and children. One of his key nonfamily political advisers, Washington consultant David Doak, argued that in the wake of Biden's 1986 soundings of political sentiment around the country, "people now look at Joe as being the third guy" in the Democratic winter-book handicapping for the 1988 nomination, behind Gary Hart and Mario Cuomo.[24]

Meanwhile, a critical event complicated Biden's political aspirations. In the 1986 midterm elections, the Democrats regained control of the Senate, elevating him to the chairmanship of the Judiciary Committee. He first tried unsuccessfully to get Ted Kennedy to take the job, which would have required Kennedy to give up the chairmanship of the Labor and Human Resources Committee. Now Biden had to ask himself whether he could undertake the rigors of a presidential campaign while taking on his new legislative responsibilities, which would include confrontations with the Republican administration over more high judicial appointments.

Many civil rights leaders were disappointed that Kennedy had declined to take the Judiciary chairmanship. They saw Biden as overcautious and protective of his reputation as an independent who was effective on defense—fighting Reagan appointments and excessively conservative initiatives—but possibly less aggressive in pressing an agenda of his own.

Shortly after the start of the new year, Biden reported he was putting up for sale his spacious home in Wilmington's Chateau country for a smaller one in the same neighborhood, for two stated reasons. First, Beau would be attending the University of Pennsylvania the next year, with Hunter entering college a year later, and the family would not need so much room. Second, Biden's new responsibility as Judiciary chairman would be more demanding of his time in Washington and he would not be able to cope with keeping up the old showplace.

He told an Associated Press reporter that "I can't say it would be preposterous" that he might give up the committee chairmanship if running for president at the same time proved to be too demanding. Notably, he did not say he would take the other option—not running—and subsequently declared "I think I can do both" if he so decided.[25]

While still not announcing his candidacy, Biden in early 1987 conscientiously played the double role, working short weeks in Washington and spending weekends in the early-voting states of Iowa and New Hampshire and also in the South. Other Democratic Judiciary members such as Ted Kennedy, Patrick Leahy of Vermont, and Howard Metzenbaum of Ohio, took up the slack in various subcommittee hearings. Meanwhile, Biden began to put together the rudiments of a national campaign—meeting party, labor, and business leaders, considering a fund-raising apparatus, more on a just-in-case basis at first. For one thing, Jill had to be convinced, but he reasoned with her that there was no harm in taking these early steps.

Biden continued to deliver his customary emotional entreaties to voters, sparking enthusiastic crowd reactions but seldom registering in the public-opinion polls. In the most recent survey of Iowa caucus voters, he stood at a dismal 1 percent, behind front-runner Gary Hart, Representative Dick Gephardt of Missouri, and other early Democratic competitors. But he was buoyed nevertheless by the crowd reactions, particularly at labor and party events.

At the California Democratic Convention in Sacramento in late January 1987 he drew such a wild standing ovation that, he wrote later, "I knew that day that there was almost nothing to stop me from making the run."[26] But in this speech, he said something that eventually contributed to just that outcome, when a speechwriter inserted a phrase from a 1967 speech by Robert Kennedy without telling Biden, who spoke it without attribution.[27]

Two weeks later, he addressed the winter meeting of the AFL-CIO in Bal Harbour, Florida, and it was more of the same. He told reporters after the speech that he was planning to run, and "only if I hit a rock between now and the time of a formal announcement would anything change my mind."[28] Like the famous story of JFK in 1959 looking around the Senate at the other prospective candidates and saying to himself, "Why not me?" Biden seemed internally to be posing the same question.

At the same time, he disregarded signals of the toll his schedule was taking on his health. He experienced severe headaches on the tiring campaign trail and took to popping Tylenols, as many as ten a day, from a large bottle an aide carried with him. Nothing like this had ever bothered him before. He brushed off suggestions from associates that his problem was stress from the pace he was maintaining. In mid-March at a Rotary Club lunch in Nashua, New Hampshire, he started to make a speech critical of Reagan's Star Wars missile-defense concept when he suffered a shooting neck pain that caused him to leave the room. The pain increased and he became nauseated. In about ten minutes he struggled back, made it haltingly through his speech, recovered, and pressed on. But the headaches persisted.[29]

To cope with a nagging impression among many in the press

corps that Biden was all glitz and insufficient substance, he undertook a series of meaty speeches on his extensive foreign policy experience, combating child poverty, and the economy. And through it all he pondered the questions Marttila had posed earlier about why he wanted the presidency and what he would do if he achieved it.[30]

More than any other outside development that might affect his prospects, however, was the sudden self-destruction of Hart's campaign, beginning in late April, when long-brewing rumors of womanizing broke open. Accounts and photographs of a dalliance with a young blonde named Donna Rice, after Hart's invitation to the press to follow him, drove Hart out of the race.

His departure encouraged the notion that there would be a clearer path for Biden. He was airing a version of the same Caddell-inspired message of generational change that he had declined to deliver in the 1984 presidential campaign but was then adopted by Hart. Inevitably, an impression continued that Biden was a Pinocchio dancing to Caddell's puppet strings, fed by an injudicious remark by Biden to Paul Taylor of the *Washington Post* that "sometimes it's hard to know where Pat's thinking stops and mine begins." It was an offhand comment that would haunt Biden long afterward.[31]

Bill Bader in fact recalled later that on one trip to Germany with Biden, the senator "spent a lot of time on how important he [Caddell] was to him and to everything he had ever done. He was a mentor and a champion, and at that particular time at least he was a godlike figure to him, and my observation was that his admiration and dependence on him was sincere and deep."[32]

In any event, as usual in the Biden political circle, a frenetic meeting of the so-called political gurus and family members occurred at the Bidens' Wilmington home just prior to the final decision on whether he would run this time. According to Caddell later, knowing Biden's personal sentiments, he urged him, counter to the advice of the other political professionals, not to announce his candidacy. And at one point late in the run-up to an announcement, reflecting on how good his private life was, Biden wrote later that he said to his wife, "I don't want to do this." But she told him: "You have to do this now. You have too many people's lives on hold."[33]

Finally, with all preparations for running in place, and about $2 million in his campaign treasury, Joe Biden on June 9, 1987, kicked off his bid for the White House at the Wilmington railroad station that had been a regular venue in his political life during his fourteen years in the U.S. Senate.

In JFK-like urgings to get America moving again after the Republican "self-aggrandizement" years of Nixon, Ford, and Reagan, he told the hometown crowd: "We must rekindle the fire of idealism in our society, for nothing suffocates the promise of America more than unbounded cynicism and indifference." And he reiterated his argument that it was time for youth to be served. "I am absolutely convinced that this generation is poised to respond to this challenge, and for my part this is the issue upon which I will stake my candidacy," he said. "The clarion call for my generation is not 'It's our turn,' but rather 'It's our moment of obligation and opportunity.' "[34]

Then the full Biden clan—his parents Joe Sr. and Jean; his wife, Jill; children Beau, Hunter, and Ashley; sister, Valerie; brothers Jimmy and Frank; and numerous other relatives—boarded a special campaign train to Washington in a symbolic reprise of his famous daily commute and swift rise to national prominence. A Delaware high school band struck up "It's a Small World After All" as the small-state candidate set out once more, as he had in 1972 by challenging the unbeatable Senator Cale Boggs, to beat the odds against him.

Delaware was all abuzz, particularly with the early declaration of former governor Pete du Pont, the man who had declined to challenge Biden for his Senate seat three years earlier, that he would at the same time be seeking the 1988 Republican presidential nomination. Delawareans with political stars in their eyes could even contemplate that the much-discussed "dream race" of Biden vs. du Pont, denied them in 1984, might yet take place after all at a much higher level.

TWELVE

A JUDICIAL INTRUSION

HARDLY HAD THE Biden-for-president campaign gotten under way when a major complication intruded. Seventeen days after the celebratory kickoff at the Wilmington train station, Justice Lewis F. Powell Jr. informed President Reagan that he was retiring from the Supreme Court. He was approaching his eightieth birthday. Immediately, as chairman of the Senate Judiciary Committee, Biden had the duty of leading consideration of the fitness of Powell's replacement and recommending confirmation or rejection by the full Senate.

Biden got the news as he and Jill, along with two aides, were flying to Los Angeles for some fund-raising events. He recognized right away the monkey wrench that had been thrown into the start-up gears of his presidential effort. The announcement was particularly unwelcome because of Powell's pivotal role on the Court. Appointed by Richard Nixon in 1969 in the expectation that he would be reliably conservative, Powell had in recent years cast the deciding vote against a series of litmus-test conservative positions involving such issues as abortion rights, affirmative action benefiting racial minorities, and separation of church and state. Biden later described Powell as "essentially the finger in the dam holding back the Supreme Court from ratifying the [Reagan] conservative social agenda."[1]

Almost at once, Biden learned of the challenge he faced. Word came quickly from the Reagan White House that the likely new nominee would be Judge Robert H. Bork of the Federal Court of Appeals. He was the same strongly conservative Republican who, as Nixon's solicitor general, had agreed to fire Watergate special prosecutor Archibald Cox after Attorney General Elliot Richardson and his deputy William Ruckelshaus resigned rather than perform that odious task. He was regarded to be of the same strict constructionist camp as Justice Antonin Scalia, appointed earlier by Reagan to the unvarnished glee of the GOP right wing.

Biden's chief counsel on Judiciary, Mark Gitenstein, informed Biden that the White House was already quoting a comment the chairman had given to the *Philadelphia Inquirer* the previous year: "Say the administration sends up Bork, and after our investigation he looks a lot like Scalia. . . . I'd have to vote for him, and if the [special-interest] groups tear me apart, that's the medicine I'll have to take. I'm not Ted Kennedy." Biden told Gitenstein he was merely saying "I wouldn't let the liberal interests groups tell me how to vote; I'd make my decision based on my own investigation."[2] Biden's reference to Kennedy implied that he believed Kennedy would be swayed by these groups.

When Rehnquist was elevated to chief justice, the appointment of Scalia had put a similarly conservative judge in the vacant seat. Bork replacing Powell would be entirely different, and if nominated and confirmed would clearly shift to the right the voting power and inclinations on the Court. Biden quickly released the following statement:

> *I hope that the President will nominate a man or a woman who is superbly qualified, who comes to the bench with an open*

> *mind—someone, in short, in the mold of Lewis Powell. A major issue upon which this nomination could turn is whether the nominee would alter significantly the balance of the Court. . . . Justice Powell has been the decisive vote in a host of decisions in the past fifteen years relating to civil rights and civil liberties. The scales of justice should not be tipped by ideological biases. I will resist any efforts by this administration to do indirectly what it has failed to do directly in the Congress—and that is impose an ideological agenda upon our jurisprudence. In light of the special role played with such distinction by Lewis Powell as the deciding vote on so many cases of tremendous importance, I will examine with special care any nominee who is predisposed to undo the long-established protections that have become part of the social fabric that binds us as a nation.*[3]

Former governor Pete du Pont was quick to accuse him of political motive. "Joe has clearly made confirmation a part of his presidential campaign and that's not right," he said. "I hate to see a job so important as this being used as a pawn in a presidential campaign."[4]

Biden understood full well what extensive hearings on a Bork nomination would do to his already taxing presidential campaign schedule. He decided to call former Republican Senate leader Howard Baker of Tennessee, who had reluctantly left Capitol Hill to become Reagan's chief of staff in the White House plagued by Iran-Contra and other scandals.

"Howard Baker had unerring political radar, and he had to know the administration was in no shape to make a tough battle," Biden wrote later.[5] He hoped he might be able to head off the Bork nomination, and Baker agreed to meet with him before any decision was made. Meanwhile, Biden went to Chicago to prepare for a critical debate among the Democratic presidential candidates in Houston in a few days.

The debate prep itself was not going well, he acknowledged, with bickering among his political and policy staffers over the campaign message and their own roles, and he kept popping Tylenols to combat the recurring headaches. Then came Howard Baker's return call. "Joe, this thing's moving pretty fast," he said. "You'd better get down here." Biden replied: "Howard, I thought we were going to talk first." Baker: "We are, but we'd better do it this afternoon."[6]

Biden and his aides caught a flight to Washington, continuing the debate preparation en route but without much focus. A police escort whisked them from National Airport to the office of Senate Majority Leader Robert Byrd, where he, Baker, and Attorney General Meese were waiting. Meese read off the names of several possible nominees and asked Biden and Byrd their reactions. When Bork was mentioned, Baker asked Byrd whether he would block the nomination. Byrd said no. Biden interrupted, he wrote later: "If you go ahead with Bork, it's going to be a long, hot summer."[7] It was left at that, and Biden flew off to Houston for the debate that night. On arrival, the press informed him that Bork had already been named. By this time, the Houston debate was totally overshadowed by the news; nothing said in the candidates' forum really mattered.

Biden's first mistake in response to the Bork nomination was in indicating publicly that, based on what he knew about Bork's views, he didn't see how he could support him. But he promised the judge would get a full and fair hearing before the Judiciary Committee. The battle lines quickly formed: conservatives who shared Bork's strict-constructionist views on one side, liberal interpreters of the Constitution including civil rights and civil liberties groups on the other. And Biden's comment of premature judgment immediately made him a target of charges of bias, unfairness, and bad politics all in one.

Doubts continued within the civil rights community about Biden's abilities to handle the fight against the Bork nomination. Indeed, Michael Pertschuk and Wendy Schaetzel of the nonprofit

Advocacy Institute wrote: "Democratic leadership of the [Senate] Judiciary Committee . . . was handicapped. . . . Joseph Biden was chairman by virtue of seniority, not choice. He had been reluctant to challenge Reagan's transformation of the federal judiciary. In particular, he had appeared to rule out ideology as a legitimate ground for challenging a Supreme Court nomination," and they cited the quote in the Philadelphia paper about having to vote for Bork if nominated.[8] Some Democratic senators confided they would have felt more confident with Kennedy leading the committee, and conservative columnist and Bork friend George Will predicted that the judge "will be more than a match for Biden in a confirmation process that is going to be easy."[9]

Ralph Neas, director of the umbrella Leadership Conference on Civil Rights (LCCR), understood the need for a broad and unified opposition to Bork. He was less concerned about Biden than he was with others in the coalition pursuing their single issues and putting politically destructive pressure on him.

One of the other leading voices, Estelle Rogers of the Federation of Women Lawyers, had already fired a shot across Biden's bow. Citing his earlier quote in the *Philadelphia Inquirer*, she demanded that Biden "retract his endorsement" and "exercise the kind of leadership expected from the chairman of the Senate Judiciary Committee." And, she went on, "if he can't, he'd be wise to think carefully about resigning the chairmanship."[10] Nan Aron of the Alliance for Justice warned of a "mass mobilization" against Bork. And Kate Michelman of the National Abortion Rights Action League, promising "an all-out frontal assault like you've never seen before" against Bork, replied when asked by a reporter whether Biden got the message: "Biden got it today. All day."[11]

As the Judiciary chairman and as a presidential candidate, Biden understood he could not afford to be seen as narrowly marching to the drum of either the civil rights or the women's movements in dealing with the Bork nomination. In a biting column Will, after labeling Biden "the incredible shrinking candidate" and also citing the *Philadelphia Inquirer* quote, had already charged he "has either changed his tune because groups were jerking his leash or, worse, to prepare for an act of preemptive capitulation."[12]

Biden concluded after long examination of past Bork rulings and speeches that the most legitimate basis for challenging him was to focus on the man's views, philosophy, and ideology, and the basic threat they posed to changing the direction of the Supreme Court.

In light of this thinking, more worrisome to Biden was Ted Kennedy's immediate, blistering attack on Bork from the Senate floor: "Robert Bork's America is a land in which women would be forced into back-alley abortions, blacks would sit at segregated lunch counters, rogue police could break down citizens' doors in midnight raids, schoolchildren could not be taught about evolution, writers and artists would be censored at the whim of government, and the doors of federal courts would be shut on the fingers of millions of citizens for whom the judiciary is often the only protector of the individual rights that are at the heart of our democracy."[13]

Kennedy's remarks, echoing the fears of special-interest groups, were the sort that Biden considered would put the confirmation of Bork on the wrong track. That approach, he wrote later, risked "quickly devolving to name-calling; Bork would be painted as a racist, a sexist, and a tool of the rich and powerful. Even if those things were true, it was the worst strategy I could imagine, and one I wanted no part of."[14]

One conservative Democrat, Senator Richard Shelby of Alabama, later to become a Republican, noted the likely reaction to Kennedy's speech. "With Senator Kennedy against him," Shelby said, "that puts a lot of southern Democrats in bed with Bork."[15]

To counter such an outcome, Biden and other anti-Bork northern senators approached the five new southern Democrats elected with overwhelming majorities of black voters in 1986. They enlisted these New South winners to support questioning Bork's ideology as a basis for opposing his confirmation.[16]

As Biden contemplated the challenge of the Bork nomination, however, he had to endure pressure from many in his presidential campaign to join Kennedy in hammering at Bork. His managers for the Iowa caucuses, especially his honorary chairman in the state, longtime Iowa Senate leader Lowell Junkins, and the principal political operatives, David Wilhelm and Mike Lux, said Democratic rivals Dick Gephardt, Michael Dukakis, and Bruce Babbitt were poised to declare against Bork at an approaching NAACP convention. But Biden was considering now the importance as Judiciary chairman of conducting a fair hearing—and, politically, of the appearance of it being so. Others in the campaign expressed concerns about how much time Biden's preparation for the confirmation hearings would take from the campaign in Iowa and in New Hampshire.

Finally, Bork himself, a few weeks after Reagan had nominated him, requested a private meeting with Biden. Accompanying the committee chairman at the meeting was his Judiciary counsel, Gitenstein, and Bork brought with him John Bolton, a former Bork student at Yale who then was an assistant attorney general for legislative affairs at the Justice Department. (Bolton much later was the controversial UN ambassador in the George W. Bush administration.)

By this time, Biden wrote of Bork later: "I'd done enough research into his avowed views on the Constitution by then to be almost certain I'd oppose his nomination. I didn't see how he could talk his way around the extreme positions he'd taken in so many published articles and speeches. Among other things, he'd refused to recognize a general right to privacy."

But face-to-face, Biden assured Bork he would get a fair hearing, there would be no filibuster, he would "get an up or down vote from the full Senate, and I gave the judge my word: 'I will engage in no personal attacks.' "[17] Bork in turn said he understood Biden was inclined to oppose him, but expressed the hope the chairman would give him a chance to change his mind in the hearings. It was a friendly meeting and Biden encouraged Bork to invite his family to attend. Bork said he would.[18]

Shortly after this meeting, Biden attended another critical and private session with Neas and other important members of the civil rights coalition on coordinating their approach to the Bork nomination. Some of them were impatient that the hearings get under way; others wanted to delay them in order to allow sufficient time to prepare. Biden responded to concerns about his presidential campaign by saying he considered the nomination of historic importance, and that he had told his campaign staff that he would sacrifice the whole presidential effort if necessary to see the nomination and hearings through. Then he told the groups he would decide the strategy and that if he was going to lead the fight it would not be a single-issue campaign—a pointed warning to abortion rights activists and others with a narrow cause to defend.

Biden found himself in the middle of pressures from all these intense anti-Bork foot soldiers, some with inflated egos and a sense of their own or their issue's importance, demanding to testify at the confirmation hearings. At the same time, he was determined not to give the Bork camp the chance to make the issue that these liberal special-interest groups were ganging up on the judge.

"The problem for the coalition," Pertschuk and Schaetzel wrote, "was that the decision *not* to testify had to be reached by a compelling consensus. . . . While it was true that Biden, committee chairman and presidential candidate, had the authority to deny any group the opportunity to testify, or to choose among the group leaders, the *political* reality was that Biden could not pick and choose without incurring the undying enmity of those who were denied. It had to be none, [or] (virtually) all."

The Advocacy Institute writers noted that "despite the tensions, no public controversy erupted to deflect attention from the hearings. No one publicly insisted upon testifying. And at the close of the hearings, Bork and his flawed constitutional vision—and not the coalition—remained at center stage." The *Washington Post*'s Mary McGrory called the decision of the various anti-Bork groups to accept this approach "the Supreme Sacrifice," and Biden had his way.[19]

Nevertheless, the anti-Bork forces formed one of the most expansive and intensive organizations to oppose a Supreme Court nomination in the nation's history, including the gathering and preparation of Bork's judicial record, writings, and speeches on which to base the case against him. The massive compilation came to be known within the coalition as "The Book of Bork" and was widely circulated among members of the Senate, the news media, and every group with an interest in rejecting the nomination. Biden himself underwent intensive preparations for his interrogation of the judge, with famed Harvard law professor Laurence Tribe grilling him and playing the role of Bork in what are called "murder boards" in the legal community.

After a telephone conversation between Biden and Strom Thurmond, the ranking Republican on the Judiciary Committee, a starting date for hearings of September 15 was agreed upon—seventy days after the Bork nomination, to allow for thorough preparation. No date was set for a committee vote, although Thurmond and some Democrats, including Kennedy, had pushed for October 1.

After the meeting, Biden met with reporters outside the room and announced the hearings' starting date, adding that the "overwhelming prospect" was that he would oppose the nomination. After having promised a fair hearing, that forecast of how he would vote was hardly an act of prudence. A *New York Times* reporter quickly broke the story of the private meeting, saying Biden was going to lead the fight against Bork, leaving the impression that the civil rights groups had successfully pressured him to take the stand they wanted.

Gitenstein and others then persuaded Biden to have his press secretary, Peter Smith, tell reporters he "intends to oppose the nomination and to lead the effort against it in the Senate." Neas told the

Times reporter that Biden "made it very clear to us that he knows what he's going to do, and that he considers the confirmation fight so important that he's willing to work on this, and not on the presidential campaign."[20]

The stories that resulted only muddled the situation for Biden. The *Washington Post* reported: "Biden told the activists, whose constituencies are important to his presidential campaign, that he will detail his reasons for opposing Bork in his upcoming statement." The *Post* editorial page hammered Biden: "While claiming that Judge Bork will have a full and fair hearing, Senator Joseph Biden this week has pledged to civil rights groups that he will lead the opposition to the confirmation. As the Queen of Hearts said to Alice, 'Sentence first—verdict afterward.' "[21]

As for Biden's promise to Bork of a fair hearing, the *Post* editorial asked: "How can he possibly get one from Senator Biden, who has already cast himself in the role of a prosecutor instead of a juror in the Judiciary Committee? If there is a strong, serious case to be argued against Judge Bork, why do so many Democrats seem unwilling to make it and afraid to listen to the other side?"[22]

Biden set to work on a major speech explaining his argument that examining Bork's ideological positions was a legitimate and not unprecedented basis for considering his nomination. In the meantime, he did not hesitate again to express his expectation that he would vote against Bork because he represented a determined effort by Reagan to change the direction of the Court toward a more dependable conservatism. In a C-SPAN interview, he said: "If I'm proven wrong—that Judge Bork is not part of the Reagan-Meese agenda on the Court—then in fact I would change my view. But I don't see any evidence of that at all. . . . The president has clearly decided on a political agenda for the Court. He and Meese have made a judgment that they will pick someone who would vote the way they want on the Court to fulfill their economic and social agenda. . . . The president has not been able to pass any of these things through the Congress."[23]

To counter the Biden argument, White House counsel A. B. Culvahouse got his counterpart in the Jimmy Carter White House, prominent Democratic lawyer Lloyd Cutler, to write a pro-Bork op-ed article in the *New York Times* suggesting that Bork's confirmation would not mean a radical change in the Court's direction. "Judge Bork," he wrote, "is neither an ideologue nor an extreme right-winger."

While it was true that Bork opposed the abortion rights law, Cutler went on, "this does not mean that he is a sure vote to overrule *Roe v. Wade;* his writings reflect a respect for precedent that would require him to weigh the cost as well as the benefits of reversing a decision deeply imbedded in our legal and social systems. . . . I predict that if Judge Bork is confirmed, the conventional wisdom of [subsequent years] will place him closer to the middle than to the right, not far from the Justice [Powell] whose chair he has been nominated to fill."[24]

With Ralph Neas and other civil rights activists clamoring for Biden to make a definitive anti-Bork speech, the senator tried to sandwich in some presidential campaigning in New Hampshire but was heckled by prolife protesters. The next weekend he flew to Cleveland for a meeting of Democratic Party state chairmen. There he added fuel to the fire by saying in a question-and-answer period that he had made "the biggest mistake of my political career in coming out against Bork the way I did."

When the comment only generated more allegations of Biden's inconsistency and of yielding to pressures by civil rights groups, he compounded the matter by saying "it was more of a public-relations mistake than a substantive mistake." He seemed unable simply to shut up. "The thing that is so galling to me," he confided to Gitenstein, "was that I've worked so hard to develop positions independent of the groups but which represent the values I believe we share. The way I'm being portrayed is as if I'm their tool. That's not me."[25]

While his campaign managers in Iowa continued to urge him to come into the state and declare himself squarely against Bork, Biden focused on his case against the judge. He told Junkins: "This was bigger than politics. I was coming off the road to prepare for Bork. The presidential thing would sort itself out."[26] Meanwhile he persuaded an uncomfortable Jill to fill in for him in Iowa.

There also was some friction between the Iowa campaign operatives and the national political team, and within them. "There was some tension, as you can imagine with those egos involved," Gitenstein recalled. "What some of them wanted and what some didn't want, and especially between the gurus and the campaign about Biden's personal time; the time he wanted to spend with the kids, he was very protective of that. Most of [the national gurus] were a lot younger than he was; they didn't have kids. They didn't understand what it was like to have children. They didn't know why it was necessary for Joe to go to Beau's soccer games when he was supposed to be in Iowa doing x, y, or z. But the real stress did not have to do so much with the kids, but with his Senate duties, with the Bork hearing." And the local guys, Wilhelm, Lux, and Junkins, wanted to know: "Why didn't he just come out against Bork?"—the popular position among Iowa Democrats.[27]

"The first thing I had to do," Biden wrote later, "was reset the table on the nomination process, which in the recent past had focused almost solely on character and qualifications. Robert Bork was a bona fide scholar . . . the way to stop Bork as I saw it was on the question of his outside-the-mainstream judicial philosophy—or ideology—and that was a long shot, too. There was no modern-day precedent for that kind of fight in the Senate. . . . The basic understanding was that as long as a Supreme Court nominee had the intellectual capacity, a breadth of experience in constitutional law, and a reasonable judicial temperament, and had committed no crimes of moral turpitude, the Senate was bound to confirm a nominee. Ideology was the third rail of Supreme Court nominations."[28]

Biden wrote that he also knew that in the toughest confirmation cases in the 1960s—LBJ's effort to elevate his friend Abe Fortas to chief justice and Nixon's failed nominations of G. Harrold Carswell and Clement Haynsworth—the nominees "were rejected for personal shortcomings, but the clear and unspoken reason was ideology. Personal attacks had merely given the opposition cover in each case. I thought it was time to take up ideology in the open and avoid personal attacks . . . [and] argue to my colleagues that if the president wanted to nominate based on Bork's stated constitutional philosophy, the Senate was bound to investigate the meaning and potential implications of that philosophy."[29]

In a major speech on the Senate floor in late July, Biden cited specific precedents in which ideology had been the deciding factor, starting with George Washington's failure to seat John Rutledge in 1795 up to FDR's court-packing scheme to preserve New Deal legislation, slapped down in the Senate. He specifically pointed out that in the Fortas fight, a senator noted that a young lawyer named William Rehnquist had urged in the Harvard Law Record restoration of the Senate practice "of thoroughly informing itself on the judicial philosophy of a Supreme Court nominee before confirming him." Biden concluded, "We are once again confronted with a popular president's determined attempt to bend the Supreme Court to his political ends. No one should dispute his right to try. But no one should dispute the Senate's duty to respond."[30]

Biden wrote later that his core argument against Bork was that he did not share, with Justice Felix Frankfurter, his belief that "there were fundamental rights that deserved the protection of courts whether or not they were specifically enumerated in the Constitution. . . . To me that was the central glowing idea that lights the path of our democracy and defines the relationship between individual citizens and government. If there were no Constitution, I believe, human beings would still have a right to marry whom they want. We would still have the right to see our biological offspring, the right to speech, the right to practice a religion. Judge Bork, however, thought we have our rights because the Constitution relinquishes them to us—and jealously. As a justice he would not recognize fundamental human rights beyond what was spelled out in the Constitution."[31]

Biden's seemingly jumping the gun on opposing Bork suffered sniping from unexpected quarters. His supposed unofficial confidant, sometimes-journalist William Schneider, was quoted in the Wilmington paper as saying: "He is ridiculed and discounted as politically maladroit and inept. People are already talking about the collapse of the Biden campaign. . . . He's made defeat of Bork more difficult, and galvanized the GOP. Joe didn't look like a statesman."[32]

But Biden dug in, joining his staff in continuing to search the Bork record and rulings, and sounding out legal experts. Together they constructed a shopping list of Bork assaults on individual rights. One issue noticeably lowballed in the anti-Bork camp strategy was abortion. The Leadership Conference on Civil Rights achieved agreement to focus on Bork's overall legal philosophy and how it would change the basic direction of the Court.

In mid-August, Biden addressed a meeting of the American Bar Association (ABA) in San Francisco in advance of its customary rating of a Supreme Court nominee's qualifications. He laid out five specific cases in which Bork had attacked decisions that defended individual privacy rights dealing with use of birth control, forced sterilization, racial covenants, even the one-man-one-vote doctrine, and concluded: "We cannot be certain that these are among the dozens of precedents that Judge Bork might vote to overturn. But we can be certain that if Judge Bork has meant what he's written for the past thirty years—and that had he been Justice Bork during the past thirty years and his view prevailed—America would be a fundamentally different place than it is today. We would live in a different America than we do now."[33] The audience of lawyers rose in sustained applause.

Retired chief justice Warren Burger followed Biden before the ABA and gave Bork a ringing endorsement: "I don't think in more than fifty years since I was in law school there has ever been a nomination of a man any better qualified than Judge Bork. . . . I don't really know what the problem is." But when the ABA rating came in later, while ten participants found him "well qualified," one voted "not opposed" and four rated him "not qualified" on the basis of judicial temperament.[34]

The Reagan White House hit back with the release of a packet labeled "Materials on Judge Bork," known internally as the Blue Book. Biden dismissed it in advance of the confirmation hearings as a "transparent effort" to paint Bork as "the apostle of judicial restraint and moderation" in the fashion of the justice he would replace, Lewis Powell. Reagan himself in a nationally televised speech before going off on August vacation touted Bork as "a brilliant scholar" who had never had a single opinion reversed by the Supreme Court on appeal and would be "an important intellectual addition" to it.

Meanwhile, Biden undertook a marathon lobbying of the nation's major newsrooms and editorial boards, presenting himself as much as his arguments. Accompanying political aides detected considerable skepticism toward the Senate Judiciary Committee chairman, and his motives as a presidential candidate. But the intensity of his presentations, and surprise about his detailed grasp of the substance of his case, seemed persuasive to many listeners. Stuart Taylor, the *New York Times* reporter covering the Supreme Court at the time, said later that the editors in New York "saw everything that Biden said as political, as part of his [presidential] campaign. [To them] Biden was just trying to make a big deal out of something that wasn't so important. . . . Sometimes New York [*Times* editors] seemed more concerned about how [what was written] affected Biden and his campaign. That tended to be more important than what Biden or the opponents of Bork were saying."[35]

Biden's often colloquial and even flippant manner could lead listeners to underestimate him, and he sometimes added to the impression by references to his own encouraged reputation as a casual student of low ranking in law school. In a hearing in 1983 for a federal appeals court nominee, J. Harvie Wilkinson III, Biden in referring to the man's strong academic record quipped "he probably would not have spoken to me in law school; he would have had to go through too many people to get down to where I was." And he often quoted a wry letter of introduction a professor allegedly wrote for him to a prospective employer: "You'll be lucky to get Biden to work for you."[36]

Bork also made the rounds of key editorial boards, but William Bradford Reynolds at the Justice Department acknowledged later that the White House was caught napping on the threat Biden posed to the Bork nomination. "We did not perceive the political debate going on in the July–August time frame as having the kind of impact that would carry over into the hearings," he said, and what was done "was too little, too late. We did not use the brunt of White House on the matter. One, because there was a miscalculation as to how serious the problem was, and, two . . . as we saw it, the strategy that was most effective was to save our big guns until later," meaning having Bork testify in his own behalf.[37]

With Jill Biden still holding the fort in Iowa in the presidential race, her husband in late August retreated to the Delaware ocean resort town of Bethany Beach to continue preparations for the Bork hearings now only weeks off. He and key staff aides pored diligently over briefing books on Bork's past judicial decisions and observations, breaking off only to catch a charter flight to Des Moines for a critical campaign debate at the Iowa State Fair, the summer's featured political affair.

The Bork hearings remained Biden's top priority, but he could not afford to miss this event, in which all the other Democratic presidential candidates would participate. What he would say there would soon prove to be an even bigger mistake than his premature declaration of opposition to the Bork nomination that had already caused him so much political angst.

THIRTEEN

DEBACLE IN IOWA

SOMETIME IN AUGUST, William Schneider, the CNN political analyst/commentator and a hanger-on in the Biden campaign, put his journalist's hat on a hook and became a collaborator. He turned over to Mark Gitenstein and Tom Donilon a videotape he had acquired of a political commercial for a British Labour Party candidate, Neil Kinnock, showing him delivering a speech in his recent unsuccessful election campaign against Prime Minister Margaret Thatcher. In it, Kinnock talked of the political opportunities that had come to him and his sturdy family forebears from coal-mining country as a result of the helping hand they had received, and given in return, along the way. Shown to Biden, it struck a chord with him, reminding him of his own fortunes.

"Why am I the first Kinnock in a thousand generations to be able to get to university?" the Labour candidate asked.

> *Why is [his wife] Glenys the first woman in her family in a thousand generations to be able to get to university? Is it because*

all of our predecessors were thick? Did they lack talent? Those people who could sing, and play, and recite, and write poetry, those people who could make wonderful, beautiful things with their hands? Those people who could dream dreams, see visions? Why didn't they get it? Was it because they were weak? Those people who could work eight hours underground and then come up and play football? Weak? Those women who could survive eleven child-bearings? Were they weak? Anybody really think they didn't get what we have because they didn't have the talent, or the strength, or the endurance, or the commitment? Of course not. It was because there was no platform upon which they could stand.

That last phrase particularly captured Biden's attention. It squared with his own view that his political party in its championing of the poor also gave voice to the broad American middle class from which he and his own family had sprung. "That was my argument at its heart," he wrote later. "It was so simple. That's what the Democratic Party should be doing for all citizens, providing 'a platform upon which they could stand.' People weren't asking for a free handout from government or a promise of fabulous outcomes. They just wanted a little support to help raise them higher. I watched the Kinnock ad once, and I never forgot it—partly because it rhymed with my own family experience. Uncle Boo-Boo never let my father forget that I had been the first Biden to go to college. I had ancestors from the coal-mining town of Scranton."

Biden liked the remarks so much that he began, in his infrequent detours from Bork preparations onto the campaign trail, to adapt them to his own situation, always with reference or credit to Kinnock. But on the flight to Des Moines for the debate at the Iowa State Fair on August 23, Biden was too occupied discussing with aides the strategy and substance for the Bork hearings to worry about what he would say. "I confess to a certain amount of arrogance in not taking some time to get ready for the State Fair debate," he wrote later. "This was about my believing I could talk my way through the most important campaign event of the summer. . . . I spent no time preparing for it. When the plane touched down in Des Moines, I hadn't prepared an open or a close [for his remarks]."

Waiting for him on arrival was his Iowa campaign director, David Wilhelm, later a Bill Clinton aide who became chairman of the Democratic National Committee. Wilhelm asked Biden whether he was ready for the debate and was told he hadn't decided on his close. Wilhelm suggested, "Why don't you use the Kinnock stuff?" and Biden readily agreed.

At the fairgrounds, he was taken to a separate holding pen and was about to start writing something when "I got an urgent summons from Jesse Jackson [also in the debate]. So I went over to see Jesse, who was just trying to help me. Each of the candidates would get to ask one question of another candidate. Reverend Jackson had drawn me, and he wanted to tip me to the question. I said he didn't have to do that, but Jesse said he was tired of how mean the little guy, Dukakis, was getting. There had been too much bickering among friends. He was trying to inject a little decency in the proceedings." By the time Biden got back to his holding pen, he recalled, the candidates were being called to the stage, and he still hadn't had time to prepare his close.

Biden later wrote what happened next: "The debate went fine, and when I got to my close, I just did Kinnock's platform thing. But it was a limited time, so I rushed: 'I started thinking, as I came over here. . . . Why is it that Joe Biden is the first in his family ever to go to university? . . .' I ran through the piece whole, from memory. 'The family of coal miners. Were they not smart? . . . Were they weak? . . . No . . . they didn't have a platform on which to stand.' The power of the sentiment was hard to miss. There was dead silence as I spoke, and I remember looking down at a lady in the front row who was, despite her best efforts, in tears."[1]

When Biden finished and came off the stage, one of his aides mentioned to him that he hadn't made the attribution to Kinnock that he usually did. His press secretary, Larry Rasky, noticed but recalled later: "There were a hundred reporters covering that debate. I figured if there was a problem, somebody would have said something to me about it." As far as Biden and his aides were concerned, that was the end of it.

Over the next couple of weeks, Biden on other occasions used the Kinnock remarks, tailored to himself, with proper attribution. A story in the *New York Times* six days later by Robin Toner reported near the end of a sixteen-paragraph story from New Hampshire that Biden had offered "a paean to the Democratic Party and its commitment to equality of opportunity improvising on a speech by Neil Kinnock, the British Labour leader."

Jack Farrell, a *Boston Globe* reporter, also heard Biden attribute Kinnock in New Hampshire and mentioned it in a story. "I followed him for three days and every time he used it, he gave Kinnock credit," he said. When later he saw the Biden speech at the Iowa State Fair on television without the attribution, Farrell said, "it was awful hard for me to criticize him for not using it once in Iowa." And in a taped interview with David Frost never aired on television, Biden again quoted Kinnock with proper attribution, in arguing "that's what this nation has to do, to build a platform on which people can stand." Biden later lamented: "All I had to say was, 'Like Kinnock,' and I didn't. It was my fault, nobody else's fault."[2]

At the time, Biden had been cruising along in the polls at about 10 percent support in Iowa, bunched with several other contenders near the top on a sort of prairie populism reflected in the Kinnock pitch. And he was gathering plenty of key Iowa political endorsements, convincing Marttila and other campaign pros that he was well positioned to break out in the kickoff caucus state. On Saturday, September 12, twenty days after Biden's Iowa State Fair remarks, the story of the borrowed usage suddenly broke simultaneously in the

New York Times and the *Des Moines Register*—a suspicious coincidence on its face. Both reports pointed out the striking parallels between the words of Kinnock and Biden. In the front page of the *Times*, under the headline DEBATE FINALE: AN ECHO FROM ABROAD, reporter Maureen Dowd wrote that Biden had "lifted Mr. Kinnock's closing speech with phrases, gestures, and typical Welsh syntax intact for his own closing speech . . . without crediting Mr. Kinnock."[3] Biden at once responded that on other occasions he had made the attribution, and Rasky called it "a tempest in a teapot."[4]

But Biden's reference to having just thought on his way to the debate of the fact that he was the first Biden to go to college—to "university" in the British manner—stuck out. So did his version of the Kinnock reference to the coal-country background and forebears coming out of the mines to kick a football around. He hadn't simply appropriated Kinnock's words; he had essentially taken on his persona in a rough and imprecise parallel.

In Dowd's story, she made no reference to any source alerting her to the similarities between the Kinnock and Biden quotes. The story the same morning in the *Des Moines Register*, by the paper's top political reporter, David Yepsen, reported that his account had been based on what he called an "attack video" of the Kinnock speech. Unlike Dowd, Yepsen wrote that "an aide to one of Biden's opponents, who spoke on the condition that he would not be named, is circulating a videotape showing a Kinnock commercial followed by Biden's closing word using some of the same lines."[5] By so doing, Yepsen let it be known that an act of political sabotage was involved.

Rasky said later the Biden strategists at first tried to fan speculation that the Kinnock disclosure had been orchestrated by the Republicans to undermine Biden's leadership of the Bork inquiry. Others in the Biden camp suspected rival Democratic presidential candidates might have played a role out of concern that Biden as committee chairman would be the chief political beneficiary of the nationally televised confirmation hearings. "What really steams me,"

Rasky told the *Washington Post*, "is that here we are on the eve of the Bork hearings and another Democratic candidate is deliberately trying to undermine Biden. . . . Somebody here has committed an outrageous act."[6]

At first, fingers pointed at the Dick Gephardt campaign, which some thought at the time would be most threatened by Biden's candidacy. Also, Gephardt strategists included two old Pat Caddell associates who had fallen out with him, Bob Shrum and David Doak, with personal animosity cited as the motive. When Michael Dukakis was asked whether any of his aides were involved, he denied it, saying he would be "very angry" and "astonished" had they been because "anybody who knows me and knows the kind of campaigns I run knows how strongly I feel about negative campaigning."[7]

But Dukakis, it turned out, had reason to be both angry and astonished. John Sasso, his longtime aide in Massachusetts and his campaign manager, had obtained the Kinnock tapes and had recognized the similarities with Biden's Iowa State Fair remarks. Just before Labor Day, Dowd had phoned him to discuss a profile on Caddell and had remarked on the similarities. Sasso told her Biden had not attributed Kinnock's remarks to him at the fair. "When Maureen raised the question," Sasso said later, "I leaped at it. I had been looking for a way to get someone to write about it."[8] He sent a copy of the videotape to her and had another delivered to Yepsen, on the basis that they would not reveal the source.

When the story broke and the speculation grew of political sabotage, and in light of Dukakis's flat denial of involvement, Sasso finally went to his boss and offered to resign. But Sasso was his closest adviser and one of his closest friends, and Dukakis did not want to part with him. He first suggested that Sasso take a leave of absence, but realized a clean break was imperative. Another key political aide, Paul Tully, who had delivered the tape to Yepsen, joined Sasso in resigning—at Dukakis's insistence.

The governor phoned Biden and apologized, telling him he had no idea that Sasso had been involved. Afterward, Biden described the conversation to my column partner, Jack Germond, and me. "Kitty and I feel very badly about this," he said Dukakis told him. "I know what it's like. I'm sorry this happened, but John's a good man." Biden said he replied: "Governor, thanks for calling."[9]

Much later, author Richard Ben Cramer, a particularly perceptive observer of both men, noted in his excellent book *What It Takes:*

> *Biden, of course, didn't believe him. Biden would have known. Hell, yes—if it was his campaign! No one could convince Biden that Michael was not involved—or that Dukakis and his minions hadn't deliberately hit him, just at the start of the hearings on Bork . . . just when he couldn't hit back.*
>
> *That's what galled Joe—the worst. Biden didn't want to talk to Dukakis. He didn't want a lot of holy explanations. He had only one point to make to the Governor:*
>
> *"Don't you assholes understand? This shit [keeping Bork off the Supreme Court] is important!"*[10]

Biden himself told us after the campaign, in gentler language, why he thought the deed was done: "I think they knew what I knew—that I'd beat them" because he had raised more money and had a better organization. His answer also reflected what many others would have conceded, that the earnestly bland Dukakis was no match for Biden in rousing a crowd. "I didn't pass a moral judgment on Sasso or Michael Dukakis," Biden said. "What angered me most was that I thought it was a cheap shot at a time when they were smart enough to know I would not be in position to really be able to respond. It was obviously extremely well and precisely timed. No one in the world can ever convince me that it just all of a sudden appeared on Friday before the Bork hearings, they put it together Saturday and put it out."[11]

Dukakis also phoned Dick Gephardt and apologized, because by this time an unfounded rumor had circulated in political circles that it had been Gephardt, even more of a Goody Two-shoes than Dukakis, who had planted the videotape, to cool off the Biden campaign in the one state in which Gephardt had been the caucus early front-runner. Joe Trippi, then a young Gephardt campaigner, said of the scuttlebutt: "There was a sense on our part that they [the Dukakis operatives] had gone for a two-fer"—making Biden look bad and getting Gephardt blamed for it.[12] Later on, Dukakis went on a sort of mea culpa tour of Iowa, reassuring voters he had not intentionally poisoned the climate for the caucuses with the Kinnock caper. He got an earful from disappointed Biden supporters and other still suspicious Iowans.

More troublesome for Biden than any particular detail was the concern raised by the Kinnock story that there might be more unattributed borrowing by him from other colorful speakers. Biden had delivered that speech earlier in the year to a Democratic state convention in California that turned out to include verbatim phrases about the virtues of the American people from another given by Robert Kennedy in 1967. A reporter for the *San Jose Mercury-News*, Phil Trounstine, compared parts of the two speeches and wrote about the similarity. In this case, Biden aides said, Biden's speech had been written by Pat Caddell without any attribution to Kennedy. "In fairness to Pat," Biden said later, "he probably assumed that I knew it was Kennedy's."[13] Some television screens showed the two versions on a split screen, a particularly damaging device.

The San Jose paper and others also weighed in with more examples of Biden using lines allegedly from other Robert Kennedy and Hubert Humphrey speeches, and Republicans were quick to join the accusers. A Ronald Reagan speechwriter and political adviser, Kenneth Khachigian, said: "What Biden did, or what his speechwriter did, is unforgivable. He has fashioned himself to be a great orator, and yet to achieve that, he is having to draw on the great language of others."[14]

A prominent labor leader, requesting anonymity, said: "It illustrates something that all of us worry about with Biden, that he doesn't seem to be in control of his mouth, or his campaign either." A leading Democratic pollster, Harrison Hickman, observed of Biden: "He has said consistently that the next president must be the one who can motivate the American people. And if you're going to make that argument, it seems a fair test to ask whether Joe Biden is a visionary or is Joe Biden [just] a good speaker?" And another pollster, Geoffrey Garin, observed: "This controversy plays into the case his opponents would like to make against him—that he is a person of style rather than substance."[15]

A former Robert Kennedy speechwriter, Adam Walinsky, said that he had written the 1967 speech for RFK from which the Biden speechwriter had borrowed without attribution. "It's a counterfeit of emotion," he said with some irritation, an odd observation for a speechwriter for hire. "If what you do is take something that is somebody else's and that somebody else's staff prepared and pass it off as your own, all that tells us is that you're good at copying."[16]

Still another blow came in the resurrection of a minor scene in New Hampshire the previous spring, when Biden replied to a heckler charging he had grossly exaggerated his academic record. Unfortunately Biden was caught by a C-SPAN camera telling his tormenter, "I think I probably have a much higher IQ than you do, I suspect." He proceeded to inform him: "I went to law school on a full scholarship, the only one in my class to have an academic scholarship. In the first year . . . I decided I didn't want to be in law school and ended up [in] the bottom two-thirds of my class and then decided I wanted to stay, went back to law school and ended up in the top half of my class. I won the international moot court competition."

Biden also claimed that at the University of Delaware he had been the outstanding political science undergraduate and had finished with three undergraduate degrees. His law school records showed, however, that in his first year at Syracuse he was ranked eightieth out of a hundred students and in his final year seventy-sixth out of eighty-five. Also, his full academic scholarship was half based on need, and rather than three undergraduate degrees, he earned one with a dual major in history and political science. Also, he had shared the moot court award with two other students.

When a *New York Times* reporter called Biden at home one night questioning him about the moot court competition, Valerie remembered a plaque on his wall, ran up to his room, and pulled it down, enabling him to read the citation to his inquisitor. "It was like we found gold," he remembered. "It was like, we're safe, we're free."

But by now, Biden recalled later of his presidential campaign, "the floodgate had opened. It was like the red tide was rolling. And all the time I was having to get up every morning, bang the gavel, know what I'm talking about with Bork, keep my committee in a posture I thought it should adopt." All the while, too, he was experiencing those splitting headaches that kept him popping those painkillers. "I couldn't figure why I was having headaches," he said later. "I could not figure out why I was living off Extra-Strength Tylenol. I'd never had a headache in my life."[17]

The Kinnock story breaking on the weekend before the scheduled start of the Bork confirmation hearings made matters worse. As the time of the opening gavel approached, Biden was caught between the pressures of gearing up for the critical confrontation and having to ward off the assaults on his personal reputation that had already dealt a severe blow to his presidential aspirations.

Back in Delaware, Joe Biden often would say, in vowing for something: "I give you my word as a Biden," and to his friends and constituents it was bankable. From childhood, his parents had drummed into him, and into his brothers and sister, what it meant to be a Biden, and the burdens it imposed in terms of loyalty and, above all, honor. As he had moved up the ladder of public life and politics, one of his most steadfast commitments had been to uphold the trustworthiness of the family name.

The worst part about the allegations of plagiarism was how they threatened to tarnish that credibility. "The alarm bells went off for Jill right away," he said later. "They [reporters] were questioning the one thing she saw as my greatest strength—and something I would never be able to defend with words alone. 'Of all the things to attack you on,' she said, almost in tears. 'Your integrity?' "[18]

All this might not have set the news media on a hunt for more holes in Biden's political armor had they not already been unleashed by the disclosures of personal behavior of another sort about Gary Hart that had forced him prematurely from the 1988 Democratic presidential race. The outbreak of "gotcha" journalism more than a decade earlier that had brought such celebrity to the Watergate sleuths, Bob Woodward and Carl Bernstein of the *Washington Post*, was now in full flower, with Biden on the receiving end.

"Challenging his integrity was really a hard blow for him to take," Marttila remembered. "He was really wounded. I was witnessing this cascade of fairly minor issues coming at him, at the time he had the responsibility of conducting the Bork hearings." And on top of all this, he said, the campaign strategists "needed to see eye to eye" in assessing the dilemma, and they "lacked strategic unity."[19]

Suddenly, as Biden took on his greatest task as a Senate committee chairman in examining the qualifications of a Supreme Court nominee whose confirmation could change the course of justice in the land, he found his "word as a Biden" very seriously open to question.

Years later, he acknowledged that "I didn't mind" having to withdraw from the president race, but the allegation against him "was a killer. I doubt if there's any person you've ever interviewed has ever questioned my integrity. And here I was, getting out of a race with my integrity being questioned. That was the killer for me. It wasn't not being president. It was the allegation. It just really ate me up. It was the core of who I thought I was, who I know I am, actually."[20]

FOURTEEN

CONCENTRATING ON BORK

ON SEPTEMBER 15, 1987, Biden finally gaveled open the Bork confirmation hearings in the historic Caucus Room of the Richard B. Russell Senate Office Building. It was the site of the announcements of presidential candidacies of John and Robert Kennedy, of the infamous Army-McCarthy hearings, and the Senate Watergate hearings that eventually led to the resignation of President Richard M. Nixon. By now, the stories of what were widely being described as Biden's plagiarisms were rampant.

"I felt I was at the edge of an abyss," he wrote later. "The White House was playing hardball politics, but I had left myself wide open. I couldn't even say it was a cheap shot. And if I let the hits on me affect the way I ran the Bork hearings, my colleagues on the Judiciary Committee would see it. The full Senate would see it. The entire watching public would see it."[1]

After brief appearances by former president Gerald Ford, Senator Bob Dole, and a couple of other Republicans as character witnesses for Bork, Chairman Biden called on several other members of the committee to make opening statements before him, with Judge Bork patiently in the witness chair. When the turn came of conservative Republican Senator Alan Simpson of Wyoming, an obvious Bork supporter, he sought to reassure the judge he would get a fair hearing, in words that might well have applied to Biden as well in his difficult circumstances.

"Obviously, we are going to be picking at a lot of old scabs," Simpson said. "Too bad. Who among us here on the panel—we in the U.S. Senate—are designated as the official scorekeepers of our fellow humans? Who does or does not judge, when we put aside the mistakes, the utterances, the errors of our earlier lives, and who in this room has not felt the rush of embarrassment or pain or a feeling of plain stupidity about a phrase previously uttered or an act long ago committed? Who of us here can pass that test. . . . It seems to be an unpleasant reality that a Supreme Court nominee has every single constitutional protection until he or she walks into this room. And once in this room, unlike a defendant in a court of law, the nominee is not guaranteed a single right analogous to the Miranda Rule or the Fifth Amendment or anything else." Then, to Bork: "I await your presentation with great anticipation. The chairman and all members of the committee will assure you that you will be treated most fairly and courteously."[2]

Biden, who had earlier given Bork his personal guarantee of an open and fair confirmation review, picked up on Simpson's assurance. "Judge Bork," he said as he raised the gavel and faced the nominee a short distance before him, "I guarantee you this little mallet is going to assure you every single right to make your views known, as long as it takes, on any grounds you wish to make them. That is a guarantee, so you do have rights in this room, and I assure you they will be protected."[3] Biden's manner was so deferential that the *Washington Post*'s television critic, Tom Shales, was moved to write, "Biden did everything but rush over to Bork's water glass with an ice-cold refill."[4]

The chairman then proceeded with his own opening statement, launching into his central argument that the issue was more than Bork himself, rather a question of how a man of his ideological views if confirmed would affect the basic direction of the whole Supreme Court. "Will we retreat from our tradition of progress," he asked, "or will we move forward, continuing to expand and envelop the rights of individuals in a changing world which is bound to have an impact upon those individuals' sense of who they are and what they can do? . . . In passing on this nomination to the Supreme Court, we must also pass judgment on whether or not your particular philosophy is an appropriate one at this time in our history. . . . My role as chairman of the Senate Judiciary Committee in my view is not to persuade but to attempt to ensure that the critical issues involved in this nomination are laid squarely before my colleagues and the American people. . . . As a matter of principle I continue to be deeply troubled by many of the things you have written . . . our differences are not personal."[5]

His concerns, Biden went on, touched on "the relationship of people of different races in our land; whether it was wrong for state courts to enforce covenants that prohibited black couples from buying homes in white neighborhoods; whether the court was wrong in not stopping the U.S. Congress from outlawing literacy tests to protect voting rights; and whether in the future as similar situations arise the Court will intervene to protect the rights of the races in the land." Biden listed other privacy rights: in marriage, in child-raising, in having private schools, in having foreign-language training, in use of birth control; above all in freedom of expression in politics and the arts.

Biden then challenged Bork's basic argument of enumerated rights in the Constitution, saying: "I believe all Americans are born with certain inalienable rights. As a child of God, I believe my rights are not derived from the Constitution. My rights are not denied by any majority. My rights are because I exist. They were given to me and each of my fellow citizens by our creator, and they represent the essence of human dignity."[6]

Throughout Biden's remarks, Bork sat calmly, occasionally nodding. Then, in his own opening statement, he replied:

> *How should a judge go about finding the law? The only legitimate way, in my opinion, is by attempting to discern what those who made the law intended. . . . Where the words are precise and the facts simple, as in the case with some of the most profound protections of our liberties—in the Bill of Rights and in the Civil War Amendments—the task is far more complex. It is to find the principle or value that was intended to be protected and see that it is protected. As I wrote in an opinion for our court, the judge's responsibility "is to discern how the framers' values, defined in the context of the world they knew, apply to the world we know." If a judge abandons intentions as his guide, there is no law available to him, and he begins to legislate a social agenda for the American people. That goes well beyond his legitimate powers. He or she then diminishes liberty instead of enhancing it. . . .*
>
> *The judge must speak with the authority of the past and yet accommodate that past to the present. The past, however, includes not only the intentions of those who first made the law, it also includes those past judges who interpreted it and applied it to prior cases. That is why a judge must have great respect for precedent. It is one thing as a legal theorist to criticize the reasoning of a prior decision, even to criticize severely as I have done. It is another and more serious thing altogether for a judge to ignore or overturn a prior decision. That requires much careful thought. . . . Respect for precedent is a part of the great tradition of our law, just as is fidelity to the intent of those who ratified the Constitution and enacted our statutes. That does not mean that constitutional law is static. It will evolve as judges modify doctrine to meet new circumstances and new technologies.*[7]

And so the Biden-Bork debate had begun. Biden opened the questioning by asking Bork to name the "dozens of cases" that in 1981 he had told Congress had been decided wrongly but had declined to identify. Biden noted that four years later in a legal magazine Bork was asked again whether he could identify them, and he curtly replied: "Yes, I can, but I won't." Biden pressed him once more and Bork replied only that he could "discuss with you the grounds upon . . . which I would reconsider them. . . . There is in fact a recognition on my part that *stare decisis*, or the theory of precedent, is important. In fact, I would say to you that anybody who believes in original intention as the means of interpreting the Constitution has to have theory of precedent, because this nation has grown in ways that do not comport with the intentions of the people who wrote the Constitution. . . ."[8]

Biden saw his opportunity. Introducing the famous case *Griswold v. Connecticut*, in which the Supreme Court ruled that a married couple had a right of privacy in the use of birth control, he noted Bork had disagreed, contending there was no such right in the Constitution. Citing one Bork article, Biden quoted him as saying "that the right of married couples to have sexual relations without fear of unwanted children is no more worthy of constitutional protection by the courts than the right of public utilities to be free of pollution control laws," or as Biden put it, "to pollute the air. Am I misstating your rationale here?"

Bork replied: "With due respect, Mr. Chairman, I think you are. I was making the point that where the Constitution does not speak—there is no provision in the Constitution that applies to the case—then a judge may not say, 'I place a higher value upon a marital relationship than I do upon an economic freedom' . . . if there is nothing in the Constitution, the judge is enforcing his own moral values, which I have objected to."[9]

A moment later, Biden said, "Then I think I do understand it, that the economic gratification of a utility company is as worthy of as much protection [from pollution control laws] as the sexual gratification of a married couple, because neither is mentioned in the Constitution." And again: "Does a state legislative body, or any legislative body, have a right to pass a law telling a married couple or anyone else that . . . behind their bedroom door, telling them they can or cannot use birth control?" Bork answered: "There's always a rationality standard in the law, Senator, and I don't know what rationale the state would or what challenge the married couple would make. I have never decided that case. If it ever comes before me, I will have to decide it."[10]

They went around like that for a few more minutes, with Biden arguing his position in the most dramatic way while Bork fell back on indicating he didn't disagree with the decision, just the rationale by which it was reached. Finally, Biden said: "As I hear you, you do not believe that there is a general right of privacy that is in the Constitution." Bork replied: "Not one derived in that fashion. There may be other arguments and I do not want to pass on those."[11]

Biden observed later of his reaction at the time: "Bork can't come up with a good reason to stop the government from intruding in my bedroom?"[12] Biden's repeated pounding on the same matter may have reached the point of boredom in the hearing room, but as it went out to the country over television, it easily won the sound-bite game.

In this first direct confrontation with Bork, Biden demonstrated a sure grasp of his material, which for weeks he and his Judiciary majority staff, led by Mark Gitenstein, had ferreted out from Bork's judicial record. They found rulings that would most effectively support Biden's case—that Bork's rigid view of interpreting and applying the Constitution promised to give Reagan and Meese the Court they wanted, predictably supportive of their conservative social agenda.

The hearings continued the next morning, with Biden presiding but this time leaving most of the questioning of Bork to other senators. Much of the interrogation concerned the judge's role as solicitor general in the firing of Watergate special prosecutor Archibald Cox. With Biden focused less on that episode to oppose Bork's confirmation than on his judicial philosophy, Democratic Senator Howard Metzenbaum of Ohio led the questioning.

He began by accusing Bork of having violated "the regulation in effect when you fired Mr. Cox" that, Metzenbaum said, flatly prohibited doing so "unless he engaged in extraordinary improprieties, which he clearly did not. That regulation had the force and effect of law. Under these circumstances, your firing of Mr. Cox was a violation of the law, was it not?" Bork replied: "No, I do not think it was, Senator." He went on to assert that the matter was mooted by the appointment of a new independent prosecutor, and in the end it was inconsequential in determining the fate of Bork's nomination.[13]

In the middle of the testimony, Biden aides Larry Rasky and Tom Donilon suddenly called the chairman out of the Caucus Room. The old story had surfaced about Biden's failed grade as a first-year law student at Syracuse for improper footnoting on a class paper on Legal Methods. As already noted, he had included several pages from a *Fordham Law Review* article without proper attribution and was obliged to take the course again, which he did and passed. The failed grade had been eliminated from Biden's sealed record, but the current dean of the law school, Craig Christensen, had seen it. At a recent dinner in Miami with some law school admissions officers, he had mentioned it and soon the story was out, feeding the latest charge of plagiarism.[14]

Biden immediately phoned Christensen, authorized him to release the file, and drafted an old friend and former law school classmate, Bob Osgood, to fly to Syracuse to pick it up. He also remembered a subsequent letter from the law school dean at the time attesting to his good character after the incident. When Osgood returned with the law school files, Biden reviewed them and again insisted he had not cheated in the long-ago paper. But Osgood noted that Biden had footnoted the law-review article only once instead of each time he had mentioned it. "Technically," Osgood told his old classmate, "they can say you plagiarized." Biden answered: "But it was an academic mistake. I wasn't trying to hide it. If I was trying to hide it, why would I cite this article that no one else in the class found? . . . I didn't cheat."[15]

Biden wrote later that his political consultant from Boston, John Marttila, told him it didn't matter. The best strategy was to cut the thing off. "Just say you did it and ask for forgiveness," he advised Biden. "Say, 'Look, it was a big mistake. It was a long time ago. I was young. I'm sorry I did this.'" Biden was adamant, though, against offering a mea culpa. He wasn't going to say he had plagiarized because in his mind he hadn't. "I didn't do it," he said. "Not what they're saying."[16]

The 1965 Biden law school paper in question had been critiqued by a fellow student named Arthur Cooper, later a New York judge. Tracked down by the Scripps Howard News Service, Cooper said of the lack of proper attribution, "maybe it was just sloppy work, very sloppy. . . . I can't call Joe a dishonest person. . . . It was a long time ago. It could have been that Joe didn't understand how to prepare a paper at that time."[17]

Biden also released a letter from the Syracuse dean for use with his application to the Delaware State Bar, written two months after his graduation. It said: "Mr. Biden is a gentleman of high moral character. His records reflect nothing whatsoever of a derogatory nature, and there is nothing to indicate the slightest question about his integrity."[18] But Marttila continued to counsel that it was better that he throw in the towel.

With the furor not abating, Biden called a private meeting of the Judiciary Committee and offered to surrender the chairmanship. Democratic and Republican members alike insisted he stay in charge. Pro-Bork Utah Republican Orrin Hatch said: "I think he is getting a bad rap. A lot of people give speeches that aren't totally original. His speeches are like everyone else's. They're written by very creative staff people."[19]

In a forty-minute news conference on Thursday, September 17, Biden released a stack of papers and school transcripts and acknowledged "I was wrong, but I was not malevolent in any way . . . I did not intentionally mislead anybody." He added: "In the marketplace of ideas in the political realm, the notion that for every thought or idea you have to go back and find and attribute to someone is frankly ludicrous."[20] In a letter he had submitted to the Syracuse Law Review panel at the time pleading that he not be thrown out of school, he wrote:

> *It has been my aspiration, for as long as I can remember, to study the law, and now that there is a possibility that this desire will never be fulfilled I am heartsick, but this I could eventually overcome. However, the indelible stain which would mark my reputation would remain with me for the rest of my life for being branded a cheat. This I could never overcome. I am aware that, in many instances, ignorance of the law is no excuse. Consequently, if you decide that this is such an instance and that I've broken the law, then any course of action on your part is justified. But please, I implore you, don't take my honor. If your decision is that I may not remain at Syracuse University College of Law, please allow me to resign, but don't label me a cheat.*[21]

It read like an early, extensive version of "my word as a Biden."

While he was at it, Biden also responded to recurring allegations that he did not, as he put it, come out of the civil rights movement of the 1960s. "I was in fact very concerned about the civil rights movement," he said. "I was not an activist. I worked at an all-black swimming pool of the east side of Wilmington. . . . I was involved, but I was not out marching. . . . I was a suburbanite kid who got a dose of what was happening to black Americans. . . ."[22]

As for not taking to the streets in protest against the Vietnam War, Biden said, "[L]ook, I was twenty-nine when I ran for the Senate, folks. Other people were marching, carrying banners . . . I'm not a joiner. . . . By the time the war movement was at its peak, when

I was at Syracuse, I was married, I was in law school. I wore sports coats. I was not part of that. . . . You're looking at a middle-class guy. I am who I am. I'm not big on flak jackets and tie-dye shirts, and, you know, that's not me."

His own antiwar credentials, he said, rested on what he did as a senator to end the American involvement in Vietnam. When the Senate Foreign Relations Committee called on President Gerald Ford, he said, "I mean, I was so nervous, asking the question. But I said, 'Mr. President . . . what is the plan?' I did more in that meeting than a lot of people did that marched. So I don't take a backseat to the notion that, somehow, I did not go on the line. Other people marched. I ran for office, got elected to the United States Senate at twenty-nine, and came down here and was one of those votes that helped stop the war. And I'm proud of it."[23]

Biden vowed he not only would press on as chairman of the Judiciary Committee leading the Bork confirmation hearings, but also that "I'm in this [presidential nomination] race to stay. I'm in this race to win. And here I come."[24] With that, he entered the Senate Caucus Room, resumed chairing the hearings, and led off the questioning. He returned to the issue of precedent and Bork's devotion to it. He cited the nominee's statement to the committee five years earlier, that "a judge ought not overturn prior decisions unless he thinks it is absolutely clear that the prior decision was wrong and perhaps pernicious." Depending on what Bork might mean as "pernicious," Biden argued, "it seems to me that the entire line of privacy decisions would be in some jeopardy."

In a reference again to *Griswold v. Connecticut*, he suggested: "It is through the right of privacy that the Supreme Court protected married couples in their decisions and found a 'marital right to privacy.'" Biden went on: "It is through the right of privacy that the Court protected the right of a grandmother to live with her grandsons in spite of an ordinance saying that you had to be a nuclear family. It is through the right of privacy that the rights of a father to see his children have been protected. What has been protected, in other words, are at least from my perspective important and fundamental liberties that, in my view, predate the Constitution. I have them because I exist, at least from my point of view."[25]

Biden finally cut through all the legal twists and turns. Noting that previous revered justices had never found the basic right of privacy written in any constitutional amendment, he asked Bork: "Do you believe that the Constitution recognizes a marital right to privacy?" Bork answered: "A marital right of privacy? I do not know. It may well. I have seen arguments to that effect, but I have never investigated that. It is certainly one that I entirely agree with. I mean, I agree with the concept, and I think it is very important that it be maintained. But I have never worked on a constitutional argument in that area."[26]

Other comments by Bork made clear he had no trouble with the concept of marital privacy; he just couldn't find a constitutional protection for it. A moment later he observed: "As to the marital right of privacy, I think it is essential to a civilized society. I do not know of any state, including Connecticut, that has ever tried to interfere with it, because even the law in Connecticut was never used. Nobody ever went in and arrested a married couple for using contraceptives or even threatened it, and I do not think it could be enforced given the Fourth Amendment and the lack of enforcement."[27]

Bork's remarks seemed to support Biden's basic position that, as he put it previously, "as a child of God, I believe my rights are not derived from the Constitution. My rights are not denied by any majority. My rights are because I exist."

Biden also challenged Bork on the scope of the First Amendment's guarantee of freedom of speech, citing articles the judge had written in which, Biden said, "you said that only speech protected was explicitly political speech." He quoted Bork as observing "there is no occasion on this rationale to throw constitutional protection around forms of expression that do not directly feed the democratic process.

It is sometimes said that works of art or indeed any form of expression are capable of influencing attitudes, but in these remote relationships to the political process, verbal and visual expression does not differ at all from other human activities such as sports or business which are also capable of affecting political attitudes, but are not on that account immune from regulation."[28]

Biden told him:

> *Now, that view, frankly, troubles me. All sorts of artistic speech—painting, hanging a nude in a gallery or your home of some great artist; dancing, the Joffrey Ballet or any other ballet that some are more expressive than others, and in [the view of] some conservatives their attire somehow is viewed as being provocative, have nothing to do with politics, that is, painting and dancing. But they are still, in my view, very important and I believe protected by the First Amendment. I do not believe that state legislatures or any government body should be able to censor or suppress expressions just because it is unrelated to expression of a political notion. I am not talking about obscene expression. I am talking about nonpolitical expression like, as I said, the American Ballet Company, Reubens's nudes, or dancing on* American Bandstand.[29]

Bork backtracked, arguing that "the core of the First Amendment is political speech," to which Biden asked: "But how about just speech for speech sake that is not pornographic but that people do not like it. It has no relationship to the political process?"

What about fiction, Biden asked. Bork said it was protected. Well, Biden asked, could fiction reach a point where it "produces only self-gratification"? Bork replied: "Produces only self-gratification? I think about the only form of fiction that does not affect ideas and attitudes or affects them in ways that are not particularly healthy is obscenity or pornography."[30] Biden let it go at that, having made his point.

The next day, Biden on several occasions broke into other senators' question periods and sparred with Bork on such differences as the deference to be observed to precedent in Court rulings and the meaning of various principles, legal phrases, and clauses such as Substantive Due Process and Equal Protection. The exchanges gave the hearings the flavor of a seminar, in which Biden held his own. Taken as a whole, he scored some telling points that should have dispelled rival impressions that he was overmatched against the learned judge.

But all the while Biden had to continue dealing with the personal distractions swirling around him. Gitenstein wrote later that Biden's strategy of focusing on Bork's philosophy of law "was going just as planned. Bork had been a profound disappointment." Nevertheless, he wrote, because of the personal allegations, "Biden's [presidential] campaign was hurtling toward disaster. By the end of the first week of the hearings, Biden and his operatives knew that the campaign was tottering on the edge of collapse. It could not take another serious allegation."[31]

The fear among Biden's aides, Gitenstein wrote, was that the accusations "were undermining Biden's performance [in the committee hearing room] and the credibility of the anti-Bork effort," and if they continued, that effort would be hurt. Another Biden campaign leader, Tim Ridley, said it was decided "to wait out the coming weekend and see if Senator Biden got any credit for his steady stewardship at the Bork hearings. He didn't. Instead, the weekend talk shows and Sunday papers rehashed, in laborious detail, the troubling events of the week—Kinnock, law school, et cetera."[32]

The second week of hearings was given over to testimony of witnesses pro and con Bork, with Biden again presiding, but only intermittently jumping in with brief questions. Leading off were prominent pro-Biden speakers including Barbara Jordan, then a professor at the University of Texas; Mayor Andrew Young of Atlanta; and Burke Marshall, the former Robert Kennedy aide at the Justice Department. They all offered strong arguments against Bork's confirmation. Taken together, they presented ample ammunition with which Biden was able to turn away pressures to testify from single-issue advocates whose appearances could undercut his broader points against the judge's whole philosophy.

But the cloud of personal misrepresentations by Biden of years before continued to hang over the hearings, making the issue in many eyes and ears Senate Judiciary Committee Chairman Joe Biden, not Judge Robert Bork.

Finally, on the night of September 22, after conversations with Ted Kaufman, Mark Gitenstein, and other Biden insiders, the chairman took the train back to Wilmington for dinner with Jill and a family meeting afterward at the house known as the Station. It included Joe's parents; Valerie and her husband, Jack; brothers Jimmy and Frank; sons Beau and Hunter; Kaufman; Gitenstein; Tom Donilon; and Larry Rasky. Ashley, now six, remembered later only that there were a lot of people there, but that was not unusual in the Biden household. Pat Caddell weighed in by phone, to the annoyance of many, pushing the candidate to get out to Iowa and up to New Hampshire to revive the campaign. " 'We can weather this,' Pat kept saying," according to Biden.[33]

Caddell, apparently trying to take proprietorship of the campaign from afar, chastised Rasky: "You people have formed a vigilante party to get my candidate out of the race." And Donilon joked: "I know this is going to be like *Poltergeist II.* Pat's going to get up from behind the bushes—he's back!"[34]

Donilon more soberly told Biden: "You've had tens of millions of dollars of negative TV thrown at you in the last week. It's time for a strategic withdrawal." And Rasky warned: "The press is in a feeding frenzy. I don't know how to stop it now."[35]

Kaufman as usual was looking out first for Biden the man and friend, not the candidate. "There's only one way to stop the sharks, and that's to pull out," he said. "Then we can catch our breath, win the Bork fight, and come back into this thing sometime later."[36] But the last part seemed unrealistic—for the 1988 cycle, anyway.

Marttila said later that Biden "was way, way further advanced in both Iowa and New Hampshire in terms of endorsements than anybody really believed, and it was all based on his personal charisma. The polls in Iowa basically showed a four- or five-man race and Joe was about to begin the Bork hearings." Then why did Biden have to withdraw?

On the eve of Bork's testimony, Marttila said later, "he was worried about that. He didn't want it [the allegations against him] to be a distraction. I think challenging his integrity was really a blow that was very, very hard for him to take. It was also a different time in American politics, when those kinds of offenses were blown into gigantic proportions. . . . I didn't try to talk him out of it. It was a kind of cascade of things that by themselves would have been okay."[37]

Biden himself wrote later: "I was torn. It really came down to family. . . . Beau and Hunter were both so angry that night; the whole thing was a cheap attack, they kept saying, and we couldn't let it stand. 'All that stuff they said, Dad—there's nothing real.' That's what Beau said in front of the group, but when my sons got me alone, I could see they weren't just angry, they were worried for me—for us. In the middle of that long evening I had wandered out of the living room and found Beau and Hunter standing together in the tiny hallway leading to our library. I tried to cheer them up. 'Don't worry about this, guys,' I said. 'But Dad, if you leave you'll never be the same,' Beau said. 'The only thing that's important is your honor,' Hunt said. 'That's what you've always taught us. Your honor.' . . . My sons were afraid I might walk away from public life, fold up my tents, and call it a career. 'You'll change, Dad,' Beau said. 'You'll never be the same.' "[38]

(Years later, Beau Biden explained his and Hunter's rationale at the time: "We both thought, as seventeen- and eighteen-year-old young men, that getting out of the race would be giving up, would not be consistent with what our grandparents taught us and my dad taught us. I think what we both missed was it wasn't about the race. The important thing was what my dad's job was in the Bork hearings at the time; he had a choice between aggressively continuing to pursue his presidential aspirations, his political career, or chairing the Bork hearings and trying to shape, in his advise and consent role, the shape of the United States Supreme Court, which was what he was elected to do. And what my dad chose was the latter.

"I think Hunt and I were focused more on the former because of all the rough stuff that was being said about him at the time. What my dad knew was two things: one, he had a job he was elected to do, and that was as chairman of the Judiciary Committee, and two, that with time and with his refusal to give up, and his resilience, I think he had faith that people would make the appropriate judgment about his character over the long haul. And he had a kind of a serene confidence in that; not just a bravado but a confidence that ultimately people would make a judgment about him and his character that was consistent with who he is.

"There's many people in public life on both sides of the political aisle who engage in the snarky or cheap shot in politics and in life. Even when my dad's been hit the hardest at any point over thirty-six years, he's rarely if ever engaged in the cheap shot that is the give-and-take of politics. That's just not what he is, and never has been."[39]

Hunter later also reflected on the situation at the time. "I believed in my dad and I felt it was incredibly unfair what happened. But I don't remember being bitter about it. It was one of the times I was most proud of my dad. As my grandmother always says, 'Everything happens for a reason' "—a comment that was soon to be borne out in another crisis.)[40]

Biden walked back to the living room and noticed that his mother was quiet. "What do you think, Mom?" he asked. "I think it's time to get out," she said. With that, Biden turned to Gitenstein and Rasky, asked them to draft a withdrawal statement, and told them he'd make a final decision in the morning. Then he and Jill retired to talk it out between themselves. Which was more important now, the campaign or defeating the Bork confirmation?[41]

In the morning of Wednesday, September 23, Joe and Jill, along with Ted Kaufman and another aide, Chris Schroeder, drove to Washington; and Biden started the day taking the testimony of retired chief justice Warren Burger in support of Bork. The chairman made the most of the opportunity by quoting one of Burger's opinions on rights not enumerated in the Constitution, a key bone of contention with Bork. Biden read:

> *Notwithstanding the appropriate caution against reading into the Constitution rights not explicitly defined, the Court has acknowledged that certain unarticulated rights are implicit in enumerated guarantees. For example, the rights of association and the right of privacy, as well as the right of travel, appear nowhere in the Constitution or Bill of Rights; yet these important unenumerated and unarticulated rights have nonetheless been found to share Constitutional protection with explicit guarantees. The concerns expressed by Madison and others have thus been resolved. Fundamental rights, even though not expressly guaranteed, have been recognized by the Court as indispensable to the enjoyment of rights specifically defined.*

Biden emphasized to Burger, "Mr. Justice, that is what this debate is all about, at least with Judge Bork and I. And I wonder if you could speak with us a little bit, educate us a little bit, about these unenumerated rights—the right of privacy?"[42]

Burger, after telling Biden he hadn't come "to give a lecture on Constitutional law" of which the committee had heard "quite a lot of that in the last few days," offered: "I see no problem with that statement, and I would be astonished if Judge Bork would not subscribe to it."[43]

To that answer, Biden was the astonished one. Burger seemed not to realize that he had just differed on the key point with the nominee he was there to support.

Biden drove his case home with another question, asking the former chief justice: "Does the Ninth Amendment [on unenumerated rights] mean anything?" Burger seemed momentarily stumped. "Well, I will follow the habit that [former justice] Hugo Black had," he said. "Whenever he wanted to talk about an amendment, he wanted to look right at it and not try to remember exactly what its words were. So let us take a look at the Ninth Amendment here," and proceeded to peruse his copy of the Constitution, calling it "one of the very, very most important parts."

Burger went on, saying "persons" was the key word and segueing into a vague monologue on the *Dred Scott* decision, in which a black slave was taken from a free to a slave state and subsequently lost his suit to be declared free, on grounds he was property, not a free man. Burger concluded his answer: "It is hard to say which amendment is more important than any other amendment, but surely, this matter of 'persons' becomes terribly important."

Biden, straight-faced, replied: "I appreciate that answer. . . . I, even much more than you, Your Honor, have to look at this to make sure I have got the amendments right. And let me just read into the record." He read the twenty-one-word amendment aloud: "Amendment Number Nine. The enumeration in the Constitution of certain rights shall not be construed to deny or disparage others retained by the people." And that was all Biden said, not pointing out that the word "persons" never appeared. He thanked Burger and yielded to Senator Strom Thurmond, the committee's ranking Republican, who had no comment on Burger's previous answer, moving to safer ground for Bork.[44]

Biden could have rested his case right there, though the hearings would drag into a third week. Over the whole time, Bork testified for thirty-two hours, and was witness to the demise of his dream of serving on the Supreme Court. Biden supporters, for their part, saw the hearings as a flowering of sorts of their man's reputation as a legal scholar and fair-minded chairman.

During the lunchtime break, Biden held a news conference and announced to a crowd of reporters and cameramen outside the Judiciary Committee hearing room that he was dropping out of the presidential race. With his wife at his side, he began:

> *Hello, everybody. You know my wife, Jill. Although it's awfully clear to me what choice I have to make, I have to tell you honestly I do it with incredible reluctance—it makes me angry. I'm angry with myself for having been put in this position—for having put myself in this position of having to make this choice. And I am no less frustrated at the environment of presidential politics that makes it so difficult to let the American people measure the whole Joe Biden and not just misstatements that I have made. But, folks, be that as it may, I have concluded that I will stop being a candidate for President of the United States.*

He continued: "There will be other presidential campaigns and I'll be there, out front. There will be other opportunities. There will be other battles in other places at other times, and I'll be there. But there may not be other opportunities for me to influence President Reagan's choice for the Supreme Court."

Of his campaign, Biden said: "Quite frankly, in my zeal to rekindle idealism, I have made mistakes. . . . Now the exaggerated shadow of those mistakes has begun to obscure the essence of my candidacy and the essence of Joe Biden." He said he had to choose been salvaging his campaign by going back to Iowa and New Hampshire or fighting the Bork nomination; he couldn't do both. So his obligation was "to keep the Supreme Court from moving in a direction that I believe to be truly harmful. I intend to be deeply involved in that battle. I intend to bring it to victory."

He paused for a moment, taking in the scene of the gathered crowd of reporters, then went on: "I appreciate your consideration.

I appreciate your being here. And lest I say something that might be somewhat sarcastic, I should go to the Bork hearings. Thanks, folks. My wife and I thank you very much."[45]

Leaving the press behind, the Bidens simply turned and headed back toward the Senate Caucus Room. As they were about to enter, Jill grabbed her husband's arm. He wrote later: "She locked her eyes right into mine and then said something that sounded like profanity. Jill didn't often use profanity, but she wanted my full attention. She wanted me to understand that doing my best wasn't good enough now: 'You have to win this thing!'"[46]

Years later, Jill Biden observed with regret: "Of all things for him to get out of the race for, it was to be attacked for his character. I knew few men who had as strong a character as Joe had. [It was] a total irony that the work that he prided himself on and the one thing that Delawareans prided themselves in voting him as their senator was that he had this strength of character. That was what made it particularly difficult." Because of that circumstance, she said, her husband was determined to run for president again sometime. "I felt he really had to validate who he was, and that he was not a plagiarizer, that he was a man of good character. I think he always had that sort of in the back of his mind, that he wanted people to perceive him [as such]." He often talked about it, she said.[47]

Upon Biden's decision to withdraw, there was none of the remonstrations with the reporters of the sort that Gary Hart had made in his own campaign's demise in the wake of journalistic enterprise and snooping. Larry Rasky, the press secretary, said later that although Biden had good relations with the campaign press corps, reporters didn't know him well because of his nightly commuting, and his closeness to Caddell didn't help. "When the shit hit the fan," he said, "it was the press corps' lack of familiarity with Joe, combined with their contempt of Pat, that made people willing to believe the worst."[48]

The extremely close tie between Biden and Caddell abruptly ended with this episode, under circumstances neither chose to reveal, but it seemed unlikely that political differences were involved. Their relationship, Tom Lewis said later, "was just intense. They both believed down to their bone marrow in the basic democratic message of social justice. I think that was what really bonded them together."[49]

When Biden called the committee to order to resume the afternoon session, Kennedy asked him to yield. "I know the chairman, Senator Biden, has made an important statement in the last few moments," he said. "I, for one, certainly respect his decision. I admire his courage in making that decision, and we welcome him as the chairman of our committee, as he was before he made the decision and as he is now. We are very glad to have Senator Biden as the chairman of our committee."

Thurmond broke in. "Mr. Chairman," he said, "just want to say the Democrats have now lost their most articulate spokesman."

Biden, smiling, replied: "Well, thank you, Mr. [former] Chairman. I do not plan on moving over. But thank you. That is a very, very nice thing of you to say." Then: "Look, my business is behind us. Let us move on." And he swore in the next witness.[50] Soon after, however, two other Senate colleagues, Republican Alan Simpson and Democrat Patrick Leahy, took occasion to commend Biden's perseverance under the circumstances.

Of his withdrawal from the presidential race, Simpson said: "I am sure that was very painful for us, and I would say too that you have never turned from your duty, your obligation as a senator, in the time I have known you, as you pursued your quest for the presidency, and I hope you remain as chairman. . . . You have handled it all with good grace, and now you will move on."

Biden thanked him, saying, "I assure you, I am going to remain as chairman as long as the Democrats are in control." Simpson continued, saying of the Senate: "It is a remarkable arena. It can be savage and barbaric and yet also caring and supportive. I think we politicians move at such a pace in our lives, we really do not have time to savor either victory or anguish defeat. And that is good. But I [have a] hunch that care and support will surface now, and it always has, and so heal up quickly. We need you here in the fray so we can box each other around. I would not want to miss any of that." Biden answered: "I will be here, Senator, and thank you."[51]

Minutes later, Leahy added: "I too watched your press conference this afternoon. I am frankly sorry to see you dropping out of the race. I thought you brought a great deal to the debate, and certainly you have been forceful in your positions. . . . But I should also note that you have juggled it well in presiding over this committee. And I have heard from an awful lot of people both for and against Judge Bork who have complimented you on the fact that you have evenhandedly arranged to have witnesses on both sides, and have handled them evenhandedly."[52]

When the day's hearing ended, the Bidens took the train back to Wilmington, and at Jill's insistence they went to a favorite restaurant, Attilio's. It was crowded, and as they entered a murmur went through the room. "This, I thought, was exactly what I meant to avoid," Biden recalled later. "And all of a sudden one guy in the dining room started clapping. Then there was a smattering, and then it seemed like everybody at Attilio's was on their feet, giving me a standing ovation."[53]

FIFTEEN

A COSTLY VICTORY

JOE BIDEN WASTED no time getting back in harness in the Senate. The day after his withdrawal from the presidential race, he undertook a swift farewell swing to Iowa and New Hampshire, thanking his supporters and blaming nobody but himself for what had happened. The outcome had been particularly hard on Biden because, Larry Rasky said much later, "honor is a trait he has always tried hard to live by and motivates him, and when that's questioned, it's painful. So he poured himself into the Bork hearings."[1] It was back to chairing them, which he had declared his top priority in abandoning his presidential ambitions, for 1988 anyway.

The first two weeks of hearings, for all of Biden's personal woes, had eroded the White House effort to cast the judge as the best and obvious choice. So tactics were switched to discredit Bork's critics. Reagan, who had remained largely on the sidelines, told the conservative Concerned Women for America at week's end that "it's clear now that the charges that Robert Bork is too ideological are themselves ideologically inspired, and that the criticism of him as outside the mainstream can only be held by those who are themselves so far outside the mainstream that they've long ago lost sight of the moderate center."[2]

At the same time, White House press secretary Marlin Fitzwater took aim at anti-Bork commercials by "liberal special-interest groups . . . producing slick, shrill advertising campaigns that not only purposely distort the judge's record, they play on people's emotions as only propaganda campaigns can."[3] He singled out a television ad featuring movie star Gregory Peck, in full Atticus Finch mode, the role he played in *To Kill a Mockingbird*, observing that Bork "has a strange idea of what justice is. . . . He doesn't believe the Constitution protects our right to privacy. And he thinks that freedom of speech does not apply to literature and art and music. Robert Bork could have the last word on your rights as a citizen. But the Senate has the last word on him. Please urge your senators to vote against the Bork nomination, because if Robert Bork wins a seat on the Supreme court, it will be for life—his life . . . and yours."[4] Those remarks in themselves confirmed Biden's wisdom in lowballing special-interest complaints in favor of going after Bork's broad judicial philosophy.

On Saturday, September 26, realizing his confirmation was in jeopardy, Bork and his wife went to the White House to ask for direct intervention by Reagan. The president's two top political aides, Howard Baker and his deputy, Ken Duberstein, were not there and so they talked to Tom Griscom, a young Baker aide, proposing that the president go on national television and make a personal pitch for him. They got no satisfaction; the White House at this point seemed more interested in protecting Reagan than in rescuing Bork.

On Monday, September 28, Senate Majority Leader Robert Byrd, without consulting Biden, said the committee should send the nomination to the Senate floor without a recommendation. But at the regular weekly Democratic caucus the next day, Biden pushed for a vote in committee and a timely vote in support of the committee position on the floor.

The final three days of hearings were given over to testimony, largely from legal, civil rights, and academic witnesses, including three former attorneys general—Democrat Griffin Bell and Republicans Elliot Richardson and Richard Thornburgh, all three supportive of Bork. Biden sometimes chipped in with questions, but his position by now was exhaustively clear—that confirmation of Robert Bork would indeed alter the direction of the Court in a substantial way.

At the close of the hearings, two committee members, one supporting Bork and one opposing, addressed Biden's chairmanship, which Bork himself had charged was unfair. Democrat Howell Heflin of Alabama, anti-Bork, went first and commended the chairman "for the fairness with which you have conducted these hearings." He went on: "We have heard a lot of charges and all sorts of misrepresentations. I am referring to before the hearings started and I might say there were a lot of distortions and inaccuracies published about you and how you would conduct the hearings. I think that everybody throughout the hearings has felt that you have been completely fair. I think that everybody on the Republican side, the Democratic side and that third element, the undecided side, will say that you have been completely fair."[5] Strom Thurmond as the ranking Republican echoed Heflin's sentiments.

The fight on Bork, however, was not yet over. A Bork critic, Democratic Senator Alan Cranston of California, not a Judiciary Committee member, decided to apply pressure by releasing a vote count of 49 senators against Bork, 40 for, and 11 undecided, telling the press: "I think he's licked." John Bolton at the Justice Department countered by declaring Cranston's count "a transparent effort to show the appearance of momentum" and calling him "the worst vote-counter in the U.S. Senate today."[6]

A key figure at this juncture was Republican Senator Arlen Specter of Pennsylvania, who had deep reservations about Bork's views on First Amendment protections. Specter met with Bork and came away unsatisfied. But when Biden learned that Specter was about to make a speech on Bork on the floor, he apprehensively phoned his Republican colleague. Specter told him, much to Biden's relief, that he would declare against the judge.

On Friday, October 2, Democrat Lloyd Bentsen of Texas also took the Senate floor against confirmation, citing his own pro–civil rights record. "I question," he said, "whether very many Americans—black, white, Hispanic, or others—want to turn back the clock on what is right and just for America when it comes to civil rights. We've already answered those questions."[7]

Over the next few days, the floodgates opened, and more than a dozen more senators declared their opposition to Bork. Cranston sent Howard Baker a telegram at the White House informing him that three more southern Democrats and Specter were all going to vote against confirmation. "I'm up to 50," he wired of his own count. "What are you down to? More to kum [cq]. Yours for better counting—and for a better Supreme Court."[8]

Bolton called the Cranston count "a charade, not a fact," but on Monday, October 5, Biden phoned Baker at the White House and informed him the nomination was going to be unfavorably reported by the committee by a 9–5 vote. Did the president want to withdraw Bork's name? Baker said Reagan wanted the committee vote sent to the floor regardless.[9]

In a radio address after that committee vote, the president did not so much boost Bork as castigate his critics. "Old-time liberals here in Washington," he said, " . . . viewed the courts as a place to put judges who would further their agenda—even if it meant being soft on crime." In opposing Bork, he said, "liberal special-interest groups work to politicize the court system; to exercise a chilling effect on judges; to intimidate them into making decisions." He said the fight against Bork "has become a distorted, unseemly political campaign" in which the judge "has been subjected to a constant litany of character assassination and intentional misrepresentation" by people who "are determined to thwart the desire of the American people for judges who . . . will enforce the law and bring criminals to justice, not turn them loose and make our streets unsafe."[10]

Nearly twenty years after Richard Nixon had successfully used this basic law-and-order scare language to win the presidency, Reagan was trotting it out again in a long shot attempt to save Bork's judicial career.

As proponents and opponents of the Bork nomination filled the post-hearings communications void with rallies and radio and television ads over the next days, Bork began to consider throwing in the towel to save himself and his family further anguish. Bork later wrote in his book *The Tempting of America:* "The question of withdrawing now became a real one. Defeat was certain. We received advice from a great many people, and almost all of it was to withdraw rather than take a bad Senate vote. President Reagan said he would prefer that I stay but clearly left the decision up to me." At a meeting in the office of Senate Minority Leader Dole, Bork wrote, while other senators urged him to stay, "only Alan Simpson said he was not sure I should stay in because I would have to live with a fairly decisive defeat."

Finally, Bork and his wife agreed "getting out seemed probably the best course." On Thursday, October 8, as he sat at his desk in his chambers tentatively drawing up a withdrawal statement, Simpson phoned him and said he had changed his mind, and if Bork stayed in, the Senate Republicans "could make a permanent record in the debates [on the Senate floor] that would tell the facts for history." Bork and family reconsidered, and the next day they all went to the White House to so inform the president he would stay and fight.

"Reagan," the judge wrote later, "said that was the way he wanted it, but I thought some of the staff looked unhappy. In any event, I said I had a statement I wanted to make in the White House press room." While a White House aide took a copy of the statement to be duplicated for reporters, Bork recalled, Reagan invited Bork to bring in his family members, "and he chatted pleasantly with them until the copies were ready. Then we went down, and I walked alone into a packed press room," where he read the statement. Notably, Reagan did not accompany him.

Before offering the news of his decision to persevere, and without naming Biden, Bork unloaded on his conduct of the hearings. "In the hundred days [since his nomination], the country has witnessed an unprecedented event," he said. "The process of confirming justices for our nation's highest court has been transformed in a way that should not, and indeed must not, be permitted to occur again. The tactics and techniques of national political campaigns have been unleashed on the process of confirming judges. That is not simply disturbing. It is dangerous. Federal judges are not appointed to decide cases according to the latest opinion polls. They are appointed to decide cases impartially according to law. But when judicial nominees are assessed and treated like political candidates the effect will be to chill the climate in which judicial deliberations take place, to erode public confidence in the impartiality of judges, and to endanger the independence of the judiciary."

Getting down to cases, Bork said that "were the fate of Robert Bork the only matter at stake, I would ask the president to withdraw my nomination. The most serious and lasting injury in all this, however, is not me. . . . Rather, it is of the dignity and integrity of law and of public service in this country. I therefore wish to end the speculation. There should be a full debate and final Senate decision. In deciding on this course, I harbor no illusions. But a crucial principle at stake," he said, "is the way we select the men and women who guard the liberties of all the American people. That should not be done through public campaigns of distortion. If I withdraw now, that campaign would be seen as a success, and it would be mounted against future nominees. For the sake of the federal judiciary and the American people, that must not happen. The deliberative process must be restored."[11]

Five days passed—two weeks after the hearings had ended—

before Reagan himself took to national television to try to rescue the judge. By now, the Senate vote against Bork was such a foregone conclusion that none of the major networks expressed interest in airing the president's speech in prime time. So he had to settle for a midafternoon live slot on CNN that had a very modest audience. But at a subsequent rally in New Jersey, the usually affable Reagan angrily warned that if the liberals in the Senate rejected Bork, he would nominate a replacement that they would "object to as much as they did to this one."[12]

A week later, on October 21, the full Senate debate on confirmation finally began, and two days later Bork suffered a humiliating defeat by a vote of 58–42, the largest rejection ever delivered upon a Supreme Court nominee. Biden offered words of commiseration that certainly could not have gone down well with Bork.

"This is an individual we're speaking of, a man of honor, integrity and intellect," Biden said. "But notwithstanding that fact, I must tell you honestly, I feel sorry for him sitting home at this moment, watching this nomination go down, feeling—because we've all been there at one point in our lives—the personal loss he must feel at this moment. With all due respect, this is not about Judge Bork. It's about the Constitution." He said he had "never had any doubts at all, once the issues were framed . . . and the American people saw them," what the outcome would be.[13]

Bork, in *The Tempting of America*, written after the rejection of his nomination, had relatively little to say about Biden. He did, however, take note of the plagiarism allegations made against the chairman during the hearings. "As the damaging facts began to pile up," Bork wrote, "Biden at first tried to explain and finally had to hold a press conference at which he withdrew as a candidate for his party's nomination. Shortly after that, during a break in the hearings, Biden came over to my table and said, with every appearance of sympathy, 'You know, your situation and mine are a lot alike.' I didn't think there was any comparison but managed not to tell him so."

The judge also noted that when his testimony ended, "Biden asked me if [the hearing] had been fair. I said it had, meaning that he let me answer every accusation, but it had not been a 'hearing' at all, since nobody heard." He added that "Joe Biden and the Democrats structured the proceedings" to keep his friendly witnesses off television. Of nine former attorneys general testifying, Bork wrote, Biden arranged to have the two against him appear "during the day and kept the seven favorable ones waiting until the deadline for the evening TV news programs and the morning newspapers had passed. Nor did the media go back the next day to report what they had missed. Tactics like this were intended to, and did, give the public a greatly distorted view of the opposition and the support."[14]

Long afterward, Specter in his autobiography, *A Passion for Truth*, wrote of a taped interview with Biden in which the committee chairman remarked that Bork's comment that being on the Supreme Court would be "an intellectual feast" was the "capstone" of Bork's arrogance. "What fun—play with your lives," Biden said. "This is like a Rubik's Cube, and I'm one of the pieces? . . . I think Bork's arrogance was his undoing."[15]

Biden himself said long after: "I thought quite frankly it was [a choice] between my campaign and Bork. I loved Ted Kennedy but I knew it would be incredibly difficult if I wasn't chairing those hearings, to get those five southern senators to defeat Bork. . . . You know, for me to save that campaign, I would have had to go out full-time, just be on the ground in Iowa and reconstruct it. And I didn't see how I could do both those things. I didn't want to be a little asterisk in history that I went out to save my rear end and Bork got on the Court. Now, that's self-serving, thinking I'm the only guy who could have stopped Bork, but I don't think it was likely that Teddy could have gotten Howell Heflin and Bennett Johnston and that crew to vote against Bork."[16]

Beyond Biden's determination to see Bork's nomination rejected on grounds of what he saw as his dangerous views, the fight had given the chairman a vehicle for validating himself after the plagiarism allegations. His wife observed much later: "He needed to be vindicated. It was about Bork, it was about Bork's politics, but it was also about Joe."[17]

After his defeat, Bork went back to his seat on the federal circuit court of appeals for the District of Columbia, often called the second most important judicial bench in the nation. About two months later, on January 7, 1988, he submitted a long letter of resignation to Reagan, reiterating his view "that a judge must apply to modern circumstances the principles laid down by those who adopted our Constitution but must not invent new principles of his own." He said he was stepping down to be free to "participate in the public debate" on the role of the court. "I had considered this course in the past," he wrote, "but had not decided until the recent confirmation experience brought home to me just how misrouted the public discourse has become."

Reagan replied that he was accepting the resignation "with deep sadness," particularly because Bork epitomized the virtues of a Supreme Court justice "at their very finest—which is why I turned to you to fill Justice Lewis Powell's seat."[18] There was considerable irony in the comment, inasmuch as what had first fired Joe Biden's determination that Bork not be seated was his conviction that he conspicuously failed to meet that yardstick.

With the Bork nomination derailed, the White House moved quickly to another choice to replace Powell. A short list of prospective nominees was quickly reduced to two: Douglas Ginsberg—a colleague of Bork's on the same federal appeals court, a close friend, and a former assistant attorney general in the Reagan Justice Department; and Anthony Kennedy, member of another federal appellate court in California and former lawyer and lobbyist in Sacramento close to Meese and Reagan.

Biden in conversation with Howard Baker indicated a preference for Kennedy, but he told the committee staff to dig up all they had about Ginsberg, whose public record was relatively skimpy. Meanwhile, Meese and conservative senators like Jesse Helms of North Carolina and Charles Grassley of Iowa urged Reagan to bypass Kennedy, and six days after the Bork rejection he chose Ginsberg. In a news conference at the White House, the president introduced him with more criticism of Bork's fate and how it was accomplished by the judge's foes. "I hope we can all resolve," Reagan said stiffly, "not to permit a repetition of the campaign of pressure politics that so recently chilled the judicial selection process." Biden only said, "I assure the judge he will receive a full and fair hearing in our committee."[19]

But matters never got that far. The conservative *Human Events* read by Reagan raised questions about Ginsberg positions on abortion, gays, and pornography. Word got out that his wife had performed abortions as a medical intern, and finally that he had been a frequent user of marijuana while teaching at the Harvard Law School. Reagan Secretary of Education William Bennett, later author of the bestselling *The Book of Virtues*, persuaded Ginsberg to withdraw his name.[20]

Before announcing a new nominee, Reagan invited Biden to the Oval Office to discuss the situation, with Howard Baker present. As Biden described the scene later, the president for all the bitterness over the Bork affair was his customary cordial self. Greeting him with "Hi, Joe" and a handshake, Reagan said, "Congratulations on Bork." Biden replied, "No, Mr. President. There's no cause for congratulations. I feel bad for Judge Bork. He was a good man." To which Reagan, according to Biden, said with a smile: "Ah, he wasn't all that much."[21]

Biden was astonished at the reply. Later, he said he assumed the reason Reagan had taken that rather cavalier attitude was that "he was really angry at Bork. Remember Bork went on this one-man campaign saying the reason he lost was because of Iran-Contra, that Reagan didn't support him enough, and that had Reagan not been weakened by Iran-Contra he would have won."[22]

In any event, Reagan then asked Biden to tell him his preference for the next appointment to the Court. The Judiciary chairman had told the president only that he would be happy to give him "an honest appraisal" of the prospective nominees he had in mind. Reagan ran down his list and when he came to his fifth name, Judge Anthony Kennedy, and asked about him, Biden said: "Based on what I know, he's a mainstream conservative. He would probably pass."

Reagan pressed him. "So that means you're for him?"

Biden said again: "No. Based on what I know, he'd pass the Senate. I might vote for him, but I don't know enough."

"Right," the president said insistently. "You're for him."

Biden turned to Baker to intercede, obviously not wanting to be pinned down. But as he got up to leave, Reagan took Biden by the arm and led him over to his private study off the Oval Office. Standing there was Kennedy, and according to Biden, Reagan told him: "Tony, Joe says he's for you!" Biden wrote later: "I tried to explain to Judge Kennedy and to President Reagan that that's not exactly what I'd said, but Reagan wasn't one to get mired in messy details. He was a guy who operated in the essence of the thing. I had to hand it to Reagan—he did have a deft touch. I sort of enjoyed the well-rehearsed theater of the meeting. And the president was not without his charms. He had a way of making people want to help him do well."[23]

Four days later, on November 11, Reagan introduced Kennedy as his third choice to replace Powell. This time, rather than lashing out at the senators and special-interest groups who had brought down Bork, he struck a softer tone. "The experience of the last several months has made all of us a bit wiser," he said. "I believe the mood and the time is now right for all Americans to join together in a bipartisan effort to fulfill our constitutional obligation of restoring the Supreme Court to full strength."

In so saying, Reagan offered in his fashion an assurance that Kennedy was not like Bork. The man he was nominating, he said, "believes that our constitutional system is one of enumerated powers, [but] that it is we the people who have granted certain rights to the government, not the other way around." Biden's own argument with Bork actually involved enumerated rights of the citizen, not enumerated "powers" of the government, but Reagan's intent to extend an olive branch seemed clear. The president also noted of his previous threat to nominate somebody Bork's critics would like even less, "sometimes you make a facetious remark and somebody takes it seriously," he said, "and you wish you'd never said it, and that's one for me."[24]

In two days of hearings in mid-December, Kennedy gave Biden no grounds to reject him. "I do not have an overarching theory, a unitary theory of interpretation [of the Constitution]," the nominee said. "Many of the things we are addressing here are, for me, in the nature of exploration and not in the enunciation of some fixed or immutable ideas." And he cited as "central to the idea of the rule of law . . . that there is a zone of liberty, a zone of protection, a line that is drawn where the individual can tell the Government: Beyond this line you may not go."

Some of the more aggressive members of the civil rights coalition were unhappy that Biden had not confronted Kennedy on some of his views dealing with racial discrimination. But Biden said in response that "one of the things that holds Judge Kennedy in best stead of all . . . is that no one's fully satisfied with his philosophy or how he would rule, which means he in fact is as open-minded as I hoped he would be."[25]

On February 1, 1988, Kennedy won unanimous support of the Judiciary Committee and three days later of the full Senate by a vote of 97–0. In all, Biden's leadership of the Senate Judiciary Committee through the long and grueling exercise of filling the Supreme Court vacancy caused by Lewis Powell's retirement had offered the greatest test of his Senate career. This was particularly so because of the manner in which the failures of his presidential campaign had intruded and had defined him in the minds of so many of his fellow Americans.

In a sense, his generally fair and open conduct of the hearings, in which at the same time he diligently pursued Bork's scalp, greatly enhanced his reputation within the Senate club as he put presidential ambitions behind him, at least for the foreseeable future.

Writing in *Legal Times*, reporter Terence Moran said of Biden that "after he withdrew from the race, gracefully, with none of the Nixonian defiance that [Gary] Hart had exhibited, [he] turned his attention to the greatest constitutional debate in recent years, and acquitted himself superbly, rising not only to the intellectual challenge but also to the institutional one of steering the Judiciary Committee through a storm of criticism." He wrote that Biden had been "paradoxically strengthened by the trials and defeats he had endured."

Prematurely, however, Moran observed that "the jokes about his oratorical borrowing have already grown stale, and his own good humor about the episode seems sure to erase it as a slur on his career." But in the Bork hearings, he concluded, "Biden can take solace in the fact that he has put to rest the one truly damaging accusation against him—that he is a poor legislator, a hollow man of little substance who survives and prospers on superficial talents."[26]

Mark Gitenstein, who had been at Biden's side throughout the Bork fight, said later it was "the only time we won a big jurisdictional fight over the direction of the Supreme Court. That's what the fight was all about. Which way was the court going to go? Was the Warren Court a mistake? Did the American people not support the Warren Court? Did they believe in the notion of fundamental rights, which was the heart and soul of the Warren Court? Which led to the progeny of all those other decisions . . . and all the civil liberties decisions; *Griswold v. Connecticut;* you not only got Kennedy, we got Souter because the best Bush One took out of this was, you can't win on these issues. . . . We defeated Bork and got two moderate justices in Souter and Kennedy. Kennedy is still the swing vote. We got a whole generation of decent jurisprudence because of that fight."[27]

Anthony Lewis, the respected *New York Times* reporter and columnist who had covered the Supreme Court for many years, wrote at the Bork hearings' end: "They have confounded the cynical view that everyone in Washington has base political motives."[28] Bork himself, however, made clear in writings and speeches thereafter that Lewis's opinion was one from which he strongly dissented.

SIXTEEN

DOWN BUT NOT OUT

FOR ALL THE satisfaction of Biden's victory in the Bork fight, it did not erase the senator's sense of his own political shortcomings in the aftermath of his forced withdrawal from the 1988 presidential race. In a long interview with David Broder of the *Washington Post* at the start of the new year, Biden acknowledged that his campaign had collapsed because "I was doing all the wrong things." He could not believe in retrospect, he said, "how naïve I was" about the demands of a presidential campaign. He said he intended at some time in the future to address the charges of plagiarism that did him in, which he continued to deny, "because there are things there that have to be corrected for the record whether or not I ever run for president again."

That reference to a future candidacy seemed to convey a sense that Joe Biden was down but not out. He said that when he dropped out of the 1988 race, "I thought about leaving politics. But I have an opportunity to do something I like a lot—being a senator. And that's what I'm doing." He credited the demands of his Senate Judiciary Committee chairmanship in conducting the Bork hearings with putting him back on track. He said the task constituted a "first marker" on his new course, and said he thought his performance boosted his standing with critics "who questioned whether I could discipline my temper and my mouth."

In Biden's observations on his failed campaign, he took note of impressions left with the news media that he was a mere puppet in the hands of wily political consultants using him to convey a message they thought could win. He said he had never solved what he called "the guru problem," a reference to his close relationship particularly with pollster-adviser Pat Caddell, who had aggressively pushed the message of generational change.

Biden said the people of Delaware knew him and what he stood for, but his general pitch on the stump around the country had reduced him to a motivational speaker. "Suddenly I'm out there talking about moving a nation, moving a generation, and you guys [the national press] naturally say, 'What's this guy up to?'" Reporters, he said, saw his ability to generate enthusiasm, and "everything I was doing to move an audience made you doubt more that I was fit to lead the country or move the government." There again was the old complaint against him of style over substance.[1]

In another interview with reporter Nathan Gorenstein of his hometown *Wilmington News Journal*, Biden acknowledged that "I wasn't ready to run" for president in 1987. He reiterated that he should not have listened to those people who urged him to "talk about moving people" and about being a leader who could unite the nation "when in fact that's not how I ever campaigned in Delaware," where he spoke of concrete issues of concern there.[2]

With the heavy lifting on the Judiciary Committee behind him for a time, Biden decided to focus on his other major Senate responsibility, as the number two Democrat on the Foreign Relations Committee. In January, he embarked on a twelve-day tour of Europe to meet the leaders of Britain, France, Germany, and NATO in Brussels regarding a new Intermediate-Range Nuclear Forces (INF) Treaty with the Soviet Union. The committee chairman, Senator Claiborne Pell of Rhode Island, had appointed him a cochairman of hearings soon to be held on the treaty. At the same time, Pell named Biden chairman of a new subcommittee to take a new look at the controversial War Powers Resolution designed to limit commitment of American troops in conflicts in the wake of the Vietnam War.[3]

Unfortunately for Biden, what generated as much news back home as his high-level talks with the European bigwigs was a brief meeting in London with Neil Kinnock, the British Labour Party leader whose poetic utterances about his rise to influence had been Biden's undoing in the presidential race he had been forced to leave. They both joked about it, but the very fact that the meeting drew press attention was a painful reminder for Biden.

Upon return to Washington prior to Senate consideration of the INF Treaty, Biden decided first to test the public climate with a speechmaking swing, starting in his hometown of Scranton. Then he was scheduled to speak at two universities in Rochester, New York, and at Yale in the second week of February, just in advance of the New Hampshire primary in which he had once hoped to emerge as a strong Democratic presidential candidate. After working out in the Senate gym one day, he felt a sharp pain in his neck, and upon returning to Wilmington by train that night it worsened, attacking his neck again, numbing his right side, and making his legs heavy. Managing to get home from the train station, Biden told Jill that he might have pulled a muscle. He went to see a doctor, who said he probably had pinched a nerve lifting weights, so he went to a pain clinic. He was fitted with a neck brace and assigned some exercises to enable him to take part in the Senate ratification debate on INF, and keep his speaking engagements.[4]

In Scranton on February 9, Biden discussed the debate and the amendment he was proposing binding the Reagan administration specifically to the terms set at the time of Senate ratification. The objective was to prevent the administration from developing and testing its questionable "Star Wars" missile-defense technology without Senate consent to any late changes.

That same night, Democrats in Iowa were holding their precinct caucuses, kicking off the selection process for national convention delegates. Now a noncandidate, Biden was in Rochester, New York, for another foreign policy speech. Afterward, in addition to questions on the INF treaty, the large crowd asked him about the Bork hearings and, inevitably, his own failed presidential bid.

To Biden's relief, the warm and welcoming audience kept him answering questions for an hour and a half. He didn't get back to his hotel room until eleven o'clock, pumped up by the reception and the absence of hostile questions about the plagiarism charges. He pondered sending out for a pizza; the next thing he knew, he woke up on the floor at the foot of his bed, fully clothed.

Only then, he wrote later, did he remember having felt a "sharp stick in the back of my neck and something like lightning flashing inside my head, a powerful electric surge—and then a rip of pain like I'd never felt before. I could still feel the waves of dull ache from that first blast of pain. My neck was stiff, too, so it was hard to turn my head. I rolled over and saw red numbers. The alarm clock read 4:10—had to be the middle of the night; there was no sunlight in the windows. I'd been unconscious for five hours."[5]

A Wilmington friend who had accompanied Biden, Bob Cunningham, found him there, freezing, and got him on the plane home and into bed. By this time it was morning and Jill, at her school, was called home. She rushed him to Saint Francis Hospital in Wilmington, where a spinal tap was administered and a priest was called to administer last rites of the Catholic Church. Blood had been found in his spinal fluid. The Biden family hastily gathered and the word finally came that he had had an intercranial aneurysm, a leaking of blood just below the base of his brain. Surgery was imperative to save his life. Brother Jimmy got on the phone and was told to bring him to Walter Reed Army Medical Center outside Washington, more than a hundred miles south, where the chief of neurosurgery waited.

Snowy winter weather prevented use of a helicopter, so Biden was loaded onto an ambulance, accompanied by Jill, with Beau riding ahead in a police escort car. At Walter Reed, after a second opinion from another expert surgeon located by Jim Biden, Jill approved the surgery. Joe was conscious when the doctor explained a possible outcome might be loss of speech. "I think I laughed out loud when he said that," he wrote later. " 'I kind of wish that had happened last summer,' I said, but I don't think the doctor heard me."[6]

Beau and Hunter, were brought in, and their father simply told them he loved them and to take care of their mom and sister—the Biden way. Jill later told Lois Romano of the *Washington Post:* "At that very moment I made my peace with what happened in the presidential race. All I remember thinking to myself was, 'My God, had he stayed in the race, he'd be dead already.' "[7]

When Biden came out of the anesthesia after successful surgery, he remained at Walter Reed for ten days. A second aneurysm that had been found would have to be repaired later with more surgery. Back home in Wilmington, local doctors found a blood clot in his lung, and he was returned to Walter Reed for treatment. Upon his release, Jill instituted a strict regimen of no phone calls. Reagan called twice; she thanked him, but the rule held.

In May the second surgery was performed successfully at Walter Reed, where this time he stayed for most of the month. When he was released again, one well-wisher sent him a note: "Dear Joe: What a smart guy! Everyone always said that anyone who goes into politics ought to have his head examined. And thank God you took it seriously."[8]

For a while Biden was left with a drooping eyelid, but it passed.

Finally, he was back driving his car, hitting golf balls at a driving range, and slowly rounding into shape. One day, Ted Kennedy took the train up to see him unannounced and talked his way in.

Meanwhile, at the Democratic National Convention in Atlanta where Michael Dukakis was nominated for president, Biden received two votes from Delaware loyalists, New Castle County Democratic Chairman Eugene T. Reed Sr. and County Treasurer Bette Rash.

It was now near the end of August, and Biden showed up at the annual Sussex County Jamboree, where seven hundred Delawareans greeted him, many of them teary-eyed. He gave them a brief dose of Biden humor. "The good news is that I can do anything I did before," he said. "The bad news is that I can't do anything better."[9]

Labor Day kicked off the traditional start of the general presidential campaign season. Biden's long illness had not only kept him out of the Senate and the INF Treaty hearings, but also off the campaign trail, in the service of his party and Democratic colleagues seeking reelection. As he sat in his kitchen with his brother Jimmy, the phone rang. It was, of all people, Michael Dukakis. In spite of all that had happened that had contributed to the end of Biden's own campaign, Dukakis as the Democratic nominee asked for his help on a special matter. He wanted to recall to his troubled campaign his exiled manager, John Sasso, who had circulated the Kinnock video that had done Biden in. Dukakis wanted to know what Biden would say if Sasso was taken aboard again. In other words, was he going to raise a noisy objection?

Biden wrote later that he asked Dukakis: "What will John say about me, Governor?"

"What do you mean?" Dukakis asked in return. Biden wrote: "He had no interest in making amends to me. He just wanted to smooth the politics of bringing Sasso back onto his campaign team." Dukakis expressed his compassion—toward Sasso. "Joe, John has had a terrible year," Biden said Dukakis told him.

A few hours later, Dukakis announced that Sasso was indeed

coming back. "He's paid the price—a year is a long time," he said. His man had made a mistake, Dukakis allowed, but "there was nothing illegal about what John did." As for Biden, he "could not have been more gracious," said the man who thus showed little of it himself.[10]

Two days after Labor Day, most of the Biden clan got aboard the Amtrak train at the Wilmington station, to mark Joe's return to the Senate for the first time in seven months. The railroad folks greeted him with balloons and signs, and on arrival his Senate staff threw a surprise party for him. Then he took over the chair at a Foreign Relations subcommittee hearing, interrupted by frequent welcome-back visits from colleagues of both parties. On the Senate floor, one after another, they cheered his return. In brief (for him) thanks, Biden said "if I say more, the Irish in me will creep out and I'll become too sentimental"—and then continued. He quoted Senator Daniel Patrick Moynihan of New York, standing nearby, as telling him, "to fail to understand that life is going to knock you down is to fail to understand the Irishness of life."[11]

Biden's long convalescence had given him a great deal of time for personal reflection, not only on how fortunate he had been to survive the latest ordeal but also to set his professional course for the immediate years ahead. The aneurysms, Valerie observed later, brought about some reflections by her brother. "When you almost die," she said, "you look at life with a different perspective. It just reaffirmed what our mom would say: 'You just have to be open to the good and the bad that comes to you.' "[12]

Beau Biden, who was only eighteen at the time, said that the challenge to his father's honor was much more devastating to him than the withdrawal from the presidential campaign. The phrase "my word as a Biden," he said, meant "your word is your bond, and that's what he means and what we mean by it, and I've seen him live his life by that. Our whole family has lived by it."

In withdrawing to pursue the Bork hearings, Beau said, "he needed to go on and demonstrate that his word is his bond, and I think he did that in the face of a really tough episode. . . . Every time he has been knocked on his tail—1972, 1987, 1988, six months later with his aneurysm—never once, ever, did I see him be a different way. I always saw, and never had any doubt, that he would go on. . . . There was no denying the gravity of those situations, for all different reasons, but never did I see any hint of giving up, ever. I call this incredible resilience and I see it today."[13]

Actually, his father said later, rather than thinking then about quitting politics, "it kind of gave me a renewed purpose." Valerie, he remembered, told him: "Joe, you've got a second chance here. Make it good." The brush with death, he said, "really gave me a sort of quiet resolve. The biggest thing it changed was, prior to that I used to think that every meeting, every opportunity, was critical, and I realized that there aren't that many things that are important, that are actually critical. It really gave me a different perspective. I would have thought if you had told me about it ahead of time that it would have created this awful sense of urgency, like you've got to get this done. It had the exact opposite effect. . . . It just stayed with me, a sense of resolve, that everything will come out, go just do your job, and move along."[14]

With his recent intense focus on the Judiciary Committee, Biden's work on the Foreign Relations Committee had taken a backseat, and his long absence meant he had not participated in the INF Treaty debate. He did, however, as chairman of a special subcommittee, conduct a review of the War Powers Act, the product of a 1973 joint resolution.

Reviewing the War Powers Act was of particular interest to Biden. Enacted over the veto of President Richard Nixon, it was aimed at limiting executive war-making power in the wake of the Vietnam War, and requiring continuing consultation with Congress. The unilateral presidential power was to expire after sixty days unless extended by Congress, but ever since its enactment the legislation has been thoroughly disregarded, with critics declaring it unworkable or unconstitutional or both. In 1988 hearings were held over two months, and Biden coauthored a law review article proposing how the liabilities could be overcome, in part by providing prior and periodic consultation with congressional leadership.[15] But this effort stalled year after year.

In the run-up to the Persian Gulf War in 1990, President George H. W. Bush insisted he did not need action by Congress to use force to expel Iraq from Kuwait. He sought it only reluctantly in the end without acknowledging it was legally necessary. Biden challenged what he later called a "monarchist" view of war-making power—that a president had near-unlimited authority to act on his own to respond to any threat. And he continued to argue to no avail that the power to declare war resided constitutionally with Congress, citing Alexander Hamilton, James Madison, and other founders as sources.[16]

In 1998 Biden introduced his own "Use of Force Act" that would repeal and replace the War Powers Act, listing five distinct circumstances in which a president could initiate use of force: to repel an attack; to deal with urgent threats to American interests; to extricate imperiled American citizens; to forestall or retaliate against specific acts of terrorism; to defend substantial threats against international sea lanes or airspace. It would retain the sixty-day expiration and extension provisions, establish a congressional leadership group with which the president had to consult on use of force, and provide for limited judicial review. In introducing it, Biden said, "I have no illusions that enacting this legislation will be easy. But I am determined to try."[17] His lack of optimism was well-placed; the repeal and replacement got nowhere, and the original continued to be ignored.

On other fronts Biden would be more successful. Only three weeks after his Senate return in 1988, the major drug bill he had spent six years crafting became law. Included was the creation of a national drug czar, a key Biden objective and a job that went to Republican William Bennett. Biden vowed to be Capitol Hill's point man in pressing the new Bush administration on antidrug spending and helping Bennett navigate his way through a thorny bureaucratic thicket of multiple congressional jurisdictions.

When President Bush announced his 1989 antidrug plan, Biden showed no hesitation in criticizing him for not funding initiatives already on the books. He called for higher taxes on cigarettes and tobacco (neither of which he ever used) to pay for them.

In a speech before the National Press Club, Biden unleashed his old fire: "Mr. President, you say you want a war on drugs, but if that's what you want we need another D-Day. Instead you're giving us another Vietnam—a limited war fought on the cheap, financed on the sly, with no clear objectives, and ultimately destined for stalemate and human tragedy. . . . What has President Bush offered in response to this crisis? Only cardboard cops, cutout courts, paper prosecutors, pretend prisons, and tenuous treatment."[18] Biden for a moment sounded like the old master of alliteration, Spiro T. Agnew.

The speech infuriated White House Chief of Staff John Sununu, to the point that he accused Biden of "plagiarizing" the administration's antidrug plans—a word choice calculated to embarrass him by recalling the senator's earlier woes. Biden defended the speech as "flat-out accurate." In reference to Sununu's remark, he noted that "the way [they] respond to Biden taking the drug issue or the crime issue from us is, call him a plagiarist." His friend, Republican Senator Bill Cohen of Maine, said in Biden's defense: "He is not one of these real partisans. He can be hard-hitting. He has that capacity," yet he added: "It's kind of like the Jack Nicholson smile. He's smiling, but the smile is not in the heart, it's on the lips."[19]

For all Biden's bombast, the year 1989 was for him one of regrouping, not as a future presidential prospect but as a senator who, in his own view, had lost his way as a result of excessive political ambition, naïveté, and hubris. Not content with reaching at such an early age what he had said at the time was his life's goal, to be a member of the U.S. Senate, he had gotten well ahead of himself in running for president.

"I did all the wrong things," he said again in reflection. "I spent probably the first four years, maybe the first six years, of my Senate career treading water. I would miss votes, and get on the train and go home because my son was sick with the flu. Because for me I was wrongly, I acknowledge, but nonetheless totally, completely preoccupied with one and only one thing: putting my family back together. . . .

"I think I gave press people every good reason to believe that this guy is Flash. Here comes Flash. Boom. He's in, he's out, he asks the right questions, he goes to the nub of an issue, but no staying power—he's gone. All of that criticism was deserved. . . . I never took the time to really get a chance to enjoy what I wanted to do in the first place. I was always one step ahead. . . ."[20]

Asked often by reporters during this time whether he would run for president again, Biden said categorically he would not do so in the next election cycle in 1992, and probably not in 1996, either. He appreciated the self-repair work he had ahead of himself to come to terms with his own shortcomings, and to be taken seriously by others. Biden told a reporter: "What I found for me is that it doesn't make a lot of sense to plan many years in advance."[21] And later in the year he told another, with obvious good reason: "I have an incredibly high regard for fate. I have never been able to plan my life. Every time my personal life has been how I wanted it, something has intervened."[22]

Notably, Biden didn't say he'd never seek the White House again. In late May, an arm of the Delaware Supreme Court at Biden's request reviewed the plagiarism allegations against him at Syracuse Law School and exonerated him. He told the *Wilmington News Journal* that the court's review, and the clean bill of health he got from the doctors who had performed the brain surgeries on him, "would erase any doubt" about his fitness "when I run again."[23]

Meanwhile, Biden recognized that his committee chairmanship gave him an ample megaphone with which to establish, or he would say reestablish, himself as a figure of substance as well as style and flash. On Judiciary, he took on a controversial fight over flag desecration, and he and others successfully detoured a constitutional amendment on grounds it was an infringement on free speech. Judiciary, with other senators often in the lead, also tackled immigration, gun control, hate crime, and other issues, giving Biden sufficient platform from which to lay claim to be an effective legislator. Some of these matters languished under his chairmanship, but it was harder now to dismiss him as a silver-tongued lightweight.

Biden's commitment to Senate committee business in 1989 even cut into his storied rail commuting from and to Wilmington, obliging him from time to time to stay at a hotel for a night or two. But he remained committed to his first priority—his wife and growing children back in Delaware, at whatever cost it imposed on him as Amtrak's most famous regular patron on the eastern seaboard.

For the fourth time in eighteen years, Biden faced the voters back home in 1990, and for all the rehabilitation he had undertaken after his ignominious withdrawal from the 1988 presidential race, the plagiarism allegation was resurrected by his Republican opponent. Efforts by the Bush White House to recruit cautious former governor Pete du Pont to challenge Biden having failed despite Biden's taunting, the Delaware GOP settled on thirty-nine-year-old M. Jane Brady, a former state deputy attorney general with a reputation as a hard-nosed prosecutor. The state party's chairman, Basil Battaglia, expressed his resignation by saying of Brady, "I couldn't ask for more, unless, obviously, I got Pete du Pont."[24]

Brady at the outset pledged she would not run a negative campaign against Biden. She came out strong nevertheless in her mid-May announcement of candidacy: "The choice is clear—the same old, tired answers of Joe Biden, a Washington bureaucrat, or an opportunity for citizens to regain control of their government with me."[25] She said she would accept no political action committee money and pledged if elected to serve only two terms. And she sought to paint Biden as a tool of the banking and credit card industries that dominated the financial world of Delaware, and in some ways beyond. None of this, however, was particularly out of bounds in personal terms.

Campaigning against Biden, Brady recalled later, was more than she had counted on. "He was Joe," she said. "He was smart and informed and tenacious." She likewise was tenacious, she said, but "remarkably naïve about the political process. I really, truly had no concept of the role of the arms of the party." Biden, she said, "tried to stay above the race. He tried not to engage me, tried to minimize me as a candidate. . . . I was running against Congress—mail franking, cheap haircuts, pork-barrel spending, spending earmark funds." A mood of anti-incumbency was in vogue, she said, "and that was a big component in my strategy. He enjoyed all the perks of office that I was running against."[26]

Brady kicked off her campaign with Delaware's traditional visit to all three counties, which Biden followed in late July. He was no longer the unpolished kid candidate taking on the giant incumbent as in 1972, but now the familiar, seasoned, and worldly Senate man and national figure. Appropriately, he spoke in his declaration of candidacy of his worldview, predicting that "long before the Senate term I seek now is over, the Soviet Union as the world has known it for seventy years will cease to exist." He prophesized that it would dissolve into separate and independent nations, possibly in a loose confederation.[27] Whether or not listeners believed Biden's vision, his confident presence was more than enough to draw a telling difference between himself and his opponent. It was clear that Brady would need more than laments about congressional perks to replicate Biden's own upset over incumbent Cale Boggs in 1972.

No one understood this reality better than Don Devine, a veteran bare-knuckle political brawler sent to Delaware by the National Republican Senatorial Campaign Committee to throw Brady a lifeline. About a month before the November election, the Brady campaign produced a ten-minute video rehashing the whole Neil Kinnock episode and associated allegations about Biden, and played it for reporters in Wilmington. She said copies would be mailed to about forty thousand homes in Delaware, insisting she was not breaking her word given previously that she would not discuss the so-called character issue unless it became relevant in the campaign.

Brady cited as justification a Biden appearance on C-SPAN during the summer in which he talked about the plagiarism charges. "Biden made it relevant," she said. "I can't believe he said that," she remarked of the C-SPAN comments. "It was like he was teasing me."[28]

The video showed Kinnock and Robert F. Kennedy speaking, juxtaposed with Biden's speaking the same words without attribution. It also rehashed the allegations about Biden's alleged plagiarism at Syracuse Law School and his claiming at a New Hampshire rally that he had finished in the top half of his class, when he had graduated near the bottom.

While the production of the video at a cost of $75,000 generated considerable print, radio, and television coverage in Delaware, it brought essentially no new information to the voters about Biden's past travails. And on election day it failed to dent the longtime romance the electorate had with the local boy who had made good on the national scene, albeit short of attaining the presidency. Biden's argument that Delawareans knew him well and approvingly triumphed over the most negative campaign yet waged against him. He won 62 percent of the vote to 36 percent for Brady and 2 percent for a minor party candidate.

Later, Brady cited Delaware's Return Day tradition in which all major state candidates gather at the Sussex County seat, Georgetown, two days after the elections for a horse-drawn carriage parade featuring winners and losers riding together. She recalled seeing three women in the crowd, one of them holding up a photo of Biden. "This is why I lost," she said, in what was obviously only a partial explanation. "People have bonded with this man, the intensity is so strong."[29]

The reelection enabled Biden to focus as a member of the Senate Foreign Relations Committee on a national crisis that now competed with his Judiciary Committee responsibilities. Iraq's invasion of Kuwait in August 1990 had triggered a strong diplomatic response at the United Nations from President George H. W. Bush that segued from stronger sanctions to a U.S.-driven military buildup dubbed "Desert Shield." Bush famously declared at the time of the invasion that "this shall not stand" and would be rolled back by force unless Iraqi strongman Saddam Hussein withdrew.

Biden called for a special session of the Senate to debate the American policy, charging that Bush had "moved from a reasoned and practical response to a misguided policy in the [Persian] Gulf."[30] Meanwhile, the Foreign Relations Committee scheduled hearings on the crisis, giving Biden another major platform from which to continue his seemingly endless quest to demonstrate his substance, to counter the equally endless derision of his "style." In Delaware circles, the conflicting evaluations were dismissed by supporters with the oft-heard comment of "That's just Joe." But elsewhere, others were not so understanding about what they saw in Biden as flippancy and arrogance unbecoming a national leader, especially when it came to the matter of war and peace now at hand.

SEVENTEEN

HOLDING OUT FOR PEACE

IN EARLY SEPTEMBER 1990, Biden made clear in a Foreign Relations Committee exchange with Secretary of State James A. Baker that he was unwilling to give the administration a blank check on U.S. military action against Iraq to roll back its invasion of Kuwait. He told Baker that while he generally approved of the impressive buildup of forces to challenge Saddam Hussein's bold move, "we should make sure we are deciding what precipice we are going to look down or jump over or jump into." He told Baker application of the War Powers Act calling for continuing consultation with Congress on use of American combat had to be resolved, and paraphrased the old axiom of Senator Arthur Vandenberg: "if you want me in on the landing, I better be there for the takeoff." Baker replied that he understood.[1]

Not satisfied, Biden, in a subsequent hearing in October, confronted Baker with a resolution of his own, authorizing U.S. participation in "collective security actions" by the United Nations against

Iraq, but conditioned on congressional consultation. Biden specified that "if you are going to move on your own you must come back to Congress and seek a declaration of war."

Baker replied that "we would have some major reservations with it," but Senators Richard Lugar of Indiana, a Republican, and Paul Sarbanes of Maryland, a Democrat, agreed with Biden. Sarbanes said to do otherwise "is contrary to what the Constitution calls for."[2]

But Baker and the administration brushed aside the Biden resolution and on November 29 pushed their own Resolution 678 through the UN Security Council. It authorized member states "to use all necessary means" to force compliance with the UN demands for Iraq's withdrawal from Kuwait, with a January 15 deadline.

At yet another hearing in December, Biden expressed concern that in the wake of Republican losses in the November congressional elections, the president had changed and broadened his foreign policy goals in the Middle East. "We talk about a New World Order," he said. "A New World Order in the United Nations and collective security adds up to 'We will hold your coat, United States. You go get them; we give you the authority to do it.' That is the essence of that New World Order. That is not a New World Order I am prepared to sign on to."

Biden also questioned what American vital interest was now involved in a broadened objective, noting that the same question had been raised in the Vietnam War. "I came to the Senate in 1972," he said, "because I was so tired of hearing that we were unknowingly playing into the hands of Ho Chi Minh where we disagreed [with administration policy]. Now I hear today we are unknowingly playing into the hands of Saddam Hussein." He quoted Henry Kissinger as saying that a perception of American failure in the Persian Gulf would shake international stability and every moderate country in the region would be weakened. "Boy, oh boy," Biden said in his folksy way. "Here we go again." In another apparent analogy to Vietnam, he added that no one had "laid out clearly what our vital interests are sufficient to have ten thousand, twenty thousand, thirty thousand, forty thousand Americans killed. I have not heard that one yet."[3]

On December 12, with the January 15 deadline for implementation of the use-of-force resolution a month away, Biden convened another Foreign Relations Committee hearing on U.S. policy in the Gulf. He used it to challenge the imperative of taking military action while the UN sanctions were working.

"All evidence indicates that the world embargo against Iraq is the most comprehensive in history," he declared. "Collectively, we have imposed nothing less than the state of siege on that outlaw regime—a policy backed by the dispatch of a powerful deterrent force that was accepted by Congress and the American people. . . . Now, however, the administration appears to have grown impatient with the policy it originally championed. But serious discussion is warranted as to when and on what basis we should regard the sanctions policy as a failure."

Biden cited "expert testimony" before the committee, saying sanctions would succeed "if not in forcing Iraq to withdraw without military action, then in so weakening Iraq to make the eventual use of force far less costly." He asked: "Precisely which countries are not prepared to remain committed to the sanctions policy which, by all evidence, holds tremendous promise for bringing the Iraqi regime to its knees?"[4]

A week before the UN deadline, Biden resumed his argument for more time to allow the sanctions to break Saddam's will. In yet another hearing, he reported that "the sanctions have produced a 50 percent drop in Iraq's GNP. Iraq cannot sell oil and has no source of revenue. It can perform no international financial transaction. It cannot produce tires for its transport or operate its expensive, imported industrial infrastructure. . . . Even if sanctions alone do not force Saddam Hussein out of Kuwait, they will certainly weaken him politically and militarily. Before we ask Americans to die for the liberation of Kuwait, I would like to be sure we have tried every possible alternative. So far, this has not been the case." While it was critical to maintain resolve against the Iraqi dictator, he said, "the United States should not be the principal policeman of the world. . . . It is not enough for the UN to be willing to fight to the last American."[5]

On the same day, Biden presided over a separate Judiciary Committee hearing on the same basic question of whether a president could constitutionally declare and wage war without explicit congressional approval. Again, he offered a vigorous insistence on Congress's role. He noted that then Secretary of Defense Dick Cheney had said, "we do not believe that the president requires any additional authorization from Congress before committing U.S. forces to achieve our objective in the Gulf." Then Biden quoted Alexander Hamilton as saying the president's designation as commander in chief was "nothing more than the supreme command of the military and naval forces."[6]

Biden also disagreed with Baker's argument that in modern warfare in the nuclear age there simply was no time for extensive debate before using force, lest the element of surprise would be lost. "The long-standing claim that we cannot debate war in the modern age," Biden said, "has been shown to be a red herring by the UN resolution, at least in this case." As for such a debate being so divisive as to shatter American resolve, he said, "if a nation is divided about war, then it is better to have that division aired now, rather than later, after the bullets start to fly."

Biden readily conceded one argument to the other side. "Finally, we have been told that the congressional debate on war could tie the president's hands or limit his discretion. To this charge I have one simple response: Exactly right. Americans once lived under a system where one man had unfettered choice to decide by himself whether we could go to war or not go to war, and we launched a revolution to free ourselves from the tyranny of such a system."[7]

Biden was not alone in making this case against the joint resolution, but he was one of only ten Senate Democrats in opposition as it narrowly passed, 52–47. Long afterward, his vote made him an easy target of war hawks in both parties, especially in light of the war's swift resolution. Biden's warning of a sustained engagement costing huge American losses was not realized, further casting him as a man who cried wolf.

As a committee chairman he summoned witnesses on both sides but used his prerogative at times to load the roster with like-minded scholars and former cabinet officers. For example, on the day the joint resolution was called up and passed, Biden summoned before Judiciary one Professor William Van Alstyne, then of the Duke Law School, who bolstered the senator's view on the constitutional power of Congress to declare war. Van Alstyne had gone so far as to argue in a newspaper article that if President Bush failed to secure its authorization "and nonetheless embarks on war, it will not mark some trivial trespass on the margins of constitutional niceties. It will mark the boldest usurpation of power by a president we will have seen since Watergate, and it should be treated in the same way."

Biden asked him what he meant by that. He replied that if Bush so acted it would be "a sufficient affront to the substance of the American Constitution as to constitute an impeachable offense, and that is one of the things I meant by the comparison with Watergate. Indeed, the number of lives imperiled by this constitutional action dwarfs any possible comparison with any of the worst deeds that others have ever imputed to Richard Nixon, with regard to whom the House Judiciary Committee had prepared articles of impeachment."[8]

With former secretary of state Alexander Haig of the Reagan administration before the Foreign Relations Committee on the same day, Biden posed the same question. Haig's answer did not help Biden's argument, but he asked it anyway. Haig said certainly there could be a war without a "congressional declaration" and American combat had been so initiated more than two hundred times, with only five ever accompanied by a formal declaration. But, he noted, no war could be "sustained without congressional funding," and refusal to pay for one was the teeth Congress had to prevent or stop one. "We saw that in Vietnam," he noted.

Regarding the Constitution, Haig observed—perhaps mischievously—that "as a former White House chief of staff and secretary of state I have a passing knowledge of this document." It was Haig who, after the 1981 attempted assassination of President Reagan, declared that he was "in control," apparently under the misapprehension that in the demise or absence of the president and vice president, he as secretary of state was next in line of presidential succession.

In any event, Haig said the debate over the constitutional power to start a war had "an air of unreality" about it, adding: "No doubt it would be politically prudent for the president to have an overt expression of congressional support, especially to send Saddam Hussein a clear signal. Still, we should be under no illusions. The ultimate test of this war, if there should be one, is not to be found here in the halls of Congress or in the courts of law, but rather on the battlefield."[9]

In many of the Foreign Relations hearings, Biden participated in a relatively benign fashion, as opposed to his often more aggressive demeanor as chairman of Judiciary. In both committee settings, as he became a more senior senator, he demonstrated a wide range of interest and expertise in the subject matters discussed. In doing so, he often undermined the simplistic derision he also often encountered from the news media and rival Republicans that, as the Texas expression had it, he was all hat and no cattle.

In any event, the die was cast on the decision to drive the forces of Saddam Hussein out of Kuwait, and the day after the UN deadline for voluntary withdrawal passed, Operation Desert Storm began with a vengeance. As air assaults hammered Iraqi targets in the invaded neighbor state, Biden was in Los Angeles for a speech. "The president has made a decision," he said. "Now the nation should unite behind that decision. The most critical near-term objective is to give the young men and women in the field all and any support they need."[10]

Without endorsing the invasion he had vigorously opposed, Biden took the only position, both political and patriotic, left to him. And the following day, he cautiously warned of a long military engagement. "This was always a possibility," he said, "which holds the specter of widening this war, but it is too early to speculate."[11]

Two weeks later, as the war continued and President Bush addressed Congress in his annual State of the Union message, Biden predicted that a positive outcome would assure the president's reelection in 1992, and that was all right with him. He told CNN: "If the president successfully prosecutes the war, and I hope he does . . . that's worth losing the White House for." And when television anchor Bernard Shaw offered that Democratic National Chairman Ron Brown might "hate you" for so saying, Biden replied: "I don't think Ron will hate me." He said "ending the war with minimal loss is desired by every American, Republican or Democrat, honest to God."[12]

Biden's swift segue from opposing the invasion to expressing his concern for the troops and his hope for a successful outcome generally spared him much immediate criticism for voting against the president on going to war. With the decisive defeat of Saddam Hussein's invaders in Kuwait and their retreat back into Iraq, President Bush's firm pledge that the invasion would "not stand" was fulfilled. And his decision not to escalate the objectives and cost of the war to "liberation" of Iraq itself, and the deposing of its dictator, moved the episode to the back pages of American newspapers and off the television screens.

But Biden's vote against sending U.S. forces into armed conflict rather than spending more time and effort seeking a diplomatic solution lingered long after in the arsenal of Biden critics on foreign policy. His holdout placed him in the sights of Republican snipers who reveled in casting Democrats as incurably soft on defense, unworthy of trust in post–cold war times rife with talk of peace dividends.

Once again, however, Biden's heavy responsibilities on the Senate Judiciary Committee would bring him to the nation's political center stage. Another vacancy on the Supreme Court suddenly presented itself, and this time more than matters of law would confront the committee chairman.

EIGHTEEN

CLARENCE AND ANITA

ON JUNE 27, 1991, eighty-two-year-old Thurgood Marshall, the country's first and only African American Supreme Court justice, announced his retirement. On July 1 President George H. W. Bush shocked the legal community by declaring Federal Appellate Court Judge Clarence Thomas the individual "best qualified at this time" to replace him.

It was not surprising that Bush would nominate another African American to the vacancy. But the selection of Thomas, who had been a federal judge for only a year and was widely seen as the most conservative black jurist on any federal bench, triggered an avalanche of incredulity and dismay among liberal Democrats. Biden and other Democrats who had reluctantly confirmed Thomas for the federal judgeship had warned Republican leaders at the time that it would be an entirely different matter if Bush were to attempt to put him on the Supreme Court.

But with President Reagan's having failed to place the classically conservative Bork there, President Bush now sought to create a clear right-wing majority with Thomas, a man with nowhere near Bork's legal credentials. The nomination of a black candidate was a deft political move, in that liberal Democrats on the committee would be wary of alienating one of their party's most loyal voting blocs by rejecting Thomas.

Biden had his work cut out for him if he were going to prevent the same ideological majority on the Court that had motivated his case against Bork. He knew that opposition to Thomas would be cast by critics as racial bias, a position anathema to any professed strong supporter of civil rights. Biden had already had a taste of that in opposing school busing in Delaware. But he was not prepared for the aggressive manner in which Thomas himself would play the race card to gain his confirmation.

The array of liberal special-interest groups that had been aroused to a fever pitch by Ronald Reagan's nomination of Bork quickly rallied against Bush's selection of the equally controversial but much less seasoned Thomas. In opening his interrogation of the nominee, Biden focused—as he did in the Bork hearings—on the man's judicial philosophy. Like Bork, Thomas championed adherence to strict construction of the Constitution as penned by the Founding Fathers and, in Thomas's case, in the adherence to what Biden called the natural law philosophy "as a moral code, a set of rules saying what is right and what is wrong . . . which the Supreme Court should impose upon the country."

Biden said there were twentieth-century advocates of this philosophy who "believe that this is the job of the courts to judge the morality of all our activities, wherever they occur, paying no respect to the privacy of our homes and our bedrooms." They "call into question a wide range of personal and family rights, from reproductive freedom to each individual's choice over procreation, to the very private decision we now make about what is and what is not a family." Without uttering the words "abortion" or "same-sex marriage," Biden made his point.

He went on to say "this sort of natural law philosophy is one which I believe this nation cannot accept." He included also its use to intrude on personal freedoms and to strike down a range of labor laws by putting certain business and economic rights "into a zone of protection so high that even reasonable laws aimed at curbing corporate excesses were struck down."[1]

Biden noted that while Thomas had "made it abundantly clear that you do not subscribe to the most extreme of these views," that the judge also had said he supported the idea of "an activist Supreme Court that would strike down laws regulating economic rights."[2] Biden argued that the economic freedoms Thomas defended were subject to governmental limits in "a balanced liberty that we have come to expect our government to provide." He said he feared "adopting a natural law philosophy that upsets that balance . . . by lessening the protection given those rights falling within the zone of personal and family privacy and speech and religion," and limiting governmental regulatory powers to protect the public against corporate excesses.[3]

Biden's monologue on natural law had nonlegal heads spinning, and they continued to twirl as Thomas sought to derail the issue in his own reply. He finally told Biden that "my interest in natural rights were [*sic*] purely from a political theory standpoint and as a part-time political theorist. I was not a law professor, nor was I adjudicating cases. . . . I do not think that the natural rights or natural law has an appropriate use in constitutional adjudication."[4] In other words, apparently, he would not be applying natural law in deciding constitutional issues if seated on the Supreme Court.

Attempts by Biden to draw Thomas into the net that had ensnarled Bork—discussions on his views on rights of privacy and whether the Constitution included language that constrained them in any way—

got nowhere. When Thomas said he believed "there is a right to privacy in the Fourteenth Amendment," Biden seized the opportunity. "Well, Judge," he asked, "does that right to privacy in the Liberty Clause of the Fourteenth Amendment protect the right of a woman to decide for herself in certain instances whether or not to terminate a pregnancy?"[5]

Thomas artfully dodged. "Senator, I think that the Supreme Court has made clear that the issue of marital privacy is protected, that the state cannot infringe on that without a compelling interest, and the Supreme Court, of course, in the case of *Roe v. Wade* has found . . . as a fundamental interest a woman's right to terminate a pregnancy." And he added: "I do not think that at this time I could maintain my impartiality as a member of the judiciary and comment on that specific case."[6]

Biden quickly told Thomas he wasn't asking him "to comment specifically" on the celebrated abortion case. Then and later, the chairman avoided doing so, intent as he had done with Bork on building a broader case on whether Thomas should or should not be confirmed. During the second round of questions, Biden assured Thomas it was his thought process on arriving at decisions he was after, not a particular decision. "You will be pleased to know I don't want to know anything about abortion," he said. "I don't want to know how you think about abortion. I don't want to know whether you have ever thought about abortion. I don't want to know whether you ever even discussed it. I don't want to know whether you have talked about it in your sleep. I don't want to know anything about abortion. I mean that sincerely, because I don't want that red herring, in my case at least, to distract from what I am just trying to find out here, which is how do you think about these things." And with that, he got back on his exploration of natural law again.[7] As other senators later questioned Thomas more succinctly on *Roe v. Wade*, the nominee offered the same elusive reply.

Biden's questions, for all his assurances of fairness, struck some in the audience as grandstanding his legal erudition. "As the hearings wore on, Biden's queries were sometimes so long and convoluted that Thomas would forget what the question was," reporters Jane Mayer and Jill Abramson wrote in *Strange Justice*, their book on the Thomas nomination. "Biden had considered the Bork hearings his finest hour, a high-minded discourse that had engaged the country. Bork was defeated fairly, in Biden's view, because of his legal opinions. This time Biden's questions seemed occasionally to be a vehicle to show off his legal acumen rather than to elicit answers."[8]

At every opportunity, Biden returned to the matter of "marital and family privacy" as "a fundamental liberty." He pressed Thomas on whether "the right of privacy and the right to make decisions about procreation extend to single individuals as well as married couples," drawing another evasive answer from Thomas. Through much of the five days of questioning, the nominee came off as reluctant to commit himself, which was part of the strategy devised by White House adviser Kenneth Duberstein, who hoped to sell him on his appealing up-from-poverty biography. One result was to present Thomas as a much less informed and learned lawyer than Bork had been, raising doubts about the certainty of his confirmation.

Even when Senator Patrick Leahy peppered Thomas with questions about whether he had ever discussed or debated *Roe v. Wade* when he was at Yale Law School, the judge denied it. He cast himself as a sort of social recluse and a grind who had little time beyond studying and working to engage in such musings. "Because I was a married student and I worked," he offered to Leahy, "I did not spend a lot of time around the law school doing what the other students enjoyed so much, and that is debating all the current cases. . . . My schedule was that I went to classes and generally went to work and went home."[9]

Leahy registered incredulity: "Judge Thomas, I was a married law student who also worked, but also found time, at least between classes, that we did discuss some of the law, and I'm sure you are not suggesting that there wasn't any discussion at any time of *Roe v. Wade*?"

Thomas replied, "Senator, I cannot remember personally engaging in those discussions."

Even when Leahy asked whether the judge had ever had any discussions about the abortion case in the years since the Supreme Court ruling or had ever expressed any opinion on whether it had been properly decided, Thomas replied, "I can't recall saying one way or the other, Senator."

Leahy: "Well, was it properly decided or not?"

Thomas: "Senator, I think that that is where I just have to say what I said before; that to comment on the holding in that case would compromise my ability—"

Leahy interrupted to ask whether the judge had ever made any decision "in your own mind" whether the case was decided properly or not "without stating what that decision is." Thomas answered: "I have not made, Senator, a decision one way or the other with respect to that important decision."[10]

So it didn't matter that Chairman Biden had so determinedly, if comically, in his proclaimed fairness assured the judge that he would not mention the controversial and explosive issue of abortion. His distinguished Democratic colleague from Vermont took care of that. No matter how disingenuous Thomas may have come across in the hearing room to television viewers, he stuck to the game plan, even in the face of snide criticisms. Much later, Leahy said Thomas's responses that he had never even thought about *Roe v. Wade* as a law student were so preposterous that they cost him the senator's vote.[11]

On September 5, five days before the Thomas confirmation hearings were to begin, an aide to Democratic Senator Howard Metzenbaum of Ohio, acting on a tip from a Thomas critic, had telephoned a young black law professor at the University of Oklahoma named Anita Hill, to inquire about the judge. The call set off a chain reaction that wiped out all concerns over "natural law" and Clarence

Thomas's views on the right to privacy. It soon plunged Joe Biden, the Senate Judiciary Committee, and the politics of the nation into a melodrama seldom seen in Washington since the Watergate scandal.

A decade earlier, Anita Hill had worked for Thomas at the Equal Employment Opportunity Commission in Washington. Hill was out when the call from Metzenbaum's aide came, but she soon returned it, the first step toward her airing of shocking allegations of sexual harassment by Thomas at the EEOC. At first she declined to confirm or deny rumors of such behavior and hedged about cooperating with Senate committee aides pursuing the rumors. Metzenbaum was advised of the allegations, and he had the information passed on to Biden's Judiciary Committee staff, and thence to Biden.[12]

The chairman reacted cautiously. Hill at first said she wanted to remain anonymous in the matter and did not want Thomas to know of her involvement at all. Perhaps as one who himself had been victimized by anonymous conspirators in the allegations of plagiarism that ended his 1988 presidential ambitions, Biden was reluctant to act on that basis. Later, in an interview for the Mayer and Abramson book, he said however that "from the beginning it was clear that this had the potential to become a giant incendiary bomb. We were handling one of the most controversial nominees in this century, and this woman who says he harassed her won't let us use her name. I was focused on how to get the truth—but without conducting a Star Chamber."[13]

Word of Hill's allegations soon circulated among anti-Thomas operatives in Washington and committee staffers on Capitol Hill. Meanwhile, Hill waited in limbo in Oklahoma, uncertain of the Judiciary Committee's intentions as the confirmation hearings approached an end. Finally, Biden decided to send the FBI to talk directly with the woman at the center of it all. For several days, as a vote on Thomas neared, Hill balked. But eventually, to make sure her story was told as she wanted it told, she sat down and wrote a four-page account of her accusations against the judge. It was given to Biden with Hill's understanding it would be distributed to all committee members—including, of course, the Republicans supporting Thomas. The ranking Republican, Strom Thurmond, however, didn't have the statement circulated among his party brethren, who had to learn about it from the committee's Democrats.[14]

Senator Arlen Specter, the former prosecutor from Philadelphia, only heard of the vague complication for Thomas from Democrat Dennis DeConcini. He had to corner Biden outside the Senate dining room to find out what it was all about. Specter wrote later that Biden told him "we were about to close down the hearings on Thomas, and I got a phone call in the back room of the hearing, where the staff worked, saying that a woman called saying that she had a complaint against Clarence Thomas, [of] sexual harassment."[15]

Republican Senator John Danforth, in Thomas's corner from the start, phoned Specter that night and got his assurance that he intended to support Thomas. But after thinking the matter over, Specter phoned Danforth and asked him to arrange a meeting with Thomas before the imminent committee vote. Thomas himself didn't learn of the allegations at once, or that the FBI now planned to interview him. Specter wrote later that when he confronted Thomas, the judge told him: "No, sir, with God as my witness . . . it just didn't happen. I wouldn't do that. . . . Black men are always accused of that. . . . Never happened, absolutely not." Thomas told him, Specter wrote, "African American men are often described sexually in terms of prowess and size, and as predators. He told me how painful it was for him to hear Hill's charges, and how untrue and extraordinary they were."[16]

Biden was faced with a major dilemma. Should the allegations be made public if Hill was unwilling to make them openly? Should the committee vote on the Thomas confirmation go forward, or should it be delayed? The matter for Biden involved not simply fairness but the pursuit of justice. Somebody had to be lying. But Biden hesitated, only informing other committee members personally as he ran into them.

In all this, the Judiciary Committee rolled on like a brakeless locomotive, with neither Biden nor anyone else calling for a delay in the vote until further investigation could take place. Some members knew little or nothing about the allegations, or their veracity, and the specter of senators voting on Thomas while in the dark about this latest bombshell would clearly shatter Biden's reputation for fairness. Specter wrote much later that Biden told him that although he was clearly opposed to the Thomas confirmation and had said so, "notwithstanding that, I couldn't bring myself without [Hill] willing to confront it fairly, to bring this thing out in the open."[17] Specter said later he would simply have subpoenaed Hill, as he often had done as a prosecutor.

On Friday morning, September 27, the day the Judiciary Committee was to vote on confirmation, Biden took a phone call from Professor Laurence Tribe of Harvard telling him he had heard of the Anita Hill allegations. Tribe urged him to distribute them to all Democrats on the committee, which Biden finally did. By now Thomas's supporters were uncertain whether the judge had convinced enough senators to send the nomination to the Senate floor with a recommendation for confirmation.

With Biden the last member to declare his vote, the chairman went to the Senate floor and said his concerns over Thomas's views on the right to privacy and other issues made putting him on the Court simply too risky. As he had advised Thomas earlier, Biden also told the Senate that "for this senator, there is no question with respect to the nominee's character"—a surprising reassurance in light of the new and unresolved allegations of sexual harassment.

Biden told his fellow senators he had also advised Thomas that "I believe there are certain things that are not at issue at all. . . . This is about what he believes, not about who he is. And I know my colleagues will refrain, and I urge everyone else to refrain, from personalizing this battle."[18] The committee then cast its votes, essentially on party lines, with only Democrat DeConcini voting with the Republicans.

The resulting 7–7 tie sent the nomination to the Senate floor with no recommendation.

The White House, and particularly Senator Danforth, who was committed unrestrainedly to Thomas's confirmation, saw the train going off the tracks unless they moved quickly. The Senate Republicans pushed for an immediate vote by the full Senate that weekend. When that failed, they won agreement to hold it in ten days, with the allegations against Thomas yet to descend on much of the Senate and the public at large.

Biden as the Judiciary chairman had until the last days adhered to his reputation for fairness earned in that earlier Bork confrontation. But this new challenge, of which only a handful of participants in the drama were then aware, was now about to cast a huge shadow over the quality of Biden's committee leadership.

Over the weekend, a shroud of mystery covered the internal machinations within the committee over what was going on. Hill was still insisting she didn't want the whole story made public. It was not until the following weekend that two reporters covering the Justice Department and the Supreme Court, Nina Totenberg of National Public Radio and Timothy Phelps of *Newsday*, broke the story of the allegations.

The general press reaction was immediate. Anita Hill, besieged by reporters, denied "this is somehow a political ploy that I'm involved in." She said the Senate Judiciary Committee had approached her, not the other way around. She said she had "hoped and trusted" that Biden would distribute her statement of allegations to members of the committee before they voted on Thomas's confirmation. Expressing resentment toward attacks on her already being aired, Hill called them "an attempt to not deal with the issue itself. It is an unpleasant issue. It's an ugly issue, and people don't want to deal with it generally, and in particular in this case. But I resent the idea that people would blame the messenger for the message rather than looking at the content of the message itself, and taking a careful look at it, and fully investigating it. And I would hope that the official process would continue." As for Thomas's alleged behavior toward her, Hill said, "I believe this conduct reflects his sense of how to carry out his job."[19]

In a phone interview with the *New York Times* two nights before the reopening of the hearings, Biden again defended his leadership of the committee. "I must start off," he said, "with a presumption of giving the person accused the benefit of the doubt. I must seek the truth and I must ask straightforward and tough questions, and in my heart I know if that woman is telling the truth it will be almost unfair to her. On the other hand, if I don't ask legitimate questions, then I am doing a great injustice to someone who might be totally innocent. It's a horrible dilemma because you have two lives at stake here."[20]

Biden, meanwhile, became subject to demands that the confirmation hearings be further delayed to permit time to investigate the charges and a public airing of them. The coalition of civil rights groups argued that a proper case against Thomas based on the new allegations could not be mobilized in only a few days. But Biden said, "I see no reasons why the additional public disclosure of the allegations—but no new information about the charges themselves—should change this decision" on a quick second round of hearings.[21]

Under a hail of criticism about the chairman's resistance, Majority Leader George Mitchell rose on the Senate floor in defense of Biden's "exemplary leadership." But Senator Barbara Mikulski, the only woman in the body, declared, "what disturbs me as much as the allegations themselves is that the Senate appears not to take the charge of sexual harassment seriously."[22]

Biden phoned Hill directly that night and informed her of the new schedule and ground rules, assuring her he was committed to fair treatment for her. Meanwhile, he advised her to get a good lawyer, a suggestion that could be taken as an indication that the Democrats on the committee would not be on her side.[23]

Anita Hill, in her own book about the episode, *Speaking Truth to Power*, wrote that when Biden first phoned her to inform her that her allegations against Thomas would get a hearing in open committee, he told her: "I give you my word that I acted only to protect your confidentiality. . . . The only mistake I made, in my view, is not to realize how much pressure you were under. I should have been aware." Then, Hill wrote, when she told Biden she had not secured legal counsel, he replied: "Aw, kiddo. I feel for you. I wish I weren't the chairman, I'd come to be you lawyer." As Biden ended the call, she wrote, "I could almost see him flashing his instant smile to convince both of us that the experience would be agreeable."[24]

The call for a longer interval to permit further investigation may have had more support had the Senate known that only two days before the hearings were to resume, another former employee of Thomas complained that she too had been sexually harassed by him. A Judiciary Committee aide, Mark Schwartz, had learned that another black woman, Angela Wright, who had once been fired from the EEOC by Thomas, had so alleged. The woman at this time was an editor at the *Charlotte Observer* and, auditioning for a position on the editorial page, had written a sample for her editors' eyes only describing Thomas's purported advances toward her. It was never published, but apparently someone on the paper sent the committee a copy.

Schwartz phoned Wright, questioned her, and asked whether she would come to Washington to testify. She said no, not wanting to get involved. Schwartz quickly briefed Biden who, the aide said later, "was very cool about it. He was not excited at any time, he was just very steady. He was already fifty steps ahead of everyone else and could see where it was leading. There was a sense of nausea, or something big. Instantly, everyone knew it was serious." Biden had another key committee aide, Cynthia Hogan, phone Wright the next day and take her deposition. Federal marshals swiftly delivered Wright a subpoena instructing her to appear in Washington the following day.[25]

Meanwhile, word of "another woman" had got out and she was besieged by press calls. The Thomas defense team, led by Duberstein, began investigating Wright's job record, including her earlier firing by Thomas. Supporters said it had resulted from Wright's calling another EEOC worker "a faggot," to Thomas's ire, which she denied. On the morning of the reopened hearings, Wright boarded a plane to Washington.

Wright, it turned out, was not the only other woman who had now contacted the Judiciary Committee and talked to Schwartz. Former Reagan White House official Kaye Savage reported having gone to then bachelor Thomas's apartment in 1982 and found explicit photos of nude women on the walls. Biden was advised of her story as well.[26] He also was told of a sighting by a prominent Washington lawyer of Thomas checking out a pornographic video from a store a couple of years earlier. The chairman had to ponder what to do about these other allegations.

In a heated discussion among Biden, Danforth, and others on both sides of the confirmation fight, and with a swirl of negative allegations flying about against Hill and Thomas, an agreement finally was reached. No testimony from anyone would be permitted that dealt with anyone else's private life beyond those that dealt with the specific allegations of sexual harassment by Thomas.

The decision was a major victory for the Thomas camp. Biden said later had he permitted the allegations of pornography use, it could have done Thomas in. But, the chairman said, "it would have been impossible at that point to further postpone the hearings for more investigation into his patterns of behavior . . . and it would have been wrong."[27] The Senate had set a firm date to take up the confirmation only a few days away.

Danforth, pushing now for a quick vote, wrote later that Biden said "he did not want the proceedings to appear to be rushed, and that would be the appearance if he agreed to my demands. I was convinced that the leader of the Thomas opposition wanted as long a delay as possible in order to continue their attack. But Joe's insistence that his committee proceed deliberately had nothing to do with aiding the opposition. At no time have I doubted Joe's motives. He had an extremely difficult challenge and he was trying to be fair."[28]

Biden seemed by now to be just as concerned about preserving his own reputation for fairness as for assuring it to the new players in this expanding drama. Only a few years earlier, after the collapse of his presidential campaign over allegations of dishonesty, the public perception of his fairness in conducting the Bork confirmation hearings had proved a lifeline to him. Maintaining that reputation in the Thomas hearings could go a long way to continuing that political rehabilitation, and Biden appeared determined to have that happen as the next chapter unfolded.

NINETEEN

IN SEARCH OF TRUTH AND FAIRNESS

IN REOPENING THE Thomas hearings on October 11, Biden took pains to defend himself and his committee against charges of mishandling the whole matter. He said he had honored Hill's request that the committee investigate her charges and that they remain confidential within the committee up to that point. "Some have asked how we could have the U.S. Senate vote on Judge Thomas's nomination and leave senators in the dark about Professor Hill's charges. To this, I answer, how could we have forced Professor Hill against her will into the blinding light where you see her today?

"But I am deeply sorry that our actions in this respect have been seen by many across this country as a sign that this committee does not take the charges of sexual harassment seriously. We emphatically do."

Biden went on to say "[t]he landscape has changed," forcing Hill "against her wishes" to discuss the charges publicly. The hearing, he said, was not about sexual harassment in America but about specific allegations against a specific person. And it was not a trial with a formal verdict of guilt or innocence at the end, he said, but "a fact-finding hearing" whose purpose was to help senators decide whether Thomas should be confirmed as a Supreme Court justice.

Again reflecting his personal interest, Biden observed that "[a]chieving fairness in the atmosphere in which these hearings are being held may be the most difficult task I have ever undertaken in my close to nineteen years in the U.S. Senate." With members of the Judiciary Committee having already cast their public votes in the 7–7 tie yielding no committee recommendation, Biden said, "it will be easy and perhaps understandable for the witnesses to fear unfair treatment, but it is my job, as chairman, to ensure as best as I possibly can fair treatment, and that is what I intend to do. . . ."[1]

Senator Thurmond, introducing Thomas, took note that among more than a hundred earlier witnesses, that no witness, "even those most bitterly opposed to this nomination, had one disparaging comment to make about Clarence Thomas's moral character," that "the alleged harassment she describes took place some ten years ago," and that she "chose to publicize her allegations the day before" the full Senate was to vote on his confirmation. "While I fully intend to maintain an open mind during today's testimony," he said, "I must say the timing of these statements raises a tremendous number of questions which must be dealt with. . . ."[2]

Thomas thereupon made a full and categorical denial of all the allegations of sexual harassment, including that he "had ever attempted to date Anita Hill" or had discussed pornographic films with her. "If there is anything that I have said that has been misconstrued by Anita Hill or anyone else to be sexual harassment," he said, "then I can say I am so very sorry, and I wish I had known."

Thomas then went on in an indignant fashion to recite the hurt, pain, and agony that had been inflicted on him and his family, and had "done a grave and irreparable injustice" to them. Then, defiantly, he told the committee:

No job is worth what I have been through, no job. No horror in my life has been so debilitating. Confirm me if you want, don't confirm me if you are so led, but let this process end. Let me and my family regain our lives. I never asked to be nominated. It was an honor. Little did I know the price, but it is too high.

I enjoy and appreciate my current position, and I am comfortable with the prospect of returning to my work as a judge on the U.S. Court of Appeals for the D.C. Circuit and to my friends there.

Each of these positions is public service, and I have given at the office. I want my life and my family's life back, and I want them returned expeditiously.

Thomas concluded by saying dramatically: "I will not provide the rope for my own lynching or for further humiliation. I am not going to engage in discussions, nor will I submit to roving questions of what goes on in the most intimate parts of my private life or the sanctity of my bedroom. These are the most intimate parts of my privacy, and they will remain just that, private."[3] It may have sounded as if Thomas was withdrawing his name.

Biden quickly responded: "Thank you, Judge. . . . You will not be asked to." But he told Thomas "with respect to Professor Hill, I intend to focus on the general nature of your relationship with her, her responsibilities in your office and the environment in which she worked."

But before Biden could go any further, Republican Senator Orrin Hatch broke in: "I just want to say something for the record here. This is not the appointment of a justice of the peace. This is the nomination process of a man to become a Justice of the Supreme Court of the United States, and he has been badly maligned."[4]

There then ensued a heated argument between Hatch and Biden over whether Hill's statement submitted to the committee, parts of which by now had been in the news media, could be used in the hearing. Biden, still clinging to his agreement with Hill that she be able to present her story herself, found himself entangled in rigid procedure. Hatch got so incensed that he threatened to resign from the committee. Only when Ted Kennedy intervened and suggested a recess did Biden back off adherence to his set schedule and the matter was resolved. Biden agreed to cut off his interrogation of Thomas to make way for Hill to testify right then, with the idea of picking up questioning of Thomas later. The chairman who had been commended for so masterfully conducting the Bork hearings now found his committee leadership under scrutiny anew.

Finally, Hill before a nationwide television audience spelled out her allegations against Thomas. She recounted occasions on which she said he invited her out socially, which she declined, and other conversations of a sexual or prurient nature, referring to her own appearance, physical properties, and pornography, which she said constituted sexual harassment.

Asked directly by Biden, Hill said, "They were very ugly. They were very dirty. They were disgusting." When Biden pressed her on whether Thomas was "more graphic" in describing his "physical attributes," she replied: "Well, I can tell you that he compared his penis size, he measured his penis in terms of length, those kinds of comments."[5] Biden thanked her and proceeded to recognize others for questioning, bringing out other specifics.

Senator Specter, who took on a leading role in questioning both Thomas and Hill, pointed out inconsistencies in some of their statements but was particularly aggressive with Hill. Regarding some of her responses in her interview by the FBI, he warned her of possible perjury. Hill said the agents had omitted in their report certain lurid details to which she had in fact testified, nevertheless raising some doubts or at least confusion about what she had said.

Further seeming contradictions in Hill's testimony, and her acknowledgment of multiple occasions in which she phoned Thomas, finally led Specter that afternoon to declare: "It is my legal judgment, having had some experience in perjury prosecutions, that the testimony of Professor Hill in the morning was flat-out perjury." The charge roiled women particularly across the country, and Specter acknowledged a few days later that while "I think I was legally correct, it was emotionally perhaps too hard and, as it turns out, politically unwise."[6] But the observation doubtless shored up Republican senators ready to vote for the Thomas confirmation.

When Hill had finished testifying, Biden again offered a long monologue on how he had wished her public appearance had not been necessary at all, but that he could not prevent it. "For those who suggest that there was some way to do it differently and still honor your commitment [to tell your story yourself]," he said, "I respond that I know of no such way to guarantee your anonymity, or to guarantee you would not have to be in this place on this day." And he thanked her "from a personal standpoint" for agreeing to testify after all the fits and starts.[7]

Thomas upon Hill's last words denied all again and lashed out at the whole process. "I think that this today is a travesty," he thundered. "I think that it is disgusting. I think that this hearing should never occur in America." He repeated that "the Supreme Court is not worth it" and launched again into the charge of racism. "This is a circus. It is a national disgrace," he said. "And from my standpoint, as a black American, as far as I am concerned, it is a high-tech lynching for uppity blacks who in any way deign to think for themselves, to do for themselves, to have different ideas, and it is a message that, unless you kowtow to an old order, this is what will happen to you; you will be lynched, destroyed, caricatured by a committee of the U.S. Senate, rather than hung from a tree."[8]

Hill, watching Thomas give his testimony on her hotel room television set much later, wrote that "I could not help but see the irony in his claim of racism. In the years that I had known him, he had always chosen, both publicly and privately, to belittle those who saw racism as an obstacle. . . . Yet, now having met an obstacle to his *own* dream, he blamed racism. I wondered if he might then change his mind about the impact racism had on other lives."[9]

In this round of questioning from the committee, Thomas held steadfastly to his position of categorical denial of all of Hill's allegations. When he told Senator Howell Heflin of Alabama he had not watched her testimony on television, the senator expressed surprise and offered that Thomas had "put [himself] in an unusual position." Thomas again indignantly argued that Hill's recitation of the facts "keep changing" but his flat denials remained constant. And when Heflin suggested that "if you are on the bench and you approach a case where you appear to have a closed mind and that you are only right, doesn't it raise issues of judicial temperament?" Thomas shot back: "Senator? Senator, there is a difference between approaching a case objectively and watching yourself being lynched. There is no comparison whatever." Hatch jumped in: "I might add, he has personal knowledge of this as well, and personal justification for anger."[10]

Through all this Biden presided in uncommon silence. When the judge concluded, he seemed bound to assuage him. "Judge," he said, "just because we take harassment seriously doesn't mean we take the charges at face value. You have pointed out that when you worked with Anita Hill, and up until the charge was made available to you through an FBI agent, you thought her to be a respected, reasonable, upstanding person." When such a person "with no blemish on her record comes forward," the chairman said, "this committee has an obligation to do exactly what you would have done at EEOC—investigate the charge. You are making a mistake if you conclude that because this is being investigated before all the evidence is in, the conclusion has been reached by this committee."

Biden noted that Hill had testified for the first time that Thomas had visited her at her apartment, and urged him: "Do not in your anger refuse to tell us more tomorrow. This is not decided. . . ." Simpson, squarely in Thomas's corner, quoted Biden from his opening statement: "Fairness also means that Judge Thomas must be given a full and fair opportunity to confront these charges against him, to respond fully, to tell us his side of the story and to be given the benefit of the doubt. Now, that's what we're doing here, and if there is any doubt, it goes to Clarence Thomas, it does not go to Professor Hill."[11]

Later, Simpson, making no pretense of objectivity, used his time with Thomas to offer unspecified reports described only as "stuff over the transom about Professor Hill" from law professors and other "people who know her . . . saying, watch out for this woman." As he spoke, watching the televised hearing from her lawyer's office in Washington, the "other woman," Angela Wright, awaited a call from the committee about her own appearance. She heard Simpson say: "Angela Wright will soon be with us, we think, but now we are told that Angela Wright has what we used to call in the legal trade, cold feet. Now, if Angela Wright doesn't show up to tell her tale of your horrors, what are we to determine about Angela Wright?" And he asked Thomas: "Did you fire her and if you did, what for?"

The judge replied that he had "summarily dismissed her" because "I felt her performance was ineffective" and "the straw that broke the camel's back was a report to me from one of the members of my staff that she referred to another male member of my staff as a faggot . . . [a]nd that is inappropriate conduct and that is a slur, and I was not going to have it."

Simpson, as always ready with the snide remark, commented: "And so that was the end of Ms. Wright, who is now going to come and tell us perhaps about more parts of the anatomy. I am sure of that. And totally discredited and, we had just as well get to the nub of things here, a totally discredited witness who does have cold feet."[12]

Wright said later she sat down and drafted her response: "I have been subpoenaed to testify before the Committee. I have cooperated fully with the Committee on both sides of the aisle, and the FBI. I am ready to testify before the Committee, and I do not have cold feet."[13]

Biden closed the day's hearing with another defense of the process and a personal wish that the cup could have passed from him. "I did not sign onto this job or run for it to be a judge," he told Thomas. "If I wanted to do that, I would be a judge now in my home state. I don't want to be a judge. I hate this job.

"But all my colleagues here were telling everybody how awful the process is. Let me be completely blunt about it. It is like democracy. It is a lousy form of government except that nobody has figured out another way. . . . So I take the heat and I take the responsibility and I will continue to do it as long as I am chairman, no matter what these guys think of this process, okay?"

Biden went on in that vein, while also assuring Thomas that "my job is not to defend you or to prosecute you. It is to see to it that you get a fair shot in a system that is imperfect, but it is a good system." He cited his experience as a lawyer and a senator in dealing with abuse against women, noting "if there is not a pattern, to me that is probative. That has some dispositive weight. No one has proved a pattern here of anything. We are not finished yet. But no one has proved a pattern."

Biden even referred to his own trials with "the process," saying: "I am getting fed up with this stuff about how terrible this system is. I hear everybody talking about how terrible the [presidential] primary system is. We are big boys. I knew when I ran for president that everything was free game. Anybody who runs for the Supreme Court or who is appointed to the Supreme Court, to be more precise, should understand, this is not Boy Scouts, it is not Cub Scouts. In the case of the president and the right to be the leader of the free world, well, no one ever said it was going to be easy. And whoever goes to the Supreme Court is going to determine the fate of the country more than anybody. For the next twenty years we are going to have people scrupulous and unscrupulous respond and react. . . . And lastly, Judge, with me, from the beginning and at this moment, the presumption is with you."

Biden then offered some personal commiseration: "But Judge, everybody says, 'We know how you feel.' No one can know how you feel. That always excites me, when I hear people tell me how it feels. 'Oh, you lost family. I know how it feels.' 'Oh, you lost this. I know how it feels.'

"'You went through that, and they ruined your reputation by it. I know how it feels.' No one knows how it feels, but I hope we stop this stuff."

That said, the garrulous chairman turned philosophical. "You will not be unaffected by this, no matter what happens," he informed the judge as one who had been there. "Nobody goes through the white hot glare of this process, any level, for any reason, and comes out unaffected. But, Judge, nobody's reputation, nobody's reputation is a snapshot. It is a motion picture, and the picture is being made, and you have made the vast part of it the last forty-three years."[14]

For all of Biden's efforts to keep the hearings on a nonadversarial, fact-finding basis, inevitably the questioning often gave the impression of a trial, with intensive interrogation of Hill on details casting her as defending herself rather than as the accuser. As confusions arose as to what she had or had not said to the FBI, compared to Thomas's steadfast and categorical denials, polls showed a distinct shift against her and in favor of Thomas. And his early charge of a "lynching" kept the cloud of racism hovering over the hearings, which clearly was his intent.

A final Sunday session consisted of four panels of witnesses offering character references and other comments on the principal adversaries. As the afternoon dragged on, the committee agreed to limit the witnesses' time, but even at that, it took four more hours to finish. Wright, still waiting, heard Biden say at one point: "We will bring forward, if it is the decision of the witness to want to come forward, and that is not fully decided yet, Ms. Angela Wright. We are talking with Ms. Wright now." He added that another witness would then be called "who allegedly—I emphasize allegedly—can corroborate the testimony of Angela Wright."[15]

It was soon two o'clock on Monday morning. The committee caucused, and with Republicans strongly opposed to calling Wright and some Democrats uncertain about the impact of her testimony, it was decided not to call her after all. Biden later reportedly told his staff that he had cast the lone vote to have Wright testify. Instead, it was decided to include in the hearing record the committee staff interviews with Wright and her corroborating witness, another EEOC colleague, Rose Jourdain. Hatch said much later that it was his understanding that Wright "didn't want to come."[16]

Biden also inserted in the record his letter to Wright saying "it is my preference that you testify before the Judiciary" but because of time constraints and the members' willingness to have the interview transcripts included, he was "prepared to accede to the mutual agreement of you and the members of the committee, both Republican and Democrat, that the subpoena be vitiated. . . . I wish to make clear, however, that if you want to testify at the hearing in person I will honor that request."[17] Biden's letter reflected his care to note his reluctance to do without Wright's live testimony and his willingness to have her give it if she desired. Or at least it reflected his desire to have that willingness in the record.

The decision against hearing the testimonies of Wright and Jourdain caused much speculation and criticism from anti-Thomas quarters long thereafter. Civil rights advocates argued that the corroborative testimony of Wright, backed up by Jourdain, an older woman who would have come from her hospital bed to verify what Wright had told her about Thomas's advances, would have swung the case against the judge. But the transcripts as printed in the official hearings were hardly as explosive as the attention given their existence had suggested.

Wright said that Thomas had "consistently" pressured her to date him and at one point "made comments about my anatomy" and that of other women "quite often." She said he had come uninvited to her Washington apartment one night and tried "to move the conversation over to the prospect of my dating him." She said such talk "pretty much pops out of Clarence Thomas's mouth when he feels like saying this." But she said she didn't think his comments were overheard by anyone else. She said he once "commented on the dress I was wearing and asked me what size my boobs were." On another occasion, Wright said, "I remember specifically him saying that one woman had a big ass," and that she wasn't the only woman in the room but she couldn't remember who she was. And when asked whether she was annoyed or felt sexually harassed at such a time, she said, "Well, I think annoyed was a better term."[18]

Rose Jourdain in her transcribed interview merely reported that Wright had become "increasingly . . . uneasy" because of comments Thomas "was making concerning her figure, her body, her breasts, her legs, how she looked in certain suits and dresses." Jourdain said that on one occasion Wright had been reduced to tears in relating Thomas's conduct, but another time "[s]ometimes she laughed about it, you know. Sometimes it got on her last nerve. . . . Sometimes it had happened so much it was, like, you won't believe what this, what he said now, you know?"[19]

Taken as a whole, neither the testimony of Wright nor of Jourdain seemed in writing to be conclusively supportive of Hill, and Jourdain's was corroborative only in her saying what Wright had told her of what Thomas had said. Perhaps had the women testified directly in person their words and manner would have had more impact on the committee and television viewers. One leading civil rights advocate long after said he was convinced that had Jourdain particularly testified, an older woman in obvious ill health, the Thomas nomination would have been rejected.

Hill herself later said "we were waiting for Angela Wright's testimony, just like everyone else. Apparently something went wrong. I was as surprised as anyone that she didn't testify. Especially after a panel of women said, 'He never harassed me.' When there was someone who could testify to the same behavior—it was astonishing to me that she was never called."[20]

Biden, in an interview with Florence Graves for an Alicia Patterson Foundation review of the Thomas-Hill hearings later, said he believed Wright's testimony could have changed the outcome. He said he had been told Wright "wanted the subpoena lifted and that people who were strong Anita Hill people . . . believed that Anita Hill's testimony was so strong standing on its own that no matter how good anyone else was, it would be diluted. It would take away from it."

But Charles Ogletree, a Harvard law professor aiding Hill, interviewed for the same review, said it was "absolutely absurd for Senator Biden or anyone else to suggest that anyone associated with Anita Hill prevented Angela Wright from testifying. . . . We desperately wanted Angela Wright to testify because she was the one additional witness who was both willing and able to come forward and state that on another occasion Justice Thomas had engaged in conduct that was tantamount to sexual harassment. . . ."[21]

Specter wrote later: "One witness we certainly should have called was Angela Wright . . . [who] would have been the second woman to make such charges against Thomas. Wright's account could have been powerful, because sexual harassment is generally habitual behavior, and Hill had so far been the only one ever to accuse Thomas of it." At the same time, Specter wrote, Wright "was the sole additional witness that some of my Republican colleagues wanted to call. But they did not want her testimony out of a sense of legal thoroughness. They wanted her testimony because they thought she would make a lousy witness." Fired by Thomas in 1985, Specter wrote, "some thought Wright's firing would give her a motive for revenge against Thomas and that this could be used to discredit her."[22]

Biden was also criticized by Anita Hill supporters for not having defended her against the blatant and unsubstantiated smears by Simpson and other Republican senators. In the same interview, Biden said: "If I had done what the Republicans did, I would have made a lie of everything I think I stand for. . . . Mark me down as not joining the school that the end justifies the means." Ogletree responded that

The Biden family in Boston, 1943. *Left to right:* Grandfather Ambrose Finnegan, Grandmother Geraldine Finnegan (holding baby Joe), mother Jean Biden, and father Joseph Biden Sr.

Joe Biden at bat: young Joe shows his early batting stance outside the Finnegan family home in the Green Ridge section of Scranton, c. 1949.

Neilia Biden holding baby Naomi, with Beau sitting (*left*) and Joe holding Hunter (*right*), in Wilmington, 1972.

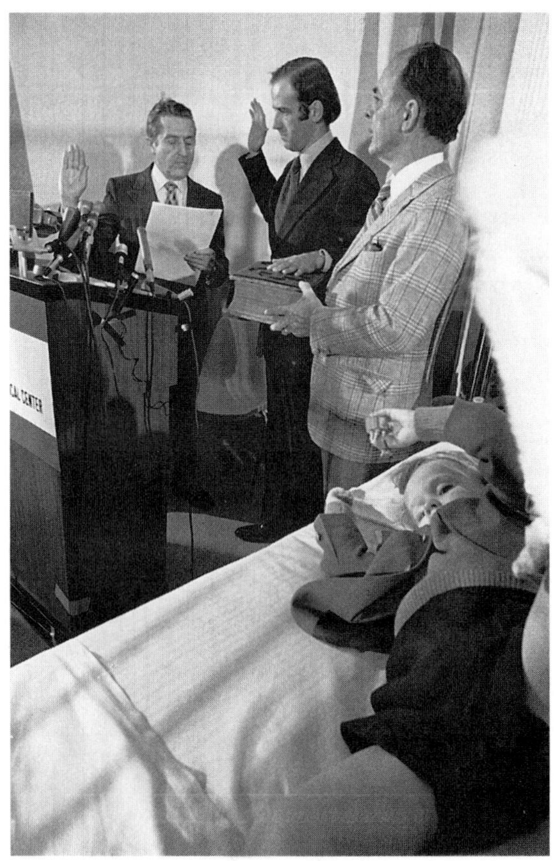

Senator-elect Joe Biden takes the oath of office from Frank Valeo, secretary of the Senate, at the bedside of Beau at Wilmington Medical Center, January 5, 1973.

Freshman U.S. Senator Biden being briefed by his chief of staff, Ted Kaufman, in his office in 1973. When Biden was elected to the vice presidency in 2008, Kaufman was appointed by Delaware's Democratic Governor Ruth Ann Minner to fill Biden's seat, pending a special election in 2010.

Senator Biden hosts his father, Joe Biden Sr., at the Senate Dining Room in the Capitol Building, Washington, D.C., March 1984.

Republican Senator Strom Thurmond of South Carolina, chairman of the Senate Judiciary Committee, discusses business with Senator Biden, ranking minority member of the committee, in the Judiciary Committee hearing room, Washington, D.C., 1987.

As chairman of the Senate Judiciary Committee, Biden led the opposition against President Ronald Reagan's nomination of Federal Appellate Judge Robert Bork to become associate justice of the Supreme Court, Washington, D.C., September 15, 1987.

President Bill Clinton and Senator Biden on Air Force One, en route to Bosnia on December 22, 1997.

Senator Ted Kennedy and Senator Biden at a Senate Judiciary Committee hearing, January 9, 2006.

As chairman of the Senate Judiciary Committee, Biden discusses procedures with Federal Judge Clarence Thomas during his Supreme Court confirmation hearings on September 13, 1991. After a stormy confrontation over allegations of sexual harassment by Anita Hill, a female lawyer, Thomas was narrowly confirmed.

Senator Biden leans on a table and shakes his fist at Serbian President Slobodan Milošević (*left*) during a meeting at the presidential palace in Belgrade on April 8, 1993. Biden later said that he called Milošević "a damn war criminal."

Joe and Jill Biden (*right*) with brother-in-law Jack and Valerie Biden Owens (*left*), celebrating Jack's birthday in Charlottesville, Virginia, May 1993.

The Biden children. *Left to right:* Beau, Ashley, and Hunter at home in Wilmington, 1984.

The Biden family in Wilmington, Christmas 2007. *Left to right:* Beau's wife, Hallie, holding daughter Natalie; Beau (*behind*) holding son Hunter; Ashley; Jill, with hands on Hunter's daughter Finnegan; Joe (*behind*); Hunter's daughter Naomi; Hunter's wife, Kathleen, behind daughter Maisy; Hunter.

The Biden brothers, Joe, Jimmy, and Frankie, with their mother, Jean, in Wilmington, 2009.

Biden and his campaign manager and sister, Valerie, celebrate in New York City, 2009.

Vice presidential nominee Biden and his mother, Jean, respond to cheers at the Democratic National Convention in Denver on August 27, 2008.

Associate Justice John Paul Stevens of the Supreme Court administers the oath of office to Vice President Biden on the steps of the U.S. Capitol while his family observes, January 20, 2009.

Vice President Biden visits Beau on duty in Iraq, July 4, 2009. Beau serves as a captain with the Delaware National Guard.

as chairman, "what the senator did regrettably was to bend over in the wrong direction, and I think he did a great disservice in the sense of civil rights by being so tolerant of [Thomas's] lack of responsiveness . . . and being so intolerant of efforts on the other side to bring out issues that may have shed light on" the judge's character."[23]

In a lengthy interview for this book in late 2009, Biden said Angela Wright said she "chose not to come. I was so concerned about that, that I wanted it in writing. My recollection is, I actually had staff draft a letter saying I wanted her to come but what was her choice, to testify or not testify and to check it off and sign it. And she said she did not want to testify. But there's a myth that's grown up that we somehow denied her. We had her in town to testify, we expected her to testify, we prepared her to testify; she chose not to testify. She had her own reasons. I don't know exactly what they were. And people say, well, why didn't you have her testify anyway? Well, that's like calling a hostile witness in a case. I strongly believed that Thomas shouldn't be confirmed. I believed that before the Anita Hill stuff broke, and here you had a witness who didn't want to testify, and God only knows what she would have said or why she didn't want to testify, so I honored her request not to testify."

Asked whether he thought the outcome would have been different had she testified, Biden said: "I don't think so, because her whole statement, we published her deposition that was taken by the staff for every member of the committee. They all had it," he said, and there was a full story in the press about it, so it wasn't that her allegations weren't known. "I don't think it would have changed anybody's mind one way or another," he said. "Some people thought it would undo the case if she turned out to be a weak witness. I remembered some senators on the Democratic side not wanting her called because they thought she'd be a weak witness and undermine the case rather than strengthen the case. That wasn't my reason. I wanted her to come because she had done a deposition, and she was on the record, and I thought she should testify."[24]

When the Thomas confirmation finally came to the Senate floor on October 15, Chairman Biden briefly defended the decision to hold open hearings on Anita Hill's allegations and again mentioned that he had never wanted to be a judge but was doing the best he could.

On one occasion, after reiterating he intended to vote against the Thomas confirmation, stepping out of his role as chairman he took the floor and defended Hill against the suggestion that since she appeared to be such a credible witness, "people were left with only one of two choices. She is credible, therefore believe her, or she is credible, therefore say she is crazy. There is absolutely not one shred of evidence to suggest that Professor Hill is fantasizing. . . . There is no shred of evidence for the garbage I hear . . . that [says] she thinks she is telling the truth; that the only answer we can come up with is that she must be fantasizing. . . . If that isn't stereotypical lynching of a black woman, what is? . . . So I hope you will drop this stereotypical malarkey."[25]

Nevertheless, when the roll call on confirmation was taken, the argument of Simpson and other Republicans that the nominee deserved the benefit of the doubt narrowly carried the day. With southern Democrats providing the margin, Thomas was confirmed, 52–48, marking the most votes ever cast against a successful nominee. Afterward, Thomas and his happy family returned to their home for a private celebration. According to Senator Danforth in his book *Resurrection*, "a gracious congratulatory message from Joe Biden left on Clarence's answering machine was greeted by a less than gracious response."[26]

As for Biden, Thomas's memorable allegation of a "high-tech lynching of uppity blacks" clearly stuck in his craw. Later, in his interview with Florence Graves for the Alicia Patterson Foundation's review of the hearings, he offered that "Thomas was the one in my view engaging in racism, and I not only mean racism in terms of playing the race card, but racism in trying to reinforce the stereotypical notion about black women. That was the sin I don't forgive the guy for, and those who were making his case." He concluded: "I think that the only reason Clarence Thomas is on the Court is because he is black. I don't believe he could have won had he been white. And the reason is, I think it was a cynical ploy by President Bush."[27]

Vince D'Anna, who was part of the Biden team on the Thomas confirmation battle, said later: "I think the thing that disarmed Joe the most, was hardest for Joe, considering Joe's beliefs and values with regard to civil rights, the history of civil rights, it was really hard for Joe to oppose a competent black running for the Supreme Court. He [Thomas] would not have had a snowball's chance [had he been white]. They would have killed him. The thing that got him approved was that the Democrats were scared to death. They just wanted this thing to be over with—'Let's be done with it; let's vote. Nothing good is going to come from it.' There were no profiles in courage on that committee"[28]

These postmortem remarks on the case made all the more remarkable Biden's relentless strivings for, and endless reassurances of, fairness throughout the hearings. Committee colleagues on both sides of the aisle fulsomely praised him for his leadership. In the next Gallup Poll, Biden received the highest approval among viewers of the hearings with 63 percent, to 48 for Arlen Specter, 41 for Alan Simpson, and only 22 for Kennedy. But after having won a clear victory in the earlier Bork hearings, it could be said that in the Thomas affair, Joe Biden may have scored, if there is such a thing, a Pyrrhic defeat.

The Thomas hearings weighed heavily on Biden long afterward. Eight months later, after long and extensive research and preparation, he requested and received a very unusual allocation of time to speak on the floor of the Senate, on the subject of reform of the judicial confirmation process. He was granted an hour and fifteen minutes and after consuming it was still not finished. He requested another fifteen minutes, as heads knowingly nodded. He apologized in advance, noting that "in my over nineteen years in the Senate, I have never sought to speak before the Senate for as long a period in morning business."

But he said he was moved to do so, because after fifteen years in which there had been only two Supreme Court nominations, there had been seven nominations to fill five vacancies in just a five-year period. In 1986, he noted, William Rehnquist was confirmed as chief justice with the highest number of votes against him in the Senate up to that time, and five years later, Clarence Thomas was confirmed with even more negative votes. The prospect for still more vacancies on an aging court, he said, demanded a reexamination of what had become one of Washington's most contentious exercises.

After a lengthy preamble citing the history of rejected and withdrawn Supreme Court nominations back to Washington's appointment of John Rutledge in 1795, Biden got down to his point, obviously driven by his experience in recent years, and particularly during the presidencies of Ronald Reagan and George H. W. Bush. Citing precedent, he noted that in three previous instances when a president was succeeded by a loyal member of his faction or party—Washington and Adams, Lincoln and Grant, and FDR and Truman—a generally ideological consensus was sustained between the executive and legislative branches, producing mostly smooth Supreme Court confirmations.

But in the Reagan-Bush period of a dozen years, Biden noted, no such consensus existed. Indeed, since 1968 the Republicans had controlled the White House for twenty of the last twenty-four years while the Democrats controlled the Senate for eighteen of them. In the Reagan-Bush era, Biden observed, half of their Supreme Court choices "have been nominated in a period of a divided government." Yet these two most recent Republican presidents, he said, had in spite of this fact "sought to remake the Supreme Court" in "an attempt to move the Court ideologically into a radical, new direction which this country does not support."

Biden segued into a stout defense of the constitutional role of the Senate to advise and consent to a presidential nomination. He cited the Founding Fathers' early resistance to giving the chief executive any role at all, to the eventual compromise of having the Senate provide an intended restraint on the president's power to choose. He contended that "the use that Presidents Reagan and Bush made of the Supreme Court nominating process in a period of divided government is without parallel in our nation's history. It is this power grab that has unleashed the powerful and diverse forces that have ravaged the confirmation process."

Furthermore, Biden charged, the discord over confirmations began with the decision of Reagan and Bush "to cede power in the nomination process to the radical right within their own administration," forcing the Senate defensively to develop "an unprecedented confirmation process." Its centerpiece, Biden said, "was a frank recognition of the legitimacy of Senate consideration of a nominee's judicial philosophy as part of the confirmation review"—obviously at the core of his own interrogations of Bork and Thomas.

Biden went on and on in what in one sense was an impressively scholarly exposition on the history and rationales of the confirmation process. At the same time, he repeatedly bolstered his central argument that the Senate had a clear constitutional role and that a nominee's ideology was a legitimate grounds for inquiry, especially when the direction of the Court was at stake.

Again at great length, Biden reviewed the standards of "character and competence," and those of "detachment and statesmanship," in evaluating a nominee's fitness for high judicial service and said they warranted due consideration if applied in "a spirit of bipartisanship." Biden said that "when the president selects nominees on the basis of their detachment and their statesmanship, with a sensitivity to the balance of the Court and the concerns of the country, then the Senate should respond in kind." But he always returned to his point that a nominee's judicial philosophy was critical in maintaining a Court that during divided government a president not be allowed to remake "in his own image."

Biden went on to allege "an unintended conspiracy of extremism between the right and the left to undermine the confirmation process, and question the legitimacy of its outcomes." The right charged, he said, that the witness list in the Bork hearings was stacked against him "and the list of excuses goes on and on. It was the camera angle, they said, the beard, the lights, the timing." The GOP conservatives "never accepted the cold fact that the Senate rejected Judge Bork because his views came to be well understood, and were considered unacceptable," he said, and for that reason "they have been on a hunt for villains ever since."

As for the left, Biden said, it "has refused to accept the fact that when one political branch is controlled by a conservative Republican and the other has its philosophical fulcrum resting on key southern Democrats . . . it is inevitable that the Court is going to become more conservative." Therefore, he said, "acceptable candidates must be found among those who straddle this ideological gulf, such as Justices [William] Kennedy and Souter, who were approved by a combined total of 188 to 9 in the Senate."

These two extremes of right and left, Biden argued, "have joined an ad hoc alliance . . . to undermine public confidence in a process aimed at moderation," and were further poisoned by "the general meanness and nastiness that pervades our political process today," as witnessed in reactions to the Bork and Thomas hearings. "A legitimate process that was built in good faith to identify and confirm consensus nominees," Biden concluded, "has been destroyed by many of the same corrosive influences that have devastated our presidential politics and our national dialog on public affairs."

In that regard, he argued that Supreme Court nominations and confirmations are particularly vulnerable to extraneous attacks in presidential election years and should be avoided. In the existing circumstances in 1992, with Republican Bush facing a serious challenge from the Democrats, it was easy to dismiss this argument as the desire to hold off filling any new vacancy until after the November election. But Biden cited the historical record of few nominations during the late stages of a presidential campaign to buttress his plea. For himself, Biden declared, he would henceforth oppose any nominee about whom genuine Senate consultation had not been sought or was clearly put forward with the intent of changing the ideological composition and direction of the Court. Nor would he vote for any nominee, he said, who declined to discuss his judicial philosophy during the confirmation hearings.

As the Judiciary Committee chairman went on at great length, he offered the Senate a long and detailed history of the evolution of the confirmation process that some colleagues and critics no doubt would write off as self-aggrandizing showboating. But those who bothered to listen could not but come away with an impression that this man so widely dismissed as a windbag knew in considerable depth what he was talking about, and was skillfully if repetitiously making his case for the Senate and for his particular focus on judicial philosophy in appraising a nominee.

At the same time, as Biden continued, his observations came off more and more as a defense of his own much-criticized chairing of the Bork and Thomas hearings, and of his struggle to convey fairness while emerging as one of the most telling interrogators of the principals. He insisted that "in most respects the Bork nomination served as an excellent model for how the contemporary nomination and conformation process and debate should be concluded and conducted." He justified his early and controversial declared opposition to Bork as galvanizing the mobilization of arguments pro and con that then were considered openly in the hearings.

Biden reiterated his earlier distinction that "conformation hearings are not trials. We are not a court; we are a legislative body. . . . Our decision on a nominee is not a neutral ruing as a judge would render. It is, as the Constitution designed it, a political choice about values and philosophy." But he acknowledged that rival outside groups also mobilized to affect the decisions of senators who did not make an early declaration of their support or opposition. And he noted there was little or no criticism of senators who made an early declaration in favor of Bork or Thomas.

Concerning the confusion over committee members' access to Anita Hill's allegations against Thomas, Biden again defended his efforts to protect her privacy, while insisting for the future on confidential senatorial briefings and closed sessions for full interrogations of all principal parties. In the end, Biden's exhaustive ninety-minute monologue conveyed a conscientious committee chairman's effort, chock-full of civics and history, to address the serious deficiencies in a critical legislative function. But it also included rationalizations and justifications for his own performance in the most contentious of confirmation hearings.[29]

For whatever else, the presentation showed Joe Biden to be a serious, well-informed, and principled legislator while confirming his acknowledged penchant for talking exhaustively and, to his critics, exhaustingly. (Three and a half years later, in a *Newsweek* interview, he indicated that assessment would be acceptable to him. "If my Achilles' heel has to be I talk too much," he said, "not that I'm a womanizer or I'm dishonest or what-not, it's fine. I set myself up for that.")[30]

Long after the Bork and Thomas hearings, and the controversy over school busing in Delaware, Professor Raymond Wolters, a civil rights scholar and critic at the University of Delaware, had harsh words to say about Biden as Judiciary chairman and senator. "I think he has played politics with the confirmation process. . . . [Particular cases] all show a certain pattern," he charged.

In the Delaware school busing controversy, in which some blacks but nearly all whites were opposed, Wolters said, "Biden recognized this, and at that time, to appease or appeal to his mostly white constituency, he came out strongly against busing, even offered legislation. And because of that, he incurred the wrath of the NAACP and the so-called civil rights community. . . .

"But when he began to get national ambitions, he needed the support of the NAACP and the civil rights community, which was by then pretty much committed to busing and quotas. So Biden developed a position of doing a political stunt. He was against busing; he was against quotas. But when it came to the judicial hearings for judges or for anybody else being put up for promotion through the Judiciary Committee, he would oppose people who were opposed to busing and quotas. He would invariably say, 'I agree with you about busing, I agree with you about quotas, but I think you are insensitive on racial matters.'" Time and again, Wolters said, "that would be his mode of operation" to defeat other Republican judicial appointees coming before the committee. "In other words, Joe is a sharp political operator."[31]

Such criticism from a Delawarean is rare in the First State, where Joe Biden is everybody's friend and a friend to everybody. Even Pete du Pont says of him, not unkindly: "Joe's a very good person to be a friend of."[32] In the Clarence Thomas confirmation hearings, Joe seemed to be trying to convince both the judge and his female adversary that he was their friend—very nearly an impossible task under the circumstances and, as it turned out, a thankless one for him.

TWENTY

COZY CORPORATE CAPITAL OF AMERICA

IN THE TIME Joe Biden had risen from lowly county councilman to chairman of the Judiciary Committee of the U.S. Senate, his home state of Delaware had seen a remarkable transformation in its business climate and stature. For many years, the First State had a reputation of being practically a wholly owned subsidiary of the du Pont family chemical empire. The state had endured economic doldrums but eventually emerged, through imaginative and daring business entrepreneurship, to claim the identification as the nation's corporate capital despite its tiny geography.

The growth had not been easy. The racial turmoil and nine-month occupation of Wilmington by the National Guard following the 1968 assassination of Dr. Martin Luther King Jr. had shaken the business community and contributed to the defeat of Democratic Governor Charles Terry by Republican Russell Peterson that year.

In 1971, under Governor Peterson's leadership, the General Assembly, in a burst of environmental responsibility, passed a Coastal

Zone Act that banned all heavy industry from Delaware's scenic and tourist-attracting Atlantic coastline. In the process, a planned refinery by the Shell Oil Company opposed by Councilman Biden was scrapped, to the dismay of the state Chamber of Commerce, and business leaders raised concerns about Delaware's pulling in its welcome mat for them.

That fact, and a mushrooming state government under Peterson that was requiring and consuming higher business taxes, cost him reelection in 1972. The new governor, moderate Democrat Sherman W. Tribbitt, bore the political brunt of the long and bitter fight over the court-ordered desegregation of the Wilmington schools. Then when Tribbitt got the General Assembly to raise the tax on oil refined in Delaware, oil magnate J. Paul Getty threatened to pull out of the state. Tribbitt blinked, and the assembly rescinded the tax.

Next, the Farmers Bank of the State of Delaware had to apply for a state and federal bailout, a further discouragement to the business community, leading to more employers' threats to pull up their Delaware stakes. In 1976 Pete du Pont easily defeated Tribbitt for the governorship. In a largely laudatory book on the du Pont tenure commissioned by the Delaware Heritage Commission, Larry Nagengast, a former *Wilmington News Journal* reporter/editor, wrote:

> *There was little indication in 1977 of how remarkable the fourth quarter of the twentieth century would be for Delaware. . . . It would be an era in which bipartisan cooperation would largely succeed political dogfights in Dover, an era in which professional managers replaced the good old boys in charge of state agencies, an era in which the financial services industry would replace chemicals as the most prominent in the state. During the du Pont administration, the state would reverse its course, transforming itself from an economic disaster into a model of prosperity that most other states could only hope to emulate, and restoring citizen confidence in the integrity and professionalism of the government's key leaders.*[1]

Before this rosy scenario played out, however, du Pont and the Democrat-controlled General Assembly engaged in a fierce budget fight, producing a du Pont veto that further alerted Delawareans to the state's financial plight. Du Pont managed to beat back a vote to override and then persuaded the Democratic legislators in Dover to join in a more cooperative atmosphere to address the state's troubles. The effort gained impetus when the president of Delaware's second largest employer, chemical giant Hercules Inc., told the Wilmington Rotary Club that the state government had grown too large and costly and his company might move its headquarters elsewhere. While the governor was in California seeking to recruit new businesses, the chairman of the DuPont Co. chimed in, telling the *Wall Street Journal* he couldn't encourage them to come into Delaware because its income-tax rate had become so high—19.8 percent.

The principal upshot of all these events was the enactment of a Financial Development Center Act (FDCA) in January 1980. It became a magnet to corporations across the country through new legislation greatly improving the business climate in Delaware, especially for banks and the burgeoning credit card industry. At the time, the largest banks in New York were being hit hard by inflation, high state corporate taxes, and state limitations on interest rates they could charge on their credit cards.

The new FDCA ended an old practice whereby the state legislature set banking interest rates, leading to constant lobbying of members. Also, du Pont said in an interview, "a new corporate law attracted companies from all over the country, because first we had a separate court system for all financial affairs. We are a much better place for corporations to come because we have a court that handles the law and they can see how well the law is all put together, so we're a great big corporate center." Another advantage, he said, "is you have very clear rules about what can and can't be done," including what corporate tax rates can be charged.[2]

Critical to the influx of corporations into Delaware was Chase Manhattan Bank in New York, whose executives approached bankers in Delaware about providing more financially hospitable conditions to companies interested in moving their headquarters. Morgan Guaranty Trust also agreed to open subsidiaries in Delaware if the terms were right. Actually, another sparsely populated and much more distant state, South Dakota, was the first to start the trend, luring giant Citicorp to move there. But that state's labor market wasn't ideal; Delaware, strategically located between New York and Washington, was more attractive.

From Delaware's point of view, greater diversification in its business base, away from the giants it already housed—DuPont, Hercules, and some General Motors and Chrysler auto assembly plants—was badly needed. A key element in the FDCA was an agreement by Delaware to tax bank profits at a considerably lower rate than those assessed in New York. Another was authority for banks to establish credit card operations in Delaware with interest-rate rules very favorable to the relocating banks. During the du Pont administration, fourteen banks created offices in the state, joined eventually by twenty-one more, until consolidations reduced the total.

By century's end, the credit card giant Maryland Bankers National Association (MBNA) moved from Maryland and became Delaware's largest private employer. Enactment of FDCA was credited with boosting Delaware's banking industry tenfold, to more than forty-three thousand jobs, or about 10 percent of the state's workforce. Taxes from credit card banks alone provided about 5 percent of the state's revenues. In 2006 Bank of America acquired MBNA and became Delaware's top private employer, MBNA America. Soon concerns were voiced of too great a concentration of banks, replacing the old talk of traditional du Pont dominance.

What did all this have to do with Joe Biden? Although the transformation of Delaware's business base and climate was achieved largely at the state level, banks and credit card companies inevitably looked to Congress to create and maintain favorable circumstances for them in terms of federal regulations and legislation. As federal legislators always attentive to and engaged in constituent service and interests, the three-man Delaware congressional delegation—Democratic Senators Biden and Tom Carper and Republican Congressman-at-large Michael Castle—looked out for their corporate as well as individual constituents. With each of them, said former Democratic state chairman Joe Farley, "he would not get elected if he ignored the financial section. We all care about the banks. That's what you're supposed to do as an elected official in Delaware."[3]

David Bakerian, president and CEO of the Delaware Bankers Association, said Biden was cooperative in the repeal of a Depression-era antitrust law that had built a "firewall" between the banking industry and businesses designed to inhibit ownership of banks by large corporations. With the later bank regulatory structure in place, he said, "there was a movement afoot for the banking industry to have one-stop shopping" for insurance, brokerage accounts, and other services. "There were restrictions on some of the products that banks could get involved with," he said, "and a national push to try and get some relief" from Congress.

"Our interaction with Joe was basically to let him know what our issues and concerns were from the state of Delaware. The overriding relationship with Senator Biden was more to keep him informed of what the major banking issues were." Biden was not on the Senate Banking Committee as Carper and Castle were, Bakerian noted, so they were more the targets of the industry's lobbying that Biden was.

"Although he wasn't on the banking committee," he said, "we have upward of thirty-five thousand people working in the banking community in Delaware, a very large employment force in a state this small, the largest nonfarming industry in Delaware. People would say Biden was close to the industry because we were such a large constituency, so whether we were banking or autos, depending on what state it was, he would listen to our concerns and he would vote with us sometimes and not vote with us other times. Joe was always fairly evenhanded. In the same vein, he would not be what I would call an overly probusiness legislator. He was supported generally by the banking industry, but that had more to do with his seniority than with his voting record."[4]

Nevertheless, charges of conflict of interest came up in Biden's 1996 bid for election to a fifth term in the Senate. The campaign of his Republican opponent, Ray Clatworthy, alleged Biden had been given a sweetheart deal on the sale of his home in Wilmington to a top MBNA executive named John R. Cochran III. About a week before the election, a woman named Cecily Cutbill at the gubernatorial campaign headquarters of Democratic nominee Tom Carper picked up a phone and was asked by a polling firm about her choice in the Biden-Clatworthy race. When she said Biden, the caller asked whether she would still vote for him if she knew he had sold his house to an MBNA executive for more than twice its value.

The caller, Cutbill said later, "didn't leave anything to your imagination," and the Biden campaign was notified.[5] The allegedly nonpartisan survey was easily recognized as what is known in the political trade as a "push poll"—airing a charge against a candidate and asking whether the voter would vote against that candidate if he or she knew that such an allegation was true.

The implication was that Biden had been given a hefty profit on his house through sale of it to the MBNA executive in exchange for, or expectation of, favorable treatment regarding legislation of interest to the huge credit card company. The Biden campaign immediately cried foul and flatly denied the allegation and implication. Details of the deal were quickly turned over to the Delaware press, including records establishing that the appraised value and the price paid by the buyer were approximately the same. Cochran had paid $1.2 million for the large wooded estate property in the fashionable Chateau country section of Greenville, just outside Wilmington.

Biden had bought the property on three acres of land, formerly part of a du Pont family estate, in 1975 for $225,000 and had made substantial improvements. Together with the intervening climb in real estate values over two decades, they more than accounted for the huge but legitimate increase over what Biden had paid for the property. Biden had obtained the appraisal by an affiliate of the Delaware Trust Co., the *Wilmington News Journal* reported, in anticipation of cosigning a loan to help pay college costs for his sons' educations.

Biden called the push poll "the underbelly of politics . . . immoral and unethical," and an MBNA official called the charge "malicious slander." Clatworthy's press secretary, Michael Flynn, said lamely that if a push poll had been taken, "it was done without the candidate's knowledge," adding "we don't know what the value of Biden's house was, and it is not an issue we're going to deal with in this campaign." But the deed was done, and the state Republican Party chairman, Basil Battaglia, said that even to raise the issue was "crazy."[6]

Clatworthy, a strong Christian candidate who had run a vigorous and essentially clean campaign against Biden, denied knowledge or involvement in the push poll. But the survey was traced to a Republican polling firm in Alexandria, Virginia, working for him. He fired the pollster involved, Guy Rogers, a former aide to Christian Coalition executive Ralph Reed.

In an appeal to the religious, antiabortion, progun right, Clatworthy had brought in movie star Charlton Heston and Iran-Contra figure Oliver North to raise money for him. As other Biden opponents before him, he tried to hang the liberal label on the incumbent, to no avail. On election day, Biden got 60 percent of the vote in a four-candidate race, to only 39 percent for Clatworthy. Biden called the outcome "the sweetest victory we have had," because it showed Delawareans didn't want "negative and dishonest campaigning" and were "able and willing to reject extremists and work in that broad middle ground" where he placed himself.[7]

(When Ray Clatworthy tried again against Biden in 2002, with the Democratic senator now supporting President George W. Bush on use of force against Iraq, the challenger fared only slightly better, losing to the incumbent by 58 percent to 41. Clatworthy's political director, Dave Crossan, explaining Biden's appeal despite occasional gaffes, said voters would tell him: "I know he's an embarrassment, but he's our embarrassment.")[8]

After the 1996 election, records at the Federal Election Commission indicated that MBNA executives and employees had given Biden $62,850, the largest bloc of contributions from any source, under another fund-raising approach known as bundling. Such contributions, ostensibly made individually and voluntarily as required by law, bore the strong suspicion of corporate direction or suggestion. But again, Biden's position as a prime provider of constituent services to state businesses and individuals made these contributions unsurprising.[9] Biden as a Senate candidate never accepted political action committee contributions and spoke out against bundling, and seemed to accept the dubious contention that such donations were made without corporate coercion.

(In 2008 the *New York Times* reported that in 2003 Biden's younger son, Hunter, after graduating from law school was hired by MBNA "as a management consultant and quickly promoted him to executive vice president." Thereafter, the *Times* said, he became a partner in a Washington lobbying firm with an annual retainer of $100,000 "to advise it on the Internet and privacy issues," and in June 2008 Senator Biden paid his son's law firm $143,000 for "legal services."[10])

In less direct ways, Biden and the other members of the Delaware congressional delegation were subject from time to time to suggestions of unhealthy coziness with the now dominant bank and credit card industries in the corporate capital of America. MBNA came to be one of the major contributors to the campaigns of many state officials and legislators, from governor on down. And in the lobbying for federal largesse, such as Urban Development Action Grants in the form of low-interest loans to spur commercial growth in depressed areas, Biden's office was an obvious port of call. Celia Cohen in

Only in Delaware wrote that "it did not hurt that Joe Biden, the first Democratic senator to back Jimmy Carter for president, was in a position to steer UDAGs in Wilmington's direction."[11]

Biden himself later said "I got the least help of anybody" in the Delaware congressional delegation from the banking and credit card industries. He estimated that out of the millions he raised for his political campaigns over the years, "MBNA associates" had given perhaps $150,000 in individual contributions, and no political action committee or bundled money of any kind. Actually, according to the Center for Responsive Politics, the amount was slightly higher, $214,050, over four senatorial elections. He voted for a major bankruptcy bill in the Senate in 2005 sought by the credit card industry that imposed a means test for protection from debt payment. But, he said, it did not include certain amendments sought by the industry that "I killed as chairman of the Judiciary Committee." He did get some heat, Biden said, for opposing use of bankruptcy to avoid paying medical bills when patients had the assets to pay them. "My view was that was a societal responsibility," he said. "The doctor in good faith provided medical services for you in a hospital and then if in fact you had assets to pay, what was the difference between a doctor getting assets to pay and a college loan or anything else? It should be resolved by national health insurance. I had never been accused of being in favor of the doctors. It became very easy to demagogue. . . . The irony was the liberals were angry with me and the conservatives were angry with me. . . . I was getting blamed by left groups that I was a supporter of the banking industry and the banking industry wasn't supporting me. If you asked anybody around Delaware, Is Joe Biden viewed as a friend of the credit card companies, [the answer would be] 'Well, he's agnostic to negative.' I don't think you'd find anybody who'd say, 'He was one of our champions.' My dad used to have an expression. He'd say, 'Joey, it's no sense in dying on a small cross. If you're going to die, make it a big one.' "[12]

Biden's personal lifestyle seemed to be the best inoculation against allegations of high living at the public trough. Required statements of personal financial holdings and worth have indicated him to be among the least financially secure of all members of the Senate. His home was his most valuable material possession, and he and Jill were living off his Senate salary of about $166,000 a year in 2006 and her smaller salary as a community college teacher.

The heavy emphasis on constituent service, both corporate and individual, stems in part from Delaware's small size, both in geography and population. As Bakerian put it, "We have a high degree of incumbency, both in the federal and state positions. Delaware is a small community. We see our politicians every day in the streets. We refer to our political representatives by their first names, and everybody knows who you're talking about. So you get to know them as people, not just as your political representatives, and therefore you'll find that the majority of our congressional and state groups are frequently reelected."

Bakerian noted that in Delaware "when you see a bumper sticker that says JOE, they don't have to say the last name. If they did that up in New York, with Schumer, if they said CHUCK, they'd say, 'Who's that?' "[13] But that is distinctly the case with Biden, who won his seventh consecutive Senate term from Delaware in 2008 even as he was being elected vice president of the United States nationwide. When he picks up a phone in Delaware, what he says is, "Hello, this is Joe," and needs no further identification. By the same token, his candor and penchant for snap observations he might later wish not to have made is greeted across the First State with the same understanding response: "That's Joe."

The ease and familiarity in Delaware toward its politicians is distinctly bipartisan. "One of the recurring themes at the end of any political year," Bakerian noted, "is the parties thanking each other for working so closely together. You hear it at any press conference, any gathering where you've got the Republicans and Democrats together. The saying is, 'We are bipartisan and we couldn't have helped the state if we didn't work together,' and they really, really mean it."

As a New Yorker originally himself, Bakerian said "entering the political arena in Delaware is a unique experience for someone coming from out of state. As such a small state and close community, they have to work together to get things done." And Biden embodies that spirit himself, the banking president said. "He once was very close to [the late] Bill Roth, the longtime Republican senator. They would ride the train together to Washington. Another reason why our congressional group has been so popular has been because of our proximity to D.C. They're not from South Dakota, they come home every night, and they ride the train together, Carper and Castle as well as Biden, as it also was with Roth in his day."

Bakerian continued. "So how come they keep getting reelected? Well, they were home almost every night, at the supermarket, at the local restaurant. And when you keep seeing these people every day of the week, you want them to represent you. Carper is our senator, he's also been our governor and he's also been our representative. Castle has been our representative and also has been our governor and may run to be our senator,"[14] which he eventually did.

The spirit of comity in politics that rules in Delaware is well illustrated by Return Day in Georgetown, the Sussex County seat. In such an atmosphere, party differences tended to blur. Jim Soles, the prominent political science professor at the University of Delaware, has labeled the condition "consensus politics."[15]

In 1999, when Democrat Tom Carper had already served ten years in the U.S. House and was finishing his second four-year term as governor, limited by law, he paid a call on Republican Mike Castle, his predecessor as governor who was winding up ten years in the U.S. House. According to Carper later, Castle said he had no intention of challenging the other Delaware senator, Republican Bill Roth, a close friend of Biden, in 2000, and suggested Carper run against him. When Carper did, Democrat Biden supported him but said nothing against Roth. Upon Roth's death in 2003, Biden delivered the eulogy.[16]

At the same time, despite his long career as Delaware's most prominent public figure, Biden by all accounts never assumed or posed to be the state's Democratic boss, attempting to dominate the party's politics there. "There wasn't room for that in Delaware," Joe Farley said.[17]

Delaware reporter Celia Cohen wrote that "although Biden never seized organizational control [of the state party], believing a senator could not run the state operation effectively, he did reorient its political compass." His first Senate election, she wrote, "would prove to be a turning point philosophically for the Delaware Democrats as a party."[18]

The tight community and political interaction is repeatedly reinforced in Delawareans' stories about how their lives are personally intertwined. One local gag is that everybody in the state "is only two degrees of separation from Joe Biden."

Bakerian recalled that in high school his oldest daughter won the Neilia Biden Community Service Award, and when his own wife died, the first floral piece to arrive at the funeral home was from Joe and Jill Biden. "I didn't know Joe that well and didn't socialize with him," Bakerian said. The next day at the funeral, a relative from New York said to him, " 'There was a guy in the back of the church who looked just like Joe Biden,' and I said, 'It *was* Joe Biden.' My relative said, 'Who the hell are you that Joe Biden shows up?' I said, 'This is Delaware. It's nothing special to us. This is what Joe does.' " About a month later, Bakerian was in Washington on banking business. "He came up the hall, gave me a big hug and said, 'Anytime you want to talk, give me a call,' because we shared a common loss," Bakerian recalled.[19]

The pride and affection in the First State for its first citizen extends among many residents to Delaware in almost a Pollyannaish fashion. Former Democratic state chairman Joe Farley said: "Delaware is so small, we all know each other. There's no real place to hide. Transparency is part of the process. There's no such thing as a secret here. What you see is what you get. Joe's career in politics is parallel with the times. You can't ignore the fact we've gone from a red state to a blue state. If you can't work together here across the aisle, the fact that we're a small state, you get along, you work together, you reach out for everyone; everyone is your constituent. One thing, once you're in, you're in forever. We don't elect people here, we anoint them. If it ain't broke, you don't need to fix it. We see each other in church, on the ball fields, bowling alleys, the golf courses, at the beach, at the coffee store, the supermarket. The environment also assures that the folks in office are good people."[20]

In some other state, Joe Biden's constituent service to the major corporate interests might have been subject to greater scrutiny and criticism. But the nature of the man, and the nature of the state, seemed to conspire against anything but a near worship of the enduring politician with the flashing grin, the photographic memory of faces and anniversary dates, the clean living and impeccable values of faith and family. No slogan more captured the relationship than the one that flew from small banners across the state in 2008, not an appeal for votes, simply: DELAWARE'S JOE BIDEN.

TWENTY-ONE

FIGHTING CRIME AND ABUSES OF POWER

JOE BIDEN HAD other reasons to lament the 1988 presidential election beyond his own premature departure from it. Michael Dukakis's loss to George H. W. Bush underscored the Democratic Party's continued vulnerability to the Republican charge that it was soft on crime. Dukakis was successfully savaged by the Willie Horton story, wherein a Massachusetts convict raped and held captive a woman in Maryland while he was on a weekend furlough that was part of a Dukakis prison policy.

Then, in a Dukakis-Bush televised presidential campaign debate in the fall, the same impression that the Massachusetts governor lacked the fortitude to combat and punish serious crime was vividly reinforced. The first question to Dukakis from the moderator, Bernard Shaw of CNN, was devastating. "Governor," Shaw pointedly inquired, "if Kitty Dukakis [his wife] were raped and murdered, would you favor an irrevocable death penalty for the killer?"

Dukakis, opposed to capital punishment, might nevertheless have been expected to react with outrage at the personal question. Instead, calmly and unemotionally, he replied, "No, I don't. . . . I don't see any evidence that it's a deterrent and I think there are better and more effective ways to deal with violent crime. We've done so in my own state. And that's one of the reasons why we have had the biggest drop in crime of any industrial state in America; why we have the lowest murder rate of an industrial state in America."[1]

Here was the Democratic nominee for president responding with cold statistics in his head—and ice water in his veins. Had it been Joe Biden, the question predictably would have gotten his Irish up and his blood boiling. But in Michael Dukakis's sober words there was no tough cop determined to wage a law-and-order assault on crime in the manner of, say, Richard Nixon. The departed Nixon had ridden the theme of law and order to the White House in 1968 and to a second term in 1972, until his own complicity in the megacrime of Watergate forced his resignation in 1974.

That latter episode in turn was critical in ushering a Democrat, Jimmy Carter, into the Oval Office in 1976, but the "soft on crime" label clung to his party. So when Biden in 1987 became chairman of the Senate Judiciary Committee, he was determined to address this public perception for which Democrats had suffered so long at the ballot box. Long before then, however, going back to the days of his youth in Scranton, Biden had been taught by his parents and grandparents never to turn the other cheek to personal abuse, whether from an individual or a mob. It was an integral aspect of his championing of civil rights. And in dealing with law and order, he defended not only the right of the individual to be protected from abuse, but also the imperative of giving the police the tools they needed to provide that protection.

While son Hunter was at Georgetown University, Biden made a speech relying on his own father's conviction that "the greatest sins on this earth are committed by people of standing and means who abuse their power." Whether it was the physical punishment of a child by his father—"it takes a small man to hit a small child," Joe Sr. used to say—or the Nazi genocide in Europe, Joe Jr. had no stomach for "might makes right." He grew up fearless, but he was no bully. And Hunter later recalled that his father "never raised a hand to us . . . enforcement came out of fear of disappointing him."[2]

At Hunter's explicit request, his father spent hours preparing the speech at Georgetown to the Jesuit Volunteer Corps, of which Hunter was a member, on how his religion had informed his public policy. "As I went through it," Joe Biden said later, "I realized that the one thread that goes through everything in my public career that is totally consistent with the Catholic social doctrine that I was taught coming up, as well as the way my parents raised me, was this notion of *the* cardinal sin of a government, an individual, a corporation, a business, a father, a parent can commit is to abuse the power they have, whether it is psychological, physical, financial, or political. . . . It was one of the reasons I ran for the Senate. I was so angry. The kicker for me was when there was a march on Washington and Nixon said those who were protesting the [Vietnam] war should not be carrying or wearing American flags. I thought to myself, that son of a so-and-so, who does he think is, defining what Americans are."[3]

Biden's personal abhorrence of the abuse of power, and his determination to share that attitude with his sons, also led him to take Beau at age fifteen to Dachau and the Berlin Wall, and then Hunter at fifteen the next year, to drive the lesson home. "At our dinner table growing up," Beau recalled later, "eating was almost incidental to our family dinners. Always imparted on us was the memory of the Holocaust, that we as a nation and a people [had to be] vigilant that it would never happen again. My dad wanted to illustrate that to us by showing us Dachau and the depravity of the Holocaust. And it made an impact. It goes to what animates my dad's public life—the necessity to stand up to the bullies, whether it's at Saint Helena's in grade school, doing the same at Archmere, at the University of Delaware, or doing the same as a United States Senator to the tyrants of the world. There is no difference between the private man and the public man. He is the same."

As a member and later chairman of the Senate Foreign Relations Committee, Hunter Biden said, his father demonstrated his contempt for the abuse of power in his deep involvement in the turmoil and 1990s breakup of Yugoslavia, and the imposition of brutal ethnic cleansing in Bosnia and Kosovo. "That was why he was so adamant about speaking out and speaking up to [Serbian dictator Slobodan] Milošević, telling him how he felt to his face. It was an illustration to me that genocides can happen and could happen again in Europe, not to mention Darfur or Rwanda. That's why he as one of a hundred senators, he was not going to stand for it."[4]

But also, from Judiciary, Biden responded at home to growing reports of police brutality on the one hand and inadequate law enforcement on the other in an era of heavy drug trafficking. Even before he became the Judiciary chairman, he had called for creation of a national drug czar to cope with the growing flood of narcotics into the American market. At the same time, he championed more community policing and spent long hours and days immersing himself in law enforcement culture, frequently attending and addressing police organization meetings.

This intensive interest in crime prevention may have been seen by skeptics as a conspicuous effort by Biden to rehabilitate himself in the wake of the collapse of his 1987 presidential bid. Rather, said Chris Putala, a key Judiciary staffer at the time, "it was like 'I'm Joe Biden and I know how to grapple with this issue.' There was no sense he was licking his wounds."[5]

For years, Biden had been pushing for the creation of a drug czar, and when Ronald Reagan appointed William Bennett as his drug czar, Biden worked with him coordinating the various governmental agency budgets dealing with narcotics. And in a pending crime bill in 1990, Putala recalled, Biden fought for more money for police departments, for a ban on assault weapons, and for tougher penalties for drug offenders; the bill was watered down by Republican opposition.

Finally, in 1994 Congress passed a $30.2 billion Violent Crime Control and Law Enforcement Act, sometimes known simply as the Biden Crime Bill, which called for one hundred thousand more police in the nation's city streets over six years. The measure won strong political support for the Democrats and President Clinton from a police community that earlier had considered the Republicans as the law-and-order party. Biden ballyhooed the Community-Oriented Policing Services (COPS) program as reducing crime for eight straight years, from 1993 through 2001.

His longest and ultimately most successful anticrime fight also came to fruition in 1994, with inclusion in the overall bill of his Violence Against Women Act. It addressed what he regarded as the most neglected abuse of power in the American society. For a time, some religious groups opposed the bill on the grounds that it was a federal intrusion into family rights. But in time many women's organizations rallied to the cause, recruiting political activists who had worked with Biden to defeat the Bork Supreme Court nomination in 1987.

In 1990 Biden became particularly interested in the legal plight of women; while perusing Bureau of Justice crime statistics, he noted that violent crimes against women had shot up even as those against men were declining. He first assumed, he wrote later, that the phenomenon represented a greater willingness of women to report domestic violence and rape against them as part of the strengthening women's movement. But he came to comprehend that it was much more than that, and that cultural attitudes played a significant role.[6]

The campaign against domestic violence did not originate with Biden. As early as 1972, a women's advocate group in St. Paul, Minnesota, created a hotline for battered women to report abuse, and a Haven House in Pasadena, California, established shelters for them. In 1974 a woman named Erin Pizzey published a book in England titled *Scream Quietly or the Neighbours Will Hear.* The National

Organization for Women announced a task force in 1976 to examine domestic violence; the same year, Pennsylvania enacted a statute providing protection for victims of domestic violence.

In 1978 the U.S. Commission on Civil Rights held a forum on battered women; in 1979 Democratic Senator Alan Cranston of California offered legislation providing $15 million in federal funds for shelters for battered women. It passed the Senate over the objections of conservatives who argued Uncle Sam had no business involving himself in family matters, but the bill died in the House.

Cranston tried again in 1980 and encountered the same opposition, with Republican Senator Orrin Hatch of Utah offering an amendment that would have limited shelter funds to married women to ensure that havens would be family-oriented. Cranston objected, asking: "Should women fleeing their homes be required to show their marriage licenses before being admitted?"[7]

In 1980 a National Coalition Against Domestic Violence held its first national conference, and subsequently the U.S. Department of Health and Human Services created an Office of Domestic Services, which was shut down two years later. In 1983 Maureen Reagan, the president's independent-minded daughter, entered the controversy, denying at an international conference that havens for battered women were mere "R and R shelters for bored housewives."

But twenty-four self-described profamily congressmen wrote Attorney General Edwin Meese complaining that a leading national organization on domestic violence was controlled by "prolesbian, proabortion . . . and radically feminist" women.[8] Nevertheless, Meese in 1984 established a task force on family violence, and the same year, Congress passed a Family Violence Prevention and Services Act earmarking federal money for domestic violence victims.

The issue got its broadest national attention in 1985 when a motion picture was made based on the fate of a victim named Tracey Thurman. She had sued the City of Torrington, Connecticut, for failure of its police department to protect her from violence by her husband. After repeated threats from him, a court order requiring him to stay away from her, numerous calls for police protection that went unanswered, and, finally, a restraining order against him, the woman was stabbed by her husband multiple times in the chest, neck, and throat and twice kicked in the head before the police came and arrested him. The woman was partially paralyzed and finally received $2.3 million in damages for violation of her equal protection rights. The case led to a mandatory arrest law in Connecticut.

In December 1989 another abuse of women case came to Biden's attention after a man walked into a college classroom in Montreal armed with a hunting rifle, separated the students by sex, and shot fourteen women, calling them "a bunch of feminists." An alert Supreme Court clerk named Lisa Heinzerling, later a law professor at Georgetown, noted that the law did not characterize such offenses pointedly against women as hate crimes, as were those involving a victim's race, ethnicity, religion, or sexual orientation. She called the omission a legal loophole "welcome to no one but the misogynist."[9]

All this happened before Biden got intensely engaged in the issue, though according to his daughter, Ashley, it was one that often was discussed at home. She recalled later that when she was in the second or third grade she wrote an essay on what she wanted to do when she grew up, in which she said "I wanted to help abused women."[10]

In fact, abuse of others was a concern unusual for someone so young. "Dad," she recalled often telling her father, "we have to save these dolphins. They're getting caught in these giant ocean floor–sweeping tuna nets." Her father introduced her to the then congresswoman Barbara Boxer of California, subsequently a senator, who shared that concern and took Ashley onto the House floor, where she pleaded with other House members to do something about what she saw as an offensive act of abuse.

"Out of it came, not just because of me but because of Barbara Boxer and others, the Dolphin-Safe Tuna bill," Biden's daughter said. "I still call her the Dolphin Lady. And that was when I realized I could be a change agent. I learned at a young age that organizing and social activism could really produce change. I learned it from Dad. Being his daughter, I kind of had it intuitively in me, but I got that from Dad, watching him." Subsequently, as a student at Tulane and then in graduate work at the University of Pennsylvania, Ashley focused on social work and went into the field.[11]

But violence against women became Joe Biden's own crusade. While working in the Senate on the omnibus crime bill, he realized it lacked that critical component. He chose a legal expert on the committee named Victoria Nourse, who was directing the research and planning for the bill, to focus on it. He recalled to her that he had worked earlier with Democratic Senator Birch Bayh on a bill involving marital rape, and that he couldn't believe the attitude that existed among some of his colleagues. He cited the comment of Senator Jeremiah Denton of Alabama that he didn't think it was a crime at all for a husband to force himself on his wife, that "damn it, when you get married, you kind of expect you're going to get a little sex."[12]

On the general proposition of the abuse of power, Biden told Nourse of one incident in which his father's employer at the car dealership he managed in Wilmington held a Christmas dinner and party for his workers. As a promotion, the company was offering a bucket of silver dollars to buying customers. As Biden related the story much later, the owner from the head table "took a bucket of silver dollars and threw them out on the floor, to watch his mechanics and salesmen reach down to pick them up. My father turned to him and said, 'I quit, God damn it!' and walked out." When his parents arrived home much earlier than expected and his father stormed up to his bedroom, he went on, "I looked at Mom and said, 'Mom, what's the matter?' And Mom said, 'I'm so proud of your father. He just quit.' And my dad didn't have any job to go to. . . . He saw that as an abuse of power. I realized it's the thing that has always pushed my button."[13]

Nourse brought abuse of power closer to home for Biden with research indicating that some states, including Delaware, not only exempted marital rape from prosecution but also reduced charges of date rape if the woman in question was "a voluntary social companion" of the man involved. In other words, Ashley Biden would be vulnerable in such a situation. Conveying that information to the senator, Nourse said later, brought "a look of horror on his face." She proceeded to draft language protecting women "from crimes of violence motivated by the victim's gender," specifically giving them the right to sue attackers in federal court under the Equal Protection Clause of the Fourteenth Amendment and the Commerce Clause. She also reached out to the NOW Legal Defense Fund to build a stronger coalition.[14]

Biden began hearings in June 1990, and had as a witness on the first day a young fashion model whose face had been slashed and seriously disfigured outside a bar in New York. The woman, Marla Hanson, testified that instead of garnering wide sympathy, she received letters and comments inferring that her behavior had brought on the attack, or that she must have been a woman of low moral character deserving of the treatment she endured. Other witnesses told of suffering date and acquaintance rape without legal recourse.[15]

"We on the staff had no clue of whether it was going to go anywhere," Nourse recalled. "We only had two or three cosponsors." But then, she said, "the hearings were extraordinary and we didn't anticipate that, either. . . . We heard from young women who were survivors, and the kinds of things they had to say were humiliating," in a way that young men who had experienced cultural change finally recognized as abuse of power.

In those days, Nourse said, there was a general comprehension of "real rape" only, for example, as "rape by a stranger with a gun." The hearings generated a much greater awareness of the other ramifications and scope of the crime, and Biden expanded the inquiry into domestic violence generally. The Tracey Thurman case, Nourse said, underscored the repetitive nature of the attacks "and the indifference of the police, the abuse of their power."[16]

By casting the abuses as a denial of a woman's equal rights to be free of any form of physical harassment, Biden was able to make the case for recourse to legal justice. He wrote a section into his bill establishing a crime of violence against women as a hate crime, and hence a civil rights offense. Doing so would give a victim the right to take civil action in a federal court if a state criminal court was not an avenue for justice. The bill also provided for sensitivity training of police, hospital workers, and others who would come in contact with domestic and sex violence victims, and called for a hotline for easier and quicker reporting of assaults.

Biden held a second hearing in August at which a female student at the University of Pennsylvania, where his son was in his senior year, testified about being violated in her dorm room by a boyfriend of one of her own friends. Biden recounted later how her residential adviser, after having heard her story, told her: "You were raped." To which the victim replied: "No, I wasn't. I knew him."[17] Such was a prevalent misunderstanding of what rape constituted.

At the same hearing, Biden told of a sense of pride that deterred some women from acknowledging the threat to them. Once, he said, when his own wife was driving to a West Chester University graduate night class, he urged her to park in an illegal but safe place. "I was pretty indignant," she recalled later. "Why did I have to take these special precautions?"[18] A lawyer-activist named Sally Goldfarb said Biden showed he understood "the lack of control that is experienced not only by women who are themselves victims but by all the women who have to constrain their daily activities to avoid being a victim." And, she said, he showed the "basic insight of the civil rights provision—that violence against women deprives women of equality."[19]

Biden's crime bill, however, didn't get to the Senate floor for a vote that year—nor the next. In the summer of 1991 Supreme Court Chief Justice William Rehnquist appointed a four-judge committee to appraise the push for a Violence Against Women Act. It challenged classifying such violence as a civil rights offense and warned that the federal caseload might be so increased as to cause "major state-federal jurisdictional problems and disruptions." Biden angrily wrote to the committee: "I will not mince words: as author of the legislation, I have stated that the bill does not federalize divorce law or domestic relations cases any more than any other civil rights law does."[20]

But Rehnquist in a 1991 report on the federal judiciary went out of his way to throw cold water on the Biden bill. Its definition, he wrote, "is so open-ended and the new private right of action so sweeping that the legislation would involve the federal courts in a host of domestic relations disputes." Biden in testimony before a House committee called Rehnquist's intrusion "outrageous." The justice, he said, "does not know what he's talking about" in saying the bill would unduly burden the federal courts with false claims to boost their alimony payments in divorce cases.[21]

Much later, Biden said, he had Victoria Nourse and other aides contact a number of female judges on the circuit "and they got mad. They turned it around from buying Rehnquist's argument."[22]

Biden continued through 1992 and into 1993 with hearings documenting the spreading growth of domestic and sex violence crimes against women. Then help came from an unexpected quarter. Republican Senator Hatch, who had been attending the hearings as a critical Judiciary Committee member, held hearings of his own in his state and was appalled at what he learned. Hatch became Biden's staunch ally and partner on the violence against women legislation and also the COPS program in a period when they alternated as committee chairman. Later they joined in pushing for the Adam Walsh Child Protection and Safety Act of 2006 against sexual predators, named after the kidnapped and murdered son of John Walsh, host of the television show *America's Most Wanted*.[23]

Biden decided for tactical reasons to fold Violence Against Women into the Clinton administration's broad crime bill, and with some legislative horse-trading it passed the Senate with $1.8 billion of funding in 1994. A key at the end was inclusion of a ban on assault weapons with strong police backing, demanded in the wake of a growing number of mass shootings.

The money allocated to combat domestic violence came from billions in deficit money achieved by a task force on government efficiency led by Vice President Al Gore. When Clinton signed the comprehensive anticrime package, Democratic Majority Leader George Mitchell singled out Biden, saying he was "the one person most responsible for passage of this bill," adding: "Joe Biden is the most underrated legislator, the most effective legislator in the Senate, bar none."[24]

Biden was able to achieve the bill's enactment, Nourse said much later, "because in fact he had by that time taken the lead on crime issues generally in the Senate. No other senator wanted to do it." Also, she said, "his whole goal was to insulate the Democratic Party from appearing uninterested in issues involving police, firefighters. It had become part of the liberal mantra that anything that police wanted was bad, and I think basically he wanted to even the playing field again; to get that voice in the room for the Democrats when people were negotiating things like habeas corpus reform or funding for cops, or death penalty restrictions.

"In fact the cops became central to actually negotiating many provisions in these bills, because they don't have what you always consider to be liberal or conservative interests. They support gun controls, for example." The police groups, she said, remarked "they didn't care about getting rid of Miranda warnings because it's easy to do. They didn't care that much. It's not a big deal to them. There are other things that are bigger deals to them. So Biden's ability to tap into, and to be respected among police officers in the various interests around the country, was extremely helpful."

Also, Nourse said, "part of the violence against women story is about women being ignored. You say you've gotten hit, but no one is arresting the guy." Police too often add to the violence by finding a rationale for it, she said, and Biden was particularly tuned into these police attitudes, to reflect them or at least see them from that police perspective.[25]

What finally assured the passage of the whole crime bill was Biden's insistence on the provision for putting one hundred thousand more cops on the streets of America, together with its cost coming from deficit reduction. It was another element that boosted Democratic stock among the men in blue. Taken together, according to Putala, the Biden initiatives produced a drop in domestic crime, "a social phenomenon that wound the clock back forty years" on the allegation of Democratic softness on crime, and on "the ability of the Republicans to wave the bloody shirt of crime." By the time of the 2000 presidential race, Putala said, crime was all but nonexistent as a major political issue.[26]

Nourse recalled how women commuters on the Wilmington-Washington train run would come up to Biden, thanking him for bringing domestic violence out to the public. But the point in the legislation, she said, was not that it was designed to give women special protection, but the same protection any citizen was warranted under the law.

In 1995, however, the Supreme Court under Chief Justice Rehnquist took aim at Biden's provision establishing domestic violence as a federal violation of civil rights under the Fourteenth Amendment, the so-called civil rights remedy. In *United States v. Lopez*, the Court declared that Congress had no jurisdiction in what was a matter constitutionally left to the states. Rehnquist observed that "punishment of intrastate violence . . . has always been the province of the State."[27]

And in 2000, in *United States v. Morrison*, while the Court let stand the bulk of the Violence Against Women Act, Rehnquist said Title III, the civil rights section, wrongly allowed Congress to regulate "family law and other areas of traditional state regulation." Furthermore, he noted, the Commerce Clause of the Fourteenth Amendment providing equal protection required an "economic" aspect, and there was none that he could see in domestic violence cases—a most arguable contention. The civil rights section was stricken by the Court in a 5–4 decision, but the rest remained.

Biden's war on domestic violence was not universally applauded. Some conservatives insisted it was also unconstitutional in establishing rape as a federal as opposed to a state issue. Raymond Wolters, the political science professor at Biden's alma mater, the University of Delaware, called him "basically a guy who thinks the federal government can do anything it wants to do. The classic example would be his Violence Against Women Act, which defines rape as a federal offense. Our Constitution nowhere authorized Congress to punish people for rape; this is a state responsibility. People like Bork and Thomas both, for slightly different reasons, do not have the expansive view of federal power that Joe Biden does."

Wolters argued that "legal distinctions are not Biden's strong point," and that he was wrong in saying "the federal government has the power, the federal government should move against rapists. Biden doesn't have much respect for the founders' original designs of dual government and states rights."[28]

But in the civil rights and women's rights communities, his leading role in enactment of the Violence Against Women Act, in writing broad anticrime legislation, and in his support of police in the war against violence and crime, won him his highest esteem. In this area, there was more reason to dismiss the old allegation that Joe Biden was a senator more of style than of substance.

TWENTY-TWO

SENATE GLOBE-TROTTER

JOE BIDEN'S FAMOUS daily commuting between Wilmington and Washington, mostly by Amtrak rail, placed his family ties and obligations always at the forefront of his agenda. The habit may have created the impression of a sheltered homebody. So casual observers probably would have been surprised to know he was among the U.S. Senate's most-traveled members abroad, sandwiching in trips spanning the globe from London and Paris to Moscow and Beijing, and scores of points between.

From the very start of his service in the Senate in 1973, Biden had hankered for a seat on the Foreign Relations Committee, a lofty goal for the chamber's youngest freshman. Against all odds, he had approached its chairman, J. William Fulbright of Arkansas, who as already noted let him down easy. But in 1975, with the assistance of Senate Majority Leader Mike Mansfield, Biden finally got the seat and was subsequently placed on the Senate Intelligence Committee and the prestigious Judiciary Committee. On Foreign Relations particularly, heavy foreign travel was involved.

Very early on, Hubert Humphrey brought him along to a conference in England. In time the young Delaware senator developed a circle of important foreign officials of his own. As related in his memoir, he had a special relationship with German Chancellor Helmut Schmidt that was unusual for a freshman American senator. In their first encounter, Biden wrote, "we got off on the wrong foot. After a long trip to Bonn, I'd overslept and was late to our meeting. Schmidt came right after me: 'It's no wonder the world is in such bad shape. You young people don't know anything.' I didn't back off. 'Well, Mr. Chancellor,' I told him, 'we can't mess up this world any more than your generation has.' "[1]

Apparently, this dose of Biden brashness went down well. Schmidt went on to express freely his concerns that new American President Carter's arms-limitation talks with the Soviet Union might imperil Western Europe. So when Biden was planning another European trip six months later, he scheduled a visit with Schmidt. But when Senator Gary Hart asked him to speak at a major Democratic dinner in his state of Colorado, Biden had to postpone it. The very next day, Biden wrote, the German embassy urgently called him to say Schmidt wanted him to come, and Secretary of State Edmund Muskie, his old Senate colleague, sent Biden a hand-delivered note asking him to call immediately.

Schmidt, Muskie said, was about to meet with Soviet General Secretary Leonid Brezhnev on arms limitations and, miffed at Carter, wouldn't talk to him, to Muskie, or to the American ambassador. He would talk only to Biden. So he flew to Bonn, where, according to Biden, Schmidt laid out what he intended to say to Brezhnev. Pounding the table, Schmidt told him: "Joe, you just don't understand. Every time America sneezes, Europe catches a cold. I think presidents have to understand that words matter. They matter."[2]

In the summer of 1979, with Senate ratification of the SALT II

Treaty signed by Carter and Brezhnev in peril, the president asked Biden to lead a delegation of six young senators to Moscow seeking assurances that the Soviet leaders would abide by new conditions adopted by the Senate. For three hours in the Kremlin, he matched wits with Brezhnev and Premier Alexei Kosygin.

"At one point," Biden wrote later, "the Soviet premier tried to engage me in a conversation about the number of American and Soviet forces in Europe. Mutual balanced reductions was also a topic of discussion. I'd been studying the balance of both conventional forces and nuclear forces, so when he gave me a number for Soviet tanks that was laughably low, I didn't let it pass. 'Mr. Kosygin,' I said, 'they have an expression where I come from: Don't bullshit a bullshitter.' He seemed to like that." The translator, however, sanitized the remark somewhat, Biden found out.[3] But once again, Joe was being Joe, whatever the occasion and wherever the setting.

In the long exchange, Biden wrote, "I did manage to get an unspoken assurance from him that the Soviets would likely accede to the treaty modifications the Senate had under consideration. The Soviets wanted the treaty passed, too."[4] In the end, however, after Biden's best efforts to sell ratification, it fell short in the Senate. Nevertheless, when Ronald Reagan won the White House in 1980, his administration informally complied with the SALT II limits. When in 1986 he threatened to abandon them, Biden along with Republican Senator William Cohen of Maine proposed legislation to prevent him from doing so.[5]

On the other side of the globe, Biden in a visit to China for the Senate Intelligence Committee in 1979 created a bit of a stir by suggesting to Vice Premier Deng Xiaoping that two American electronic spy installations be built on the Chinese border with the Soviet Union.[6] In 1981 NBC correspondent Marvin Kalb cited State Department sources in reporting that Biden was "desperately seeking a substitute" for two such sites the United States had operated in Iran until early 1979, when they were closed after the overthrow of the

Shah. Biden said he had made the proposal to spy on Soviet nuclear and rocket tests because "the far right was questioning our ability to verify [the SALT II] treaty after we lost two listening posts in Iran."[7]

In 1980 Biden went to Turkey, Greece, and Italy on an international drug-trafficking investigation for the Senate Foreign Relations and Judiciary committees, looking into what was called "the Sicilian Connection." The flourishing trade in heroin into the United States had by this time spread from opium fields in Turkey into Laos, Thailand, and Burma, then to Mexico and finally to Afghanistan and Pakistan. The trade route ran westward from the Khyber Pass through Turkey and Greece to Sicily for shipment to America.

From this trip, Biden's interest in the problem intensified and led to his prominent role in the eventual creation of an American drug czar to coordinate the activities of the various federal agencies engaged in combating the menace of the international drug trade.[8] Much later, when coca and opium production in Colombia had further broadened the illegal narcotic traffic into the United States, Biden went there in support of President Clinton's congressional bid for nearly a billion dollars to fight the murderous drug trade.[9]

Through all this time, Biden kept abreast of the economic and political developments in the Communist world, especially in the Soviet Union. Asked in 1989 whether the drug problem was his most important issue, he replied: "It's not the issue that I care most about. If I could wipe the table clean [of his senatorial responsibilities] . . . and pick one issue . . . I would focus on U.S.-Soviet relations, arms control, and Eastern Europe."

That and, of course, his obligations to Delaware. In 1984 his Republican colleague, Bill Cohen of Maine, by this time a senator, was deeply involved in arms control. He was working to convince the Russians, amid their suspicions of President Ronald Reagan's intentions, of a Cohen proposal for a gradual "build-down" of nuclear weapons. On short notice, he asked Biden to fly to Moscow with him to provide bipartisan reassurance. Biden said he would go, Cohen recalled later, "but I have to be back for a big dinner in Delaware." So they flew to Moscow and met with the Russians; then Biden got back on a plane and returned home the same day.

"He was willing to travel all the way to Moscow just for one day," Cohen remembered, "to say, 'Look, I'm a Democrat, and I believe in what Cohen is trying to do; this is not just a Reagan idea.' There are not many people that would do that."[10]

Cohen recalled, too, Biden's easy informality with the Russian bigwigs. In a serious discussion with Foreign Minister Yevgeny Primakov, Biden interrupted him by saying, as he had with Kosygin, " 'Yevgeny, don't bullshit a bullshitter.' I was kind of taken aback," Cohen remembered. "It was typical Joe."[11]

Biden continued to follow the Soviet Union with a sharp eye to internal developments, and intensified his role as a veteran member of Senate Foreign Relations Committee, focusing particularly on the political disintegration of Yugoslavia. As early as 1991, as chairman of the Subcommittee on European Affairs, he had conducted hearings that convinced him the country's ethnic and religious factions were headed for a bloody catastrophe.

Biden's interest had been whetted a dozen years earlier in a trip to Yugoslavia for a state funeral, when he accompanied Averell Harriman to a rare meeting with Marshal Tito. Along with the then U.S. ambassador Lawrence Eagleburger, they were treated to a personal review of the Balkans' political landscape and how the Yugoslav dictator had managed to keep his patchwork country together.[12]

With Tito's death the following year, Yugoslavia fell under the control of Serbian Communist leader Slobodan Milošević, who was bent on gobbling up all of it into a Greater Serbia. After the United Nations imposed an arms embargo on Yugoslavia, disintegration followed, first with Slovenia declaring its independence, then Bosnia and Herzegovina in a March 1992 referendum. Milošević and Bosnian Serb henchman Radovan Karadzic kicked off an ethnic cleansing against Croats and Muslims in Bosnia, and the bloodbath was on in earnest. The UN arms embargo left the Serbs, the overwhelming faction in the Yugoslav army, in the driver's seat. As accounts of shocking atrocities came to Biden's attention, including reports of massive concentration and murder camps, he pressed the George H. W. Bush administration to take action. Meanwhile, Milošević insisted what was going on was a civil war and no outsider's business.

With Clinton assuming the presidency in early 1993, Biden from the Senate floor pushed for lifting the arms embargo to give the Bosnians the means of defending themselves, as they had a right as citizens of a recognized independent state. He also called for NATO air strikes on Serb positions encircling Bosnian cities. But the full Foreign Relations Committee balked at supporting him. One day, the Yugoslav ambassador appeared in Biden's office and passed on an invitation from Milošević for a face-to-face meeting in Belgrade. Wary of being drawn into a propaganda or publicity trap, Biden finally agreed on condition that they meet in private, with no press coverage at all. Beyond that, Biden wrote later in his memoir, "I promised myself I would not break bread with a mass murderer."[13]

With the approval of the Foreign Relations Committee chairman, Senator Claiborne Pell of Rhode Island, Biden—as chairman of the European subcommittee—three staff aides, and a military attaché flew to Yugoslavia in April for a quick view of the situation on the ground. They visited schools and a refugee camp in Bosnia, where, Biden said later, he got an earful of first-person accounts of brutal atrocities against the Bosnian population.

He wrote of arriving at his hotel in Belgrade and turning on the local television station for "the state-controlled news. Milošević's TV commentators were baldly asserting that Bosnian Muslims were massacring Serbs, killing Serbian babies, and hanging them on hooks like they'd hang chickens. This was the big public relations campaign in Belgrade," Biden wrote, "and it tracked with one of the State Department reports we'd read. Young Muslim internees had been forced by heavily armed Serb television reporters to confess on television to killing Serbian babies. They threatened to kill parents, wives, and siblings if the men refused to perform for the camera."

But even in a meeting with declared dissidents in Belgrade arranged by the U.S. embassy, Biden found remarkable myopia about what was happening. "They hated Milošević, but they were still Serb nationalists," he wrote. "They were feeding on state-controlled television and had a full belly. They never once made the argument that Milošević was a bad guy because he was killing Muslims. He was a bad guy because he was denying his own people, the Serbs, freedom of speech. . . . Either they didn't know what was going on, or they didn't care. There was room for only one set of victims in Belgrade, and that was the Serbs," who in terms of European history were being denied their place in the sun.[14]

The meeting with Milošević took place in his office in the presidential palace. Biden likened him to "a cherubic, self-satisfied banker at the coasting end of a long and lucrative career. . . . He came at me with his hand extended; he seemed happy to see us. When I refused to shake his hand, he nodded calmly, unperturbed."

According to Biden, Milošević began by informing his visitor that he had the wrong idea of what was going on in his country. He spread maps of the old Yugoslavia on a conference table and pointed out places where the Serbs were being attacked by the Muslims and Croats. "I told him the whole world knew who was doing the attacking, and it was up to him to stop it," Biden wrote. "He was still calm. And he lied to my face." When Biden confronted him with what he had heard about the sieges of Sarajevo and other towns and the atrocities, he quoted the dictator as saying the Muslims were shelling their own people, that there was no genocide going on, and if there was any ethnic cleansing, it was against the Serbs.

At this point, Biden wrote in his memoir, "Milošević could tell I had just about had it with his lies, and at one point he looked up from the maps and said, without any emotion, 'What do you think of me?' " Biden said he replied: "I think you're a damn war criminal and you should be tried as one." Milošević greeted the charge with "not the slightest twitch in his face," Biden wrote.[15]

When Biden asked his host for an escort to see the damage done to the recently bombarded Bosnian town of Srebrenica, Milošević told him it wasn't safe to go into a war zone. And when Biden said to him if that was so, then he needed to let UN forces intercede, Milošević replied that such a decision was up to the Bosnian Serbs, implying he had no control over them.

Biden's subsequent report to the Foreign Relations Committee noted that he had "urged Milošević specifically to endorse use of NATO forces, under United Nations authorization, to (a) eliminate any heavy weapons not turned over to UN control; and (b) police disputed areas if a final map could not be agreed among the three sides. Milošević declined to do so, unless the Bosnian Serbs agreed first."[16]

Milošević then surprised Biden by asking whether he wanted to talk to Karadzic, the Bosnian Serb strongman who Milošević pretended was an independent player in the drama. When Biden said yes, Milošević picked up a phone, and within several minutes a hard-breathing Karadzic burst into the room, telling Milošević he had come as fast as he could.

For more than two more hours, Biden wrote, he wrangled with the two leaders over letting UN forces in to achieve a cease-fire that would hold. He warned them that with a new American president they could be facing an end to the arms embargo, which would change the military equation, and tighter economic sanctions. Also, he told them, "the United States hadn't ruled out using force" and the Gulf War had demonstrated what that could mean. But the two Yugoslavs didn't budge. When Milošević suggested that they all have drinks and dinner, Biden declined, noting his host "didn't act offended that I refused his offers of food and drink."[17]

Long afterward, Biden critics expressed doubts that he actually had called Milošević a war criminal to his face, implying the claim was another Biden self-aggrandizing invention. Years later, a

Foreign Relations Committee staffer who was present, John Ritchie, told the *Washington Post* Biden had made the allegation more gently, "but Biden certainly introduced into the conversation the concept that Milošević was a war criminal," and that "Milošević reacted with aplomb."[18] Biden's closest aide, Ted Kaufman, who also was present, vouched for the authenticity of Biden's recollection and produced a photo showing Biden leaning over the conference table and gesturing vigorously at the Yugoslav leader.

On the same trip, a visit to Sarajevo by Biden and his party reinforced his view that the United States should back a policy of lifting the arms embargo and supporting NATO air strikes. So did a side trip to Naples, where the American commander of the Allied Forces Southern Command, Admiral Mike Boorda, agreed. But on return to Washington, Biden was unable to convince President Clinton, who balked at acting alone, without European support. He did accede, however, to Biden's plea to send Secretary of State Warren Christopher to sound out the British and French on a strategy to "lift [the embargo] and strike" the Serb forces, but Christopher was summarily rebuffed.

On his return, Christopher appeared before Biden and the Foreign Relations Committee. European foreign ministers had just met in Brussels and stories appeared saying they had argued, according to Biden, "that the United States lacked standing to call for lift and strike because we had no troops on the ground in Bosnia."[19]

Biden, livid, gave Christopher an earful. "Let me put it plainly, Mr. Secretary," he said to the rather stiff-necked chief diplomat.

> *You are required to speak diplomatically; I am not. I cannot even begin to express to you my contempt for a European policy that is now asking us to participate in what amounts to a codification of the Serbian victory. . . . What you have encountered is a discouraging mosaic of indifference, timidity, self-delusion, and hypocrisy. . . . After they held our coats on Kuwait and*

Somalia, they are asking us to put in a few thousand troops on the ground in order to have the right to speak. . . . Let's not mince words. European policy is based on cultural and religious indifference, if not bigotry. And I think it is fair to say that this would be an entirely different situation if the Muslims were doing what the Serbs have done, if this was Muslim aggression instead of Serbian aggression.[20]

Biden joined his Delaware Republican colleague, Senator William Roth, in calling on NATO to bomb the Serbian heavy artillery pounding Sarajevo, and on the United Nations to lift the arms embargo. "These idle threats and talk have caused a significant portion of the carnage that has occurred," he said. "If the Serbs knew we meant it, you would have had this settled ten months ago."[21]

In June 1994, in France for the fiftieth anniversary of the D-Day invasion, Biden and Republican Senator Bob Dole took a side trip to Sarajevo and saw the result of further Serbian shelling and the price of Western inaction. A year later, after Bosnian government forces had agreed to give up their weapons in exchange for UN protection, the Serbs overran Srebrenica with the United Nations basically standing by. Biden was mortified. "The United Nations had defaulted on its honor," he wrote later. "It had disgraced itself. And it felt like I had personally failed the Bosnians; whatever I'd done, it hadn't been enough."[22]

On the floor of the Senate, Biden painted the killings and atrocities in cataclysmic terms: "[W]e have stood by—we, the world—and watched in the twilight moments of the twentieth century something that no one thought would ever happen again in Europe. It is happening now."

The following day, three years after Biden had called for action, the Senate finally voted to lift the arms embargo in accordance with a bill principally sponsored by Republican Senator Dole and

Democrats Joe Lieberman and Joe Biden. NATO planes subsequently hit Serb positions in Bosnia. Only then did Milošević and Karadzic agree to peace terms hammered out in Dayton, Ohio, in late November 1995, including the recognition of the independence of Bosnia and Herzegovina and safe return home of war refugees. Richard Holbrooke, the chief American negotiator, said later that Biden "in no uncertain terms made it clear to me that the policy on Bosnia had to change, and he would make sure it did. He believed in action, and history proved him right."[23]

But to Biden's chagrin, Milošević remained Yugoslavia's supreme leader. He turned his attentions to the ethnic cleansing of the neighboring state of Kosovo, and Biden urged Clinton to bomb Serb positions there. In late 1997 Biden called for "creation of a new, well-armed European paramilitary police force" with a greater share of the cost borne by countries of the region and for the arrest and imprisonment of Radovan Karadzic and Ratko Mladic.

In March 1999 Biden introduced a resolution in the Senate authorizing Clinton to take action against the ethnic cleansing of Kosovo, and with Clinton's approval NATO air strikes began. At one point Clinton told Biden he was considering halting the bombing, but Biden urged him otherwise; shortly afterward, Milošević pulled his forces out of Kosovo.

Biden continued to make trips to the Balkans, insisting that Milošević be seized and turned over to the international court at The Hague for trial as the war criminal Biden had proclaimed him to be. It finally occurred in 2001. In February 2002, Milošević was tried in a long procedure that only ended in March 2006 when he was found in his jail cell, dead of a heart attack.[24]

Meanwhile, Biden had turned to the opportunity and challenge of NATO enlargement—encouraging and approving the applications of European countries recently liberated from the Soviet Union yoke or newly declared independent states. In March 1997, when Republican Senator Jesse Helms of North Carolina was still chairman of the committee, Biden on the heels of an American-Russian summit in Helsinki, Finland, went to Moscow to sound out the Kremlin on enlargement, and then to four countries seeking admission to NATO—Poland, the Czech Republic, Hungary, and Slovenia. The first three had been members of the Soviet Union's Warsaw Pact, the cold war counterpart to NATO.

Biden was accompanied on the trip by his chief European adviser on the Senate Foreign Relations Committee, Michael Haltzel. Biden's preparations, Haltzel said later, were prodigious. The man, he said, had "a real intellectual curiosity. He had the best of the lawyer's methodology. He had a wonderful ability to spot the soft underbelly of your argument and go for it. And then you had to go and defend it. He was not one of these people who would just sign his morning toast if somebody put it under his nose. That's what they used to say about [German President Paul von] Hindenburg in the 1930s, that he would just sign his morning toast."[25]

In a speech at Warsaw University, Biden told the audience, "I am in favor of enlargement in principle, but there are several questions that need answering to my satisfaction before I would vote to admit any candidate country." Earlier, he had spotted some anti-Semitic graffiti on a Warsaw wall.[26] In his speech, while making a point of praising Poland's progress in democratic reform, he emphasized the need for respect of minority rights. He particularly mentioned vigilance on protection of Jews from physical persecution in the wake of a recent firebombing of a Warsaw synagogue. In successive talks in the other three candidate countries, he raised other questions and in all four emphasized the necessary willingness to bear the appropriate costs for their parts in the enlargement and NATO obligations.

In a subsequent committee report, Biden wrote that during the cold war, "NATO provided the security umbrella under which former enemies in Western Europe were able to cooperate and build highly successful societies. The enlargement of the alliance can now serve to move the zone of stability eastward to Central Europe, thereby preventing a 1930s-type renationalization of that historically volatile region. For the United States, this translates into investing today in a modernized, enlarged NATO in order to avoid, once again, having to spill incalculably more blood and expend more resources to settle conflict tomorrow."[27]

NATO enlargement was to be considered by the member countries at a meeting in Madrid in July, and in advance Biden had gone to Moscow to "seek a broader and deeper relationship with Russia" in light of this development. He reported that in conversations with key leaders, "no Russian politician with whom I met believed that NATO enlargement posed a security threat to Russia. Rather," he wrote, "their opposition to enlargement reflected a deeper psychological problem of coming to grips with the loss of empire and a fear of Moscow's being marginalized in the changed world of the twenty-first century."

Biden wrote, "there is much we can do to allay their misgivings. By stabilizing Central and Eastern Europe, NATO enlargement can induce Moscow to reorient its political and economic policies westward toward Europe and the United States." Also, he optimistically wrote, "through intensified trade, investment, and technical assistance, the United States and Western Europe can help Russia overcome the real threats to her security—crime, corruption, environmental degradation, and loosely guarded nuclear, chemical, and biological weapons and material."[28]

The report laid out an eight-point plan of action: start a national debate on the enlargement; call on all NATO members to fund the move; support the admission of the four candidate countries at Madrid; encourage other prospective candidate countries; push members on more forces to Bosnia; strengthen relations with Russia; continue support for security and eventual destruction of nuclear weapons in all former Soviet states; and maintain support of Radio Free Europe/Radio Liberty.[29]

Biden expressed concerns over the willingness of Congress and other member countries to bear the financial costs of enlargement, and warned that planned withdrawal of all U.S. troops from Bosnia and Herzegovina might also trigger similar pullouts of European forces, destabilizing the still volatile Balkans. In a separate article, he wrote of the importance of American public support for the enlargement. "Understandably, after four decades of the cold war, the American people may be ready to turn inward. Congress and President Clinton share the duty of convincing the public that engagement with Europe and Russia is critical to securing the gains of our cold war victory."[30]

In October and early November, Jesse Helms held hearings on the enlargement proposals for the three former Warsaw Pact countries: Poland, the Czech Republic, and Hungary. By this time, Slovenia, never a member of the Warsaw Pact as a part of nonaligned Yugoslavia, had been put on hold.

Biden had been convinced by now of inclusion of the three, again vowing for the acquiescence of Russia. He warned once more that if NATO were to fail to enlarge its membership, the countries between Germany and Russia would inevitably seek other security arrangements themselves, "creating bilateral or multilateral alliances as they did in the 1930s with, I predict, similar results." Also, he said, "there is a powerful moral argument for enlargement: redeeming our pledge to former captive nations to rejoin the West."[31] At the same time, he said, existing NATO partners—particularly France and Britain—needed to agree to concrete, long-term plans assuring the survival of Bosnia, with firm commitments on sharing the costs and burdens of the enlargement.

Finally, on April 30, 1998, the U.S. Senate solemnly ratified by roll-call vote the admission to NATO of three of the four applicants—Poland, the Czech Republic, and Hungary. The accomplishment of bringing into NATO three former allies of the Soviet Union who had been members of the old Eastern Bloc created specifically to be a military challenge to NATO was a remarkable historical milestone signaling the interment of the cold war. In a rare observation fitting the occasion, senators sat at their desks and rose individually to cast their vote when their names were called. Formal acceptance of the three was voted by NATO the next year.

Biden continued to monitor the Balkans situation. In January 2001 he returned to Kosovo, Serbia, and Bosnia and called for continued American troop and reconstruction efforts. Writing in the *New York Times*, he said "we must make clear that our security umbrella and economic assistance will continue only if Bosnia breaks free from the stranglehold of its three nationalist parties. . . . The fact is, nation-building, if done well, can prevent vastly more expensive full-scale military actions."[32]

In another interview on his return, Biden reported on shakiness about the American commitment under the prospect of a new president. "Everywhere you go in Europe, whether it's a high-level meeting in Paris or walking down the street in Pristine [Kosovo]," he said, "the question is the same: Are the Americans determined to go it alone? The French say it with a kind of false bravado: 'Go ahead and leave, we can handle things on our own.' But they don't mean it. In Kosovo, I had shopkeepers stop me in the street and ask if the Americans were leaving. Because if we are, they don't think the [other] Europeans will stay."[33]

In his report to Committee Chairman Helms, Biden pushed for the NATO admission of Slovenia, arguing that "the peoples of the Balkans must not believe that they are seen as congenitally incapable of joining the trans-Atlantic community. If one of their number fulfills the alliance's stringent requirements for membership, as Slovenia manifestly has, then it should be welcomed forthwith as a full-fledged partner."[34]

In 2002 Slovenia finally was admitted, along with six other Central and Eastern European countries: Bulgaria; the Baltic states of Estonia, Latvia, and Lithuania; Romania; and Slovakia—all former members of the Warsaw Pact. In a sense, the cold war world had been dumped into the same ash bin of history into which Soviet Premier Nikita Khrushchev had once assigned American democracy. And Joe Biden could claim to have given it a robust shove himself.

In this whole range of engagements abroad, Biden came off as impressively well versed in the history and the complications of the world of international diplomacy and conflict. And in that world as at home, along with the respect he achieved, his reputation for long-windedness and his folksy, informal style sometimes led to the underestimation of his knowledge and capability.

Regarding this impression, one Republican member of the Senate Foreign Relations Committee recalled a meeting in Mexico City in the early 2000s between Biden and his colleagues and their Mexican counterparts on the subject of illegal drug trafficking. "We were having dinner at the home of a Mexican business philanthropist," former senator Chuck Hagel of Nebraska said, "and there must have been twenty or twenty-five people around a huge round table. It was a pretty tough conversation. A lot of the Mexicans were blaming America for producing the market because of our demand, and I think Joe Biden stunned everybody there at the depth of his knowledge about all these issues."

When one of the Mexicans challenged him on a point he was making, Hagel recalled, "Joe said, 'No, you're wrong,' and this guy said, 'No, you're wrong, Senator.' Biden looked across the table at him, and said: 'Sir, let me explain something to you. Every current law on the books in America today regarding illicit drugs—drug abuse, drug sentencing, the courts—I wrote it, as chairman of the Judiciary Committee, as a member for thirty years. There's no one knows more about it than I do. Now, what is it you want me to do to show you that what I say is correct?' "

Biden, Hagel said, "was unbelievable. He put on a display that night of his knowledge of all these laws, the depth of these laws; he was reciting them by sections. That's the kind of Joe Biden that is real. You get into these kinds of discussions on this stuff, you may not agree with him, but he knows what he's talking about."

Hagel went on: "A lot of the Mexicans didn't like it, and one of the Mexicans who was sitting next to me said, 'So this is your famous Joe Biden who constantly talks.' Well, they didn't want to hear what Joe had to say, and Joe did dominate much of the conversation, and that was because he did know more about the issue than anybody sitting around the table. That's kind of good and bad. . . . Why does this reputation hang over him, that he's a motormouth? He probably could have handled that a little differently, not said quite as much, let other people talk a little more. But he had such a depth of knowledge [about the drug traffic] that there was nobody in the room who was even in his universe. So it's both a blessing and a curse."[35]

Through all of Biden's globe-trotting, he continued to display his comprehensive command of the issues that confronted him wherever he found himself—along with his particular compulsion to convey it with a passion and at lengths that would subject him to criticism and even ridicule. Back home in the dawn of a new century, all these mixed blessings would now come into play in an infinitely more critical foreign policy challenge to Joe Biden, the Senate Foreign Relations Committee, and indeed the whole country. A new president, voted into the White House in late 2000 in the most controversial election in the nation's history, was about to encounter the worst terrorist attack ever on American soil. The ultimate responses of the new man in the Oval Office would launch a radical, essentially unilateral foreign policy that would shatter America's reputation and standing in the world community for nearly a decade.

TWENTY-THREE

WARS OF NECESSITY AND CHOICE

THE WHITE HOUSE sex scandal that brought House impeachment of Bill Clinton in 1998 did not drive him from office, thanks to the votes for acquittal from Joe Biden and other Democratic senators. They held their noses, kept the vote under the two-thirds majority required for conviction, and thus saved their party leader. But in 2000 that outcome did contribute indirectly to the presidential election defeat of the Democratic nominee, Vice President Al Gore, who chose not to make optimum use of the still popular Clinton in his campaign.

In probably the most contentious presidential election in history, the U.S. Supreme Court stepped in and decided a farcical Florida ballot controversy in favor of Republican nominee George W. Bush. The decision was blatantly political, with the Republican majority on the highest court ignoring its precedent of nonintervention in state political matters while stipulating that its own intervention not be taken as a court precedent in the future. It was a fateful ruling, the ramifications of which eventually and disastrously were to go far beyond that election.

Only five months after Bush had been sworn in, Joe Biden suddenly found himself chairman of the narrowly divided Senate Foreign Relations Committee. A moderate Republican member, Senator James Jeffords of Vermont, became so disillusioned with the new president and his party that he abruptly switched allegiance, becoming an independent and choosing to vote with the Democrats for purposes of organizing the Senate. His action elevated Biden as the ranking Democrat on the committee to the chairmanship.

Soon after, Biden got a phone call at home from the new president, asking him for help in preparing for his first trip to Europe—not just as president, but ever. Biden sized up the call as presidential massaging, and he went along with the exercise. But he was disturbed by what he saw as a growing foreign policy dichotomy within the new administration between the new secretary of state, General Colin L. Powell, and a conservative triumvirate of Vice President Dick Cheney, Secretary of Defense Donald Rumsfeld, and Rumsfeld's chief deputy at the Pentagon, Paul Wolfowitz. The latter three, Biden wrote later, "seemed to think our nation was so powerful that we could simply impose our will on the rest of the world." They seemed "so intent on overturning President Clinton's foreign policy initiatives that they were losing sight of the bigger goal, which was keeping Americans safe at home and engaging in doing good in the world."[1]

In Bush's first defense budget, Biden became concerned about evidence that Bush was ready to abandon the existing Anti-Ballistic Missile Treaty with the Russians and to pour billions of dollars into the unproved missile-defense system dubbed "Star Wars" by highly skeptical critics. To the new chairman's alarm, the administration also was contemplating cuts in funding the program to help the Russians in destroying their existing nuclear, chemical, and biological weapons. Biden thereupon called a committee hearing to inquire into this threat to getting rid of the "loose nukes."[2]

At the same time, Biden intensified an already hectic globe-trotting schedule. Beyond his focus on the Balkans and the NATO countries, he went to China in August, discussing with President Jiang Zemin the Bush intentions for scrapping the ABM Treaty and building the contemplated missile shield. On his return, Biden stepped up his own opposition.

On September 10, 2001, at the National Press Club in Washington, he delivered a stern warning of the peril in those notions. He ridiculed the Star Wars defense as no more than a "technological" concept and asked: "Are we willing to end four decades of arms control agreements and go it alone—a kind of bully nation, sometimes a little wrongheaded but ready to make unilateral decisions in what we perceived to be our self-interest, and to hell with our treaties, our commitments, and the world?"

Biden noted that the Joint Chiefs of Staff had said "a strategic nuclear attack is less likely than a regional conflict, a major theater of war, terrorist attacks at home or abroad, or any other number of real issues." He warned of the diversion of money "to address the least likely threat, while the real threat comes to this country in the hold of a ship, the belly of a plane, or smuggled into a city in the middle of the night in a vial in a backpack."[3] That night he returned to Wilmington, as was his daily routine.

The next morning, a clear, bright sunny day, Biden boarded the 8:35 train back to Washington for some routine meetings and a committee hearing. In about half an hour, word spread through the car via cell phones of some kind of aircraft crashing into a New York skyscraper, and presently Biden's own phone rang. It was Jill Biden at her school, watching television in the lounge with astonishment as an airliner was seen crashing into one of the soaring towers of the World Trade Center in lower Manhattan. As her husband remembered later, Jill suddenly said, "Oh, my God. Oh, my God. . . . Oh, my God. . . . Another plane . . . the other tower!"[4]

As soon as the train pulled into Union Station in Washington at the foot of Capitol Hill, Biden wrote later, "I could see a brown haze of smoke hanging in the otherwise crystal clear sky beyond the Capitol dome." It came from the direction of the Pentagon, across the Potomac River in northern Virginia, where a third plane had plowed into it. Biden encountered a colleague who told him the Capitol police had ordered the halls of Congress evacuated. Another hijacked plane reportedly was headed their way.

As Biden recalled later, he nevertheless started running toward the steps of the Senate, the side of the Capitol closest to Union Station. "I was really insistent about getting in," he said later, "because I thought it was awfully important that the Senate be in session. That people see us. That they could turn on their TV and see where we were."

Standing in the park across the way, a staff worker for a Republican senator approached him, declaring this terrifying moment was why the country needed missile defense. Another Capitol cop ran up to him, yelling, "Senator, you've got to get out of here. Another hijacked plane, eighteen minutes out, heading our way." Biden's daughter, Ashley, called. "She was crying," he recalled. "'Daddy, you have to leave Washington right now! They're going to do something in Washington!'"[5] But another cop reported that a fourth plane had crashed on a field in eastern Pennsylvania, presumably on another mission of death against Congress or the White House.

From the park, Biden gave a television reporter a brief interview, calling for calm. He said Congress would be going back into session soon and he had heard that the president, who had received the news at a school in Florida, was "coming back to Washington, and I applaud him for that."[6] Then, unable to get into the Capitol, the senator and his brother Jimmy, who had been in Washington and had located him there, hitched a ride back to Wilmington with Democratic Congressman Bob Brady, who lived in Philadelphia.

As they were approaching Baltimore, Biden's cell phone rang again. It was President Bush, saying he had just seen him on television and commending him for "saying the right things." Biden, recalling the conversation later, said he had asked Bush where he was, and was told: "I'm on Air Force One, heading to an undisclosed location in the Midwest." Biden said he urged the president to return to Washington. "I said, 'Mr. President, don't do that. Come home. Land at Andrews [Air Force Base in nearby Maryland]. Get a helicopter. Get on the front lawn. Let everybody see you.'" But Bush said his security people would not permit it.[7]

Later, Biden said he liked to tell the story of how, upon the liberation of Paris in World War II, General Charles de Gaulle was marching down the Champs-Élysées toward the Arc de Triomphe when suddenly a sniper started firing at him. Everyone ducked or ran, as Biden later recounted the scene, "except Charles de Gaulle. He kept marching, head erect and high. He did not flinch. He did not move, even though he was being shot at. That one defiant act rallied a nation."[8]

Biden said he had tried to get the Senate leaders to call the senators back into session to make the same demonstration, to no avail. As it was, he wrote later, it wasn't until early evening that Bush finally returned to the White House and addressed the American people. Biden wrote that "given what I had seen on September 11, I was less sure that President Bush could provide the wisdom and judgment this new reality demanded."[9]

Meanwhile, Biden pushed back against an inevitable public animosity against all things Muslim. Days later, he spoke to a packed crowd at the Muslim mosque in Newark, Delaware. With al-Qaeda leader Osama bin Laden now recognized as head of the perpetrators of the 9/11 attacks, Biden told the assemblage: "Whoever is guilty of these terrible crimes, whether it is Osama bin Laden or somebody else, one thing is certain: by his very actions, by killing thousands of innocent people, he proved that he is no true Muslim."[10]

In his memoir, Biden wrote that although there was no evidence that Iraqi dictator Saddam Hussein had anything to do with the 9/11 attacks, neoconservatives were already plotting to tie him to them and justify an invasion of Iraq. Biden wrote that he had a chance encounter with Richard Perle, then chairman of the Defense Policy Board and a leading anti-Iraq hawk, who mentioned a "decapitation plan" to parachute American forces into Baghdad to capture or kill the Iraqi strongman, using pursuit of the Taliban in Afghanistan as a feint. Biden said he hoped Bush was listening more to Colin Powell than to the Pentagon triad in Perle's orbit.

As for Bush, Biden gave him high marks for the preparations to respond to the attacks, telling the Council on Foreign Relations "the president's done a pretty darn good job" assembling a multinational force to go after the Taliban protectors of bin Laden and "resisting what were very strong entrees [*sic*] from parts of the administration to bypass Afghanistan and go straight to Iraq."[11] Obviously, responding against the perpetrators was clearly an act of self-defense in keeping with the UN Charter.

Meanwhile, never one to neglect his Delaware constituency in a time of public unease, Biden went to the University of Delaware in Newark several days after the terrorist assaults and delivered a defiant speech to twenty-seven hundred students and faculty jamming the campus arena. "Don't make these guys bigger than they are," he said. "They did a horrible thing and got lucky, but they are not some great juggernaut. Put it in perspective. Don't let yourself get carried away. It will not, cannot, must not change our way of life."

At the same time, he sought to reassure the student body. "You are worried about a war, all this talk of war," he said. "But this is not a war in the traditional, conventional sense. You won't see a return of the draft."[12] He was right about that, and that was only half of it. The only mobilization Bush called on the American people to undertake was to keep shopping.

Meanwhile, the Foreign Relations Committee chairman moved quickly to demonstrate his solidarity with the Bush administration in its response to the challenge of 9/11. In October he called Secretary of State Powell before the committee and lauded him and the president for "the way you have put together a coalition of the willing here and some of the timid, it seems to me . . . [including] a coalition of our friends in the Muslim world." He praised both of them for "keeping Muslim leaders on board who are experiencing demonstrations and protests about U.S. bombings and accusations that we are attacking the whole Islamic world, which is simply not true." He thanked them for keeping him informed and promised he had no intention of being a "Monday morning quarterback" in second-guessing their strategy and actions.[13]

Once the military mission against the Taliban got under way, working in conjunction with the indigenous Northern Alliance of local tribes in Afghanistan, the ruling regime was ousted and a new Afghan government was installed under handpicked Hamid Karzai.

As mopping-up operations continued, Biden brushed aside Pentagon resistance to on-the-ground congressional examination of the situation, and in early January he flew to the war zone. With the assistance of Colin Powell; his chief deputy at State, Richard Armitage; and CIA director George Tenet, Biden got clearance for himself, three staff, and two military aides. Then, leaving behind the military aides, who were ordered by their Pentagon superiors not to accompany the Biden party farther, they flew on in a commercial plane and then UN aircraft to Pakistan and on to Bagram Air Base, a former Soviet installation in Afghanistan.

From Bagram, the chairman and his aides were driven under tight security to Kabul for a meeting with Karzai. En route, Biden got out and talked to refugees who, he later wrote, told him they were not going to return to their village. They feared the Taliban militias would merely have melted into the population or the mountains, and would be back with a vengeance when the "liberators" went home. The party also visited a ramshackle school before going on to meet Karzai and some of the tribal leaders.

What Biden heard from all quarters, he wrote later, were pleas for more of everything—money, troops, security—and a commitment for the U.S. presence to remain, at least until circumstances greatly improved. Karzai urged that reconstruction and other aid be routed through him to the tribal leaders as an inducement for them to work with the new central government.

On leaving Kabul, Biden wrote, "I was a lot more hopeful about the possibility of securing and rebuilding not only Kabul, but large parts of the rest of Afghanistan." He credited Karzai with "wisdom, strength and integrity," and expressed his belief that "if we supported Karzai and listened to what he needed, we could make this country safe and whole again."[14] At the same time, he emphasized that in his view U.S. participation in a multinational military force was essential to restore order against the Taliban, not simply as a peacekeeper but "with orders to shoot to kill. Absent that," he said, "I don't see any hope for this country."[15]

Biden's departure from Afghanistan ended as it began. When bad weather shut down the UN flights out of the country and he asked for space on the next military cargo plane leaving, he was told orders had come from Washington forbidding it. Only a phone call to Secretary of State Powell, who fingered Rumsfeld as the culprit and called Wolfowitz to back off, got Biden clearance for the military transportation.[16]

Biden returned conveying a plea for urgent help, and Powell joined it, but while Bush "was agreeable and still willing to listen, he was also noncommittal," Biden wrote later. Though Bush talked of a Marshall Plan for Afghanistan, he had other ideas, and "was already scrumptiously" giving Cheney and Rumsfeld "the force and resources they requested for a new target"—Iraq.[17]

By now it was becoming increasingly clear to Biden that a critical pivot was under way from the unfinished business in Afghanistan to the neoconservatives' vision of spreading democracy throughout the Middle East, starting with deposing Saddam Hussein. When Biden complained to Bush that he was reneging on his promise to keep the congressional leaders informed, Bush instructed National

Security Adviser Condoleezza Rice to meet weekly with him. But she didn't, and when they did meet, Biden wrote later, she brushed him off. He and Republican Senator Chuck Hagel of Nebraska introduced a bill providing more money for Afghanistan, but the administration opposed it.

Biden, whose heavy and eclectic reading ran counter to his critics' image of him as an intellectual lightweight, continued to bone up on neoconservative writings that preached the rightful role of a lone superpower to work its will militarily, with of course the noblest of intentions. With Bush targeting the "Axis of Evil" of Iraq, North Korea, and Iran, Biden saw Cheney, Rumsfeld, and Wolfowitz convincing Bush, "an optimistic, ambitious, but woefully unprepared and uninformed president," of their vision of world dominance through power. As Biden put it in his memoir, "What was the use of being the world's only superpower if political leaders were shy about demonstrating and exploiting our overwhelming force capabilities? 'Shock and awe' was more than just a slogan. They believed they could scare rogue states into subjection."[18]

As talk of "regime change" in Iraq spread through Washington, Bush in a commencement speech at West Point on June 2, 2002, argued that the policy of deterrence and containment that had kept the country safe through the long and perilous cold war with the Soviet Union was no longer reliable. In the new era of terrorism, he said, deterrence through the threat of massive retaliation "means nothing against shadowy terrorist networks with no nation or citizens," and "containment is not possible when unbalanced dictators with weapons of mass destruction can deliver those weapons on missiles or secretly provide them to terrorist allies." The United States, he told the military graduates, must "be ready for preemptive action when necessary to defend our liberty and to defend our lives."[19]

At the same time, the Bush propaganda machine was working overtime spreading speculation of Saddam Hussein as an "imminent threat" who planned to use the weapons of mass destruction he supposedly had—though UN inspectors continued to look and couldn't find them. In response to this talk justifying preemptive action, Biden held full Foreign Relations Committee hearings on July 31 and August 1 to explore the stakes in such a response. At the same time, he and the committee's ranking Republican, Dick Lugar, wrote an opinion article asking what it would cost, "what level of support we are likely to get from allies in the Middle East and Europe," and "when Saddam Hussein is gone, what would be our responsibilities?" They observed, prophetically: "This question has not been explored but may be the most critical." They concluded: "We need to know everything possible about the risks of action and of inaction. Ignoring these factors could lead us into something for which the American public is wholly unprepared."[20]

Biden in opening the hearings seemed to accept the idea that the Iraqi dictator had the fearful weapons. Noting he had used them before, Biden said, "These weapons must be dislodged from Saddam Hussein, or Saddam Hussein must be dislodged from power. . . . If that course is pursued . . . it matters profoundly how we do it. . . ." A foreign policy involving the use of force "cannot be sustained in America without the informed consent of the American people," so a "national dialog on Iraq that sheds more light than heat" was imperative.

In a bipartisan approach, he said, the committee had coordinated the hearings closely with the White House and was "honoring the administration's desire not to testify at this time"—a rather soft way of saying it was kissing the committee off. Biden said he would "expect" the White House to later "send representatives to explain their thinking once it has been clarified and determined." In the meantime, he said, the committee would set out to determine what the threat was from Iraq, what the appropriate response would be, how Iraq's neighbors saw the threat, and "most important, if we participate in Saddam's departure, what are our responsibilities the day after?"

In anticipation of what proved to be the case, Biden said the committee had to "ask whether resources can be shifted to a major military enterprise in Iraq without compromising the war on terror in other parts of the world," and what the cost in American treasure would be for what outcome. "It would be a tragedy," he said, "if we removed a tyrant in Iraq, only to leave chaos in its wake." From these comments it was clear that Biden was not going blindly over a cliff in assessing his own position.[21]

Tony Blinken, his staff director on the Foreign Relations Committee who traveled ten times to Iraq with Biden, said the July–August 2002 hearings "were really the opening round in the entire national debate on Iraq. Until then there had been lots of rumblings; there was at that time of year a bit of a news vacuum and they dominated. They got a lot of coverage. The hearings were not prejudging anything. They asked the basic questions: What is the challenge posed by Iraq to the United States and our interests? What are the possible courses of action to deal with it? What are all the alternatives, from doing nothing to doing everything, and what are the implications of those different courses of action?"

The hearings, Blinken said, were "a very objective effort to look at the problem and look at the possible answers. It wasn't prejudging whether we should or shouldn't use force. Biden had reservations. At the very first hearing, he was extraordinarily prescient [in asking] in many ways what was the most important but least explored question: What do we do after we go in, if we go in? What do we do, not to win the war but to secure the peace? This is the hardest thing, and it's the thing we spend the least time talking about. And of course that proved to be exactly right."[22]

Biden wrote later that he hadn't asked the White House to send witnesses to the hearing "because I didn't want to force their hand. My intention in holding the hearings was to make public the disincentives to going to war in Iraq. I didn't want the president to get locked into going to war."[23] With no clear voice from the administration justifying the use of force in Iraq, no persuasive case was made for it in the hearings. Instead, they did generate more public debate over the stakes involved and the prospect for a drawn-out American involvement that should have injected a greater degree of caution.

As early as August 4, Biden himself was anticipating war, saying on NBC News's *Meet the Press* that "the only question is, is it alone, is it with others, and how long and how costly will it be?" He went on: "We're talking about the United States preemptively moving upon a country with tens of thousands [of armed forces]. The American people must be brought along. And ultimately that is going to be a responsibility that rests with the president, to be able to make that case."[24] Others were already pushing for a resolution requiring congressional authorization for use of force.

Notably, two respected veterans of the first Bush administration, Brent Scowcroft, its national security adviser, and James A. Baker, Bush's secretary of state, questioned the imperative of such force, especially unilaterally, raising speculation about whether they were channeling the current president's father. Meanwhile, Cheney resumed beating the drums, categorically telling the Veterans of Foreign Wars there was "no doubt that Saddam Hussein now has weapons of mass destruction" and was "amassing them to use against our friends, against our allies, and against us."[25]

Biden, Lugar, and Hagel joined in a Senate resolution that would give Bush authority to use military force to disarm Saddam Hussein only if he refused to do so himself as demanded by the United Nations, and then only after all other options had failed. It drew about seventy supporters and called on the White House to report its diplomatic efforts at the United Nations to Congress, and to make the case that his weapons were indeed a "grave threat" to the United States. "Mr. Bush ought to accept it," the *Washington Post* said editorially. "It would unite Congress behind him and offer a responsible way forward."

But the White House was having none of it, according to then Senate Republican leader Trent Lott. In his autobiography, *Herding*

Cats: A Life in Politics, he wrote that "Bush and Cheney made a special request: They wanted me to see what I could do to kill the Lugar-Biden-Hagel bill, and make certain it didn't pollute the strong language [for invasion] they required." Lott said that resolution "would have limited the authorization of military force to the destruction of WMD installations in Iraq," would have called for another round of inspections, and if the United Nations would not do so, "would have required the president to demonstrate to Congress that the danger posed by Iraq's WMD programs was such that only military action was adequate to the task."

Lott reported Bush's response: "My question is: What's changed? Why would Congress want to weaken a resolution? I don't want a resolution such as this—that ties my hands." Lott went on: "Bush's order to me grew more emphatic: 'Derail the Biden legislation, and make sure its language never sees the light of day again.'"[26]

Lugar urged Biden to get the full Foreign Relations Committee to approve their resolution, but of all people, outspoken antiwar Democratic Senator Paul Wellstone of Minnesota balked, as a matter, he said, "of principle." Biden wrote later of Wellstone's refusal: "Spare me the lectures. I thought our objective is to do all we can to avoid an unnecessary war."[27] Opposition to the president collapsed when House Democratic Leader Dick Gephardt persuaded his Senate counterpart, Tom Daschle, to urge fellow Democrats to vote for the Bush use-of-force resolution in October, providing a congressional green light for the eventual invasion of Iraq.

Gephardt at the time was a prospective 2004 Democratic presidential candidate. Senator John McCain, who had been supporting the Biden-Lugar resolution, phoned Biden and told him, according to a close Biden aide: "Joe, it's over. I can't be seen as being to the left of Dick Gephardt."[28]

With the midterm elections approaching, Gephardt argued that the party's chances to pick up seats might be enhanced by taking the war issue off the table. Then, the argument went, the case for

Democratic gains could be made more effectively on the basis of a weak economy—a futile and failed strategy of deplorable consequences.

Biden finally dropped plans to offer the alternative and, on the Senate floor on the day before the vote, announced he would back the Bush resolution, offering this rationale: "We should support compelling Iraq to make good on its obligations to the United Nations—because while illegal weapons of mass destruction do not pose an imminent threat to our national security, they will if left unfettered, and because a strong vote in Congress increases the prospects for a tough new UN resolution on weapons inspections, which in turn decreases the prospect of war."

Biden pointedly insisted that his vote was "not a blank check to use force against Iraq for any reason. It is an authorization to use force, if necessary, to compel Iraq to disarm, as it promised to do after the Gulf War." Biden said the resolution "does not make Saddam's removal its explicit goal" because doing so could alienate countries "who do not share that goal" and "weaken our hand at the United Nations." He said the United States still had the right to act alone but doing so "must be a last resort," and that support of Bush would increase the chances of getting from the Security Council "a tough new resolution that gives the weapons inspectors the authority they need to get the job done."[29]

On October 10, 2002, the Bush war resolution passed the House, 296–133, and the next day the Senate, 77–23, with Biden voting for it. Like his vote against the Gulf War in 1991, this decision in support of authorizing force against Iraq would haunt him politically long afterward. His observation that the United States had the right to act alone flew in the face of the UN Charter that authorized military action against another member state only in self-defense when attacked. Biden thus joined Senator John Kerry of Massachusetts, the party's eventual 2004 presidential nominee, and Senator Hillary Clinton of New York, who would seek the nomination in 2008, in voting for the Bush resolution on the same dubious grounds.

Biden meanwhile was running for his sixth term in the Senate, challenged again by Delaware businessman Ray Clatworthy, whose first try in 1996 had been marred by a political hired gun's effort to smear the senator with allegations of receiving a sweetheart deal from an MBNA executive on the sale of his house. This time around, the Clatworthy campaign seized on an impolitic Biden comment in October that the United States risked being called "a high-tech bully" in the Islamic world for its bombing of Taliban targets in Afghanistan. Biden never opposed the bombing but the rival campaign tried to make the most of the remark as another example of Joe Biden thoughtlessly mouthing off.

Clatworthy characterized the incident as Biden being the first U.S. senator to attack Bush's "conduct of the war on terrorism." He wrote that "the bombing continued and Senator Biden was widely repudiated for irresponsible speech." Clatworthy also charged that the Biden-Lugar resolution, which never was voted on, sought to "severely limit the president by all but forcing him to get UN approval to act against Iraq."[30]

Biden countered with the standard Democratic attack line of the time that Clatworthy favored the "privatization" of Social Security, through the president's proposed option of investing part of old-age insurance payroll taxes in the stock market. By whatever name, the option proved to be a political dud for Biden's opponent. The only issue for Delaware voters was whether they wanted to keep Joe in the Senate, and it was no contest. On election night, Biden won easily, 56 percent to 41. The notion of making Democratic gains in Congress by getting the war issue off the table also failed. The Democrats lost two seats in the Senate and seven in the House.

With strong congressional backing now in place for Bush's war resolution, the UN Security Council on November 8 passed by a vote of 15–0 another resolution calling on Saddam Hussein to readmit with dispatch the UN and International Atomic Energy Agency (IAEA) weapons inspection teams. Though no traces of nuclear, chemical, or biological weapons were found, the Bush administration continued to press the allegations against the Iraqi regime, with impatient warnings of military action steadily mounting.

Finally, Bush sent Secretary of State Colin Powell before the Security Council on February 5, 2003. Armed with information and pictorial support from the CIA and other U.S. governmental sources, he gave a lengthy and forceful case that Saddam Hussein had such weapons, and that he had one last chance to come clean or be forcibly disarmed. Biden was convinced that the weapons existed, especially after the Powell presentation. Powell himself, however, later regretfully labeled his own speech "the worst blot" on his career record, as it turned out he had been misled by the CIA and there were no such weapons to be found.

Biden recalled long after that when Powell's deputy Richard Armitage had appeared before the Senate Foreign Relations Committee, Biden asked him to convey to Powell "my strong view not to say anything you do not know for certain." Armitage replied, Biden said: "Mr. Chairman, message received." And, Biden said, Powell himself expressed hope that in going back to the United Nations "we may yet be able to avoid this war, and how bad would that be?"[31]

Biden mused long afterward, however, that Bush had apparently agreed to send Powell to the United Nations one more time with the idea of getting the weapons inspectors back into Iraq. But he would meanwhile continue to let Rumsfeld build up U.S. forces in the Gulf region to the point where they would have to be used before bad weather set in.

As Bush adamantly insisted there were no plans set for the invasion of Iraq, they went forward along with more talk of "regime change." When a report from UN and IAEA inspectors that no WMDs were located stiffened opposition to the invasion by key allies, Powell, encouraged by Biden and others in the Senate did persuade Bush to go back to the Security Council in one last bid to win explicit sanction for the invasion.

In an impassioned plea for seeking a second UN resolution, Biden in a *Washington Post* article on March 10 wrote it was imperative for reasons of allied solidarity. "France, Russia, and Germany are engaged in a game of dangerous brinksmanship at the United Nations," he said. "Together, they threaten to drive the interests of our countries over a cliff. There is still time to pull back from the precipice and disarm Iraq without dividing the Atlantic alliance and debilitating the Security Council."

But he charged at the same time that "for some in the administration, not going to war has never been an option, no matter what Iraq does," pointing to the calls for "regime change"—the ouster of Saddam Hussein that was not being universally demanded. He noted that it had not been mentioned in the UN resolution recently passed, and that insisting on it "has produced an unprecedented anger with our allies that is bound to corrode cooperation beyond Iraq, including cooperation in the war on terror."

While a second resolution was not legally required, Biden wrote, it was "a strategic one [that] would give political cover" for the resisting allies. "Invading and occupying Iraq under a UN rather than a U.S. flag," he wrote, "would minimize resentment" toward the United States.[32] But when it was clear a vote on a second war resolution would fail, Bush withdrew it and decided to proceed without it.

Asked in 2009 for this book why, considering he was aware neoconservative administration figures were bent on taking the country to war in Iraq with or without UN approval or acquiescence, he supported the invasion, Biden replied: "I thought Bush meant what he said. What everybody kind of forgets was that [when] 9/11 occurred, Bush responded with both reason, patience, and dispatch in going after al-Qaeda in Afghanistan. He didn't just jump the gun. He consulted everyone; he consulted with our allies, he reached out. Here was the question: Had Bush become an internationalist? Had Bush changed? Because he had so responsibly acted in making the decision on Afghanistan. So the idea that on into that spring and summer he was the cowboy, he was being the opposite. . . .

"Now, I was aware of the neocon view, but I was also aware that the administration had a San Andreas Fault running right down the center. For every neocon there was a traditional realpolitik Republican, starting with Colin Powell. Powell had told me, Lugar had told me, a number of people had told me Bush had not made up his mind in this struggle. And Bush personally told me that it was not his intention to go to war; it was not his intention to invade Iraq, and that he believed the only way that we could get a resolution from Saddam Hussein that he would begin to 'act responsibly' was if there was total unity in the American position, to demonstrate the resolve of America, then he would cooperate."

Biden said he had met with Bush three or four times, "constantly" with Powell, and with "a number of the generals who weren't anxious to go to war, at the time weren't looking to invade Iraq," as well as Powell's deputy at the time, Richard Armitage, and Senators Lugar and Hagel. "We all really thought Bush was in play, [that] he hadn't made up his mind. So I thought it would give him a stronger hand to get Saddam Hussein to act responsibly, and it was a very bad bet I made."[33]

Despite Biden's professed serious doubts, having signed on to support the president he did what he had done in 1991 regarding the Gulf War. He put aside those reservations and watched as Secretary of Defense Rumsfeld's "shock and awe" blitzkrieg against the forces of Saddam Hussein methodically pulverized them and brought about the downfall and flight of the Iraqi strongman.

When the assault opened with an air attack on the presidential palace in Baghdad on March 18, Biden was recovering from emergency surgery just a week earlier for removal of his gall bladder. He, Jill, and Ashley had been visiting his brother Jimmy's vacation home in Fort Myers, Florida, when the senator experienced abdominal pains and was rushed to a local hospital. He was released the next day. After his earlier health problems, including the brain aneurysm that had been life-threatening only fifteen years earlier, the episode was another unnerving one for the family, but he was soon back to work in Washington.[34]

Almost at once, as the "liberation" of Baghdad led to widespread looting and other chaos on the ground, Biden saw reasons to believe his preinvasion warnings that more U.S. force would be required to cope with the aftermath. A committee aide came upon a secret office established by Rumsfeld deputy Paul Wolfowitz in the Crystal City complex near the Pentagon. A makeshift government-in-exile was being set up under the notorious self-promoting Iraqi exile Ahmed Chalabi, who had predicted an easy transition amid a welcoming citizenry. The supposed shadow government, Biden wrote later, was composed of expatriates who "didn't seem to know much about the functions of the various ministries to which they were going to be assigned in Baghdad. And none of them had the foggiest notion about the conditions on the ground in Iraq—either pre- or postinvasion," even as those ministries were burning, being looted, and destroyed.[35]

Biden's concerns mounted when he received a visit from Paul Bremer, the man Rumsfeld chose as head of the Coalition Authority to replace the first U.S. official in charge in Iraq after the overthrow of Saddam Hussein. Bremer came across as so ill-informed that Biden, Lugar, and Hagel decided to go to Iraq themselves to assess the situation.

On arrival, Bremer told them he was planning to sell off some state-owned industries, a scheme that, Biden wrote later, "seemed to me a benighted, ideologically based decision that turned Iraq into an instant laboratory of Reaganomics in the Middle East. This made no sense at all. Iraq was little suited to this approach, and I thought the decision was apt to swell unemployment and provide more potential recruits for the insurgency."[36]

Bremer told his visitors that thirty thousand local police had already been trained, but a tour of the police training academy in Baghdad left Biden, who had seen similar places in Bosnia and

Kosovo, appalled at the ragtag troops and facilities he saw. "It would have been funny in a Keystone Kops sort of way had it not been so tragic," he wrote.[37] Upon return, Biden from the Senate floor gave his colleagues a further candid assessment that signaled his most serious break with Bush on the war up to then.*[38]

Conceding that the situation in Iraq was complicated, Biden said it had been made "much more difficult, frankly, by the wrong assumptions that were made by the administration. This is not second-guessing. These are things that, for a year before, many of us argued with [the administration] about. I supported us taking out that tyrant, but there seems to be a tone deafness right now, and that is that the administration thought building the peace would be built on three assumptions they had. . . . [T]hey expected to find a fully functioning bureaucracy when they got to Iraq, a literate country that would have in place each of their departments . . . that all we had to do was go in and decapitate the Baathists. . . . [W]e would have this infrastructure ready to take over the running of their country. But it melted away. It is not there."

The second assumption that melted away, he said, was for a

---

* The night before, the oldest member of the Senate, Strom Thurmond of South Carolina, had passed away at the age of one hundred. Biden preceded his Iraq remarks with an extemporaneous preview of his eulogy that, to the surprise of many, Thurmond, through his wife, had asked Biden to deliver at his funeral. Biden now spoke touchingly—and, of course, at length—of how the Senate "produces relationships that are a consequence of you looking at the best in your opponent, the best in the people with whom you serve, the best about their nature." These relationships, he said, "actually grow . . . because of the diverse views that are here and the different geography represented; if you are here long enough, it rubs against you. It sort of polishes you. Not in the way of polish meaning smooth, but polishes you in the sense of taking off the edges and understanding the other man's perspective."

Reciting how Thurmond's views on race had changed over the years, including voting for the Voting Rights Act and for the Martin Luther King Jr. national holiday, Biden said he believed in Thurmond's conversion not just because "there is a chemistry that happens in this body [but] because I believe basically in the goodness of human nature, and it will win out, and I think it did in Strom."

standing army that had instead turned into an insurgency. And the third was that similarly there would be a functioning police force to maintain public order. The Iraqis themselves had to perform these functions, Biden said, and until they could do so "we the international community should be filling the gaps, not we the United States alone. What is worse is we should have known better. We had extensive experience in the Balkans. We had considerable experience in Afghanistan, which is a failure in my view." He was not speaking out to say, "[L]ook at the mistake you made, you did not listen," he said. "It is to say, let's get over this. Now that we realize and the whole world understands these infrastructures do not exist, it is time to internationalize the effort."[39]

In October Biden made the same plea for greater international involvement in Afghanistan. "With our attention focused on Iraq," he wrote in the *New York Times*, "we run the risk of overlooking the alarming deterioration of security" there. He said the eleven thousand American troops in the country were not intended as peacekeepers, and the UN security forces should be expanded.[40] In early November Biden called for making "a U-turn away from the unilateral model we've been following for securing and rebuilding Iraq, by making the country "a NATO mission" with the American commander at the time, General John Abizaid, wearing both hats of command.[41]

It was clear at year's end in 2003, a little more than nine months after the invasion of Iraq, that Joe Biden had come to believe he had made a serious mistake in voting for George W. Bush's war of choice. On December 13 Saddam Hussein was captured, ignominiously hiding in a hole in the ground, but the insurgency had spread like wildfire, with Uncle Sam the chief firefighter. Gradually, it was dawning on Democrats that their own country might need a regime change of its own to end the war, and for Biden to consider what his role should be in achieving it.

TWENTY-FOUR

REASSESSING A QUAGMIRE

THE IDEA OF making another try for the presidency had never escaped Joe Biden's musings. As dissatisfied as he had been with the outcome of the disputed 2000 election that had put George W. Bush in the White House, he seemed in the wake of the 9/11 terrorist attacks to be content for a time to give him the benefit of doubt.

Although by mid-2002 Bush had already been moving toward a unilateral response to Saddam Hussein's defiance of UN sanctions, Biden said that he saw no "fundamental mistakes" in his Iraqi policy to cause him to challenge his reelection in 2004.

But when Bush invaded Iraq in March 2003 with inadequate plans to secure the occupied country, Biden could no longer say his differences with him were not fundamental. Still, even as admirers continued to urge a second try for the White House, Biden declared in mid-August that he would not run in 2004, gauging that he would be "too much of a long shot."[1]

Ever since he had been forced from the earlier presidential campaign by the allegations of plagiarism, Biden had, as Larry Rasky later put it, "worked very hard to rebuild his reputation of being thought of as the kind of person he thought of himself, as an honorable and substantive senator. By 2004 he had all that back." If he were to run again then, Rasky said, "there was a lot to put at risk—his quality of life, his stature, and Jill was against him taking all that risk."[2]

Instead, as the election year approached, a classic underdog candidate emerged in Governor Howard Dean, a Vermont medical doctor who burst onto the Democratic stage as a fiery antiwar critic. Dean proved to be a Roman candle, however, flaming out in the early Iowa caucuses, replaced as the party front-runner by Senator John Kerry of Massachusetts, Biden's colleague on the Senate Foreign Relations Committee. Speculation spread that Joe Biden would make an ideal secretary of state for him. That wasn't the presidency, but it would be a decent alternative, if offered.

When Kerry quickly wrapped up the Democratic nomination in March, Biden two days later proposed that he choose Republican Senator John McCain as his running mate on a unity ticket. The idea got some initial traction because Biden and Kerry often talked on matters of foreign policy. But McCain, after first saying he would "entertain" the idea, observed that he couldn't imagine the Democrats accepting "a prolife, nonprotectionist, deficit hawk" and the matter went no further.[3]

Kerry, who also had voted for the Bush war resolution, like Biden had now become an outspoken critic of the war of choice. Both men had come to regret deeply their 2002 votes for the war resolution. It was not simply the failure of inspectors to find the weapons of mass destruction that were Bush's prime rationale for the invasion. In Biden's case particularly, the deliberate deceptions that he had failed to see through early enough gnawed at him.

He wrote in his memoir that he had "vastly underestimated" the influence, disingenuousness, and incompetence of Cheney, Rumsfeld, and "the rest of the neocons" in selling Bush on the war without seeking "the informed consent of the American people."[4] Actually, it was their *competence* in doing so that deserved the credit, and Biden in so saying seemed to cast the president of the United States as an unknowing or unthinking puppet in the hands of his subordinates.

In any event, in Biden's role as the former chairman and now ranking Democrat on the Senate Foreign Relations Committee, he set out with his customary intensity and zeal to salvage what he could from the fiasco. As the administration pressed on with the war as a fully owned subsidiary of Uncle Sam, Biden used committee hearings to goad Bush and advisers to internationalize the burden.

The designated Coalition Provisional Authority was now staggering toward a promised June 30 date for restoration of Iraqi sovereignty. In three committee sessions in April, Biden hammered at reducing if not removing the American face on the reconstruction. And at subsequent confirmation hearings of Bush's choice to be ambassador to Iraq, veteran diplomat John Negroponte, Biden argued that only if the effort to put Iraq on its feet was seen as the will of its people, and not imposed by the United States, could it succeed.

"We cannot want a representative government for the people of Iraq more than the Iraqi people want a representative government," he began. "Furthermore, the very institution, the very entity most vital to Iraqi success, the United States, is going to be seen as something that cannot be embraced. They are not going to kiss us in public." In Biden's penchant for restating the same point, he told Negroponte that "the higher your profile . . . the more they're going to want to keep you at arm's length so they don't appear to be doing the bidding, whether they are or not, of the United States."

Citing a recent European trip he and Hagel had made, Biden said the disaffection there was obvious. He asked rhetorically: "Why would anybody want to help [the reconstruction]? They didn't like what we did, they didn't like the way we did it, they don't like the way we're doing it now, so why would they possibly come along and help?"[5]

Sometimes this repetitious manner of argument irritated other committee members, feeding the old complaint that Biden talked too much and too long. But he certainly made himself clear with his conversational pitches, often peppered with colloquialisms. At one point, he asked Negroponte "are we going to go from the image of Clark Kent to Superman? You know, we're not wearing our 'S' in this coalition. You're [going to be] the Super Ambassador." It might be more useful for Negroponte, he said, to be seen as part of a "much broader coalition of countries with the most at stake in Iraq, and a representative who speaks in their name, [to have] that foil at least, so that it is not just you, not just the United States. . . . Otherwise, we will be continued to be viewed and blamed for everything that goes wrong. We will continue to be viewed as the target of every malcontent of the country."[6] There never was any doubt where Joe Biden stood when he finally stopped talking.

After Negroponte was confirmed, the Foreign Relations Committee held another hearing in anticipation of the planned transfer of sovereignty. Biden zeroed in on security of the population, questioning administration officials about what he saw as the inadequacy of the training of Iraqi police forces. The exchange was a prime example of Biden's talent for weaving in aspects of his own experience to buttress the points he was making.

"Those of us who do foreign policy as a major part of our occupation, we like to make things sound really complicated all the time to people," he began. He noted that as a member of the Senate Judiciary Committee he spent as much time "dealing with criminal justice issues—the mob here in the United States of America, the drug cartels. . . . It seems to me the situation, whether it's Samara, whether it's Baghdad, whether it's Basra, Mosul, wherever it is, is the same exact thing that exists in any large city in the United States of America dealing with a major drug cartel, the mob, and/or a crime wave," he said. "There is not a single difference between someone in Baghdad and someone living in South Philly, if they believe they can't walk outside their house without fear of something very bad happening to them."

The Bush administration, Biden told the officials, didn't seem to grasp that the situation was the same in Iraq. The authorities in both cases, he lectured, had "to flood the zone with cops who get to know the people on the block. That's what community policing is all about." And the police had to be sufficiently armed, he said, to get their job done in the face of mob firepower.

In Biden fashion, he concluded by telling the administration security officials of the occasion he was trying to squire a tough anticrime bill through the Senate. He was riding late one night in a Chicago police car, he said, and the driver told him: "Last night I was coming home and I got a call there was a major drug deal apparently going down on one of the piers on Lake Michigan. So we drove into this alley, came through, opened up, got out of our cars in the usual form, like Starsky and Hutch. . . . All of a sudden, 'Everybody freeze!,' we shined the lights on them, they popped open their trunk, took out high-caliber weapons that could literally blow our car away. We said, 'No problem,' got in the car and we backed out. You got to have the same firearms. You got to have the same capability."[7]

It was Joe Biden at his storytelling best—obvious and perhaps too long-winded and overdramatized, but the message was unmistakable. As his sister, Valerie, and other defenders of the Biden style often remarked, audiences listened when Joe spoke, and seldom walked out no matter how long he went on or how repetitious he was. He had the Irishman's natural knack, whether in normal conversation, making a speech, or engaging in a heavy disputation at a committee hearing.

At the Democratic National Convention in Boston in late July, Biden gave Kerry what was for him an unusually brief but ringing endorsement. He said Kerry could be depended on to keep America strong and safe, but he never would have gone into Iraq accepting the brunt of the cost and casualties. Diminishing Bush's "coalition of the willing," Biden declared: "Instead of walking alone, we must lead. It's only leadership if others follow. . . . When John Kerry is president, military preemption will remain as it always has been an option, when we face a genuine imminent threat. But John Kerry will build a true preemption strategy—to defuse dangers long before the only choice is war."[8]

In debating Bush in the fall, Kerry often offered views that coincided with Biden's own, sometimes in similar words, suggesting a collaboration that led a *Wilmington News Journal* reporter, Sean O'Sullivan, to speculate that Biden was on Kerry's vice presidential short list.

For example, Biden, alluding to Bush's deception in launching his war of choice in Iraq, had said in his convention speech: "Forty years ago, during the Cuban missile crisis, President Kennedy sent former secretary of state Dean Acheson to Europe to seek support. And Acheson explained the situation to French President de Gaulle. He then offered to show President de Gaulle classified information as proof of what he said. And you know what de Gaulle did? He raised his hand and said, quote, 'That is not necessary. I know President Kennedy, and I know he would never mislead me on a question of war and peace.' I ask you, would a single foreign leader react the same way today?"

Kerry in the September 30 presidential debate said: "I mean, we can remember when President Kennedy in the Cuban missile crisis sent his secretary of state to Paris to meet with de Gaulle. And in the middle of the discussion, to tell them about the missiles in Cuba, he said, 'Here, let me show you the photos.' And de Gaulle waved them off and said, 'No, no, no, no. The word of the President of the United States is good enough for me.' "[9]

The Wilmington reporter matched other Kerry remarks in the debate with observations Biden had made. They suggested that Biden had Kerry's ear on foreign policy and might have been the 2004 Democratic nominee's choice for his running mate, or for secretary of state, in a Kerry administration. Instead, Kerry chose Senator

John Edwards of South Carolina to run with him, and he never had a chance to pick a secretary of state.

In mid-September, once the transfer of national sovereignty had occurred in Iraq, the Foreign Relations Committee met again to assess progress in stepping up U.S. aid to the reconstruction. Biden, almost beside himself, noted that the count of American dead had just passed one thousand, and he decried Bush's minimizing of his administration's failures as "miscalculations."

"I don't see any learning curve from the repeated mistakes in judgment we've been making," he said, noting that only about $1 billion of $18.4 billion appropriated for reconstruction had been spent in the last year. He cited a nightmare of city streets flooded with sewage, constant power outages, insufficient police and training, failure of burden-sharing among UN partners, and above all no exit strategy.

As he had done in the past, Biden talked of his special impatience coming from Delaware, where "[t]he Dover Air Force Base is the place that every single coffin out of Afghanistan and Iraq sets on U.S. soil first. We owe it to those young men and women to get this right. We owe it to them to have a plan," he lamented. For once, he apologized to his committee colleagues for going on at such length with what he had said before and for his frustration. "In different ways, I'm sure you all share it. You are just better at being able to articulate it than I am," he said. "But I am really frustrated, because I think we're at the last piece of that rope. We're hanging on. We can still climb that rope, but man, there's not many more handholds on that rope, and we had better get it right."[10]

By this time, one potential handhold for pulling not only Iraq but also the United States out of the quagmire—the presidential election at home—was itself slipping away. The Bush strategists had once again come up with a disingenuous but effective smear against Kerry, a decorated skipper of an armed river vessel in the Vietnam War. A group calling itself Swift Boat Veterans for Truth started running a television ad challenging his right to his combat decorations. Kerry instead of angrily defending himself did not sufficiently respond, and the ad took its toll. In the end Bush, capitalizing on voters' fear and misplaced sense of patriotism regarding his war of choice, celebrated election to a second term, assuring a continuation of that war.

This being so, Biden was back in Iraq as the new year began, to assess the results of the election in which a new Transitional Assembly was elected. With Bush's reelection came a modest remaking of his administration, and with it further watchdog responsibilities for the Senate Foreign Relations Committee. The most significant change was the jettisoning of Colin Powell as secretary of state, the loser to Cheney and Rumsfeld in winning Bush's ear and his collaboration in pursuit of the war.

In the confirmation hearings of Powell's successor, Condoleezza Rice, moving up from her first-term post as Bush's national security adviser, Biden gave her a hard time as a defender of the war, but then voted to confirm her. Afterward, when he was asked why in an interview with *Rolling Stone* magazine, he said "it's not because I have confidence in Condi to execute foreign policy. I don't. But the president is entitled to choose the people he wants to surround himself with, even if I think his views are cockamamie."

It was a rationale he had not used in voting against Bork and Thomas, but neither of them, theoretically at least, was going to be working for the president. Biden, however, expressed his fears of what the second term would bring. "It may be worse," he told the magazine. "The moderate voices are now gone. The president pointedly got rid of Colin Powell and the other centrists. The only people left are the neocons."[11]

That certainty in turn convinced Biden that the time had come for him to seek the presidency again. In March 2005, there was another gathering of the clan—the family and Biden's closest political advisers—in Wilmington. According to Larry Rasky, who was present, "Jill was very shaken by John Kerry's loss, and that George W. Bush had been reelected. We were inexorably moving toward running in 2008. Joe as chairman of the Senate Foreign Relations Committee was heading the Democratic foreign policy apparatus, like a shadow government, a government in absentia."[12] According to John Marttila, Biden held off as long as his friend Kerry might run again, and Marttila told him that since he had been with Kerry in the 2004 campaign, he felt obliged to stay with him if he ran again. But when Kerry decided he would not, Biden, dismayed as he was about Bush's disastrous foreign policy course, called Marttila and told him he was ready to try for the presidency again in 2008 and wanted him aboard.[13]

Although the next presidential election was more than three years away, Biden in an appearance on CBS News's *Face the Nation* in June 2005 said, "it is my intention to seek the nomination. I know I'm supposed to be more coy with you. I know I'm supposed to tell you, you know, that I'm not sure. But if, in fact, I think that I have a clear shot at winning the nomination by this November or December, then I'm going to seek the nomination."[14]

The statement reflected not only Biden's candor but also his awareness, after 1987, that this time around he would need a much more deliberate approach, with a better appreciation now of what the undertaking would require of him. Actually, he said, he had been proceeding since the election the previous November "as if I were going to run," assessing his chances, and over the next six months would be out around the country gauging his political strength.

Although Biden continued to support the war in Iraq, he made clear that the president's conduct of it would be in his sights as a candidate. "I think the administration figures they've got to paint a rosy picture in order to keep the American people in the game," he said, "and the exact opposite is happening." He predicted that "if nothing changes here, we're going to be out of Iraq by the end of 2006 as a nation that has been viewed by the rest of the jihadists in the world as having been pushed, which is a very bad thing for us."[15]

At the same time, Biden resisted the notion, raised in the fall by a prominent military hawk, Representative John Murtha of Pennsylvania, for an immediate drawdown of American forces in Iraq. In a speech to the Council on Foreign Relations in New York, Biden said he shared Murtha's frustration but was "not there yet." While he had already abandoned an earlier call for even more troops, he said "the hard truth is that our large military presence in Iraq is necessary." However, he said, Bush should work harder at diplomacy, bringing together the ethnic rivals in the country and training more Iraqis to take over the task of providing security, instead of trying to create a model democracy in Iraq.[16]

It was a position that was not likely to win political allies among the Democratic Party's antiwar liberals who were increasingly clamoring for an end to the war. But the view was shared by the early 2008 front-runner for the nomination, Hillary Clinton. This situation did not go unnoticed in the *Nation*, the liberal antiwar magazine, where an article called both of them "enablers" of the Bush policy. "The prominence of party leaders like Biden and Clinton," it said, "and of a slew of other potential prewar candidates who support the U.S. invasion and occupation of Iraq, presents the Democrats with an odd dilemma. At a time when the American people are turning against the Iraq War and favor a withdrawal of U.S. troops, and British and American leaders are publicly discussing a partial pullback, the leading Democratic presidential candidates for '08 are unapologetic war hawks. . . . It's more than a little ironic that the people who got Iraq so wrong [by voting for the Bush war resolution] continue to tell the Democrats how to get it right."[17]

Such harangues from the left branch of his party did not, however, detour Biden from assessing whether he had the support, political and financial, to have a reasonable shot at the next Democratic presidential nomination. In the fall of 2005, he faced another confirmation task that might have deterred him, when Supreme Court Justice Sandra Day O'Connor retired. Bush appointed another federal appellate judge, John R. Roberts, to replace her, but before his confirmation hearings could be held, ailing Chief Justice William Rehnquist died in office; three days later Bush appointed Roberts to take the highest seat.

Unlike the 1987 nomination of Judge Robert Bork, which had disrupted Biden's first presidential campaign, the Roberts confirmation only gave Biden more public visibility that fall; this continued in January, when Bush's nomination to take the O'Connor seat came before the Senate Judiciary Committee.

In Roberts's confirmation for chief justice, Biden took the opportunity to engage him in the same examination of the right of privacy not enumerated in the Constitution that had served the senator from Delaware so well in the Bork hearings. He told Roberts "there is a genuine intellectual debate going on in our country today over whether the Constitution is going to expand the protections of the right to privacy, continue to empower the federal government to protect the powerless," or be arrested by the "very narrow reading of the Constitution" by so-called strict constructionists—like Bork.

"In every step we have had to struggle against those who saw the Constitution as frozen in time," he said, arguing that "our journey of progress is under attack, and it is coming from, in my view, the right. . . . [We] need to know whether you will be a Justice who believes that the constitutional journey must continue to speak to these consequential decisions [involving privacy], or that we have gone far enough in protecting against government intrusion into our autonomy, into the most personal decisions we make."

Biden then let Roberts know where he was coming from, if the nominee had any doubts. "Judge, if I look only at what you have said and written, as used to happen in the past, I would have to vote no. You dismissed the constitutional protection to privacy as 'a so-called right.' You derided agencies like the Securities and Exchange Commission as 'constitutional anomalies.' And you dismissed gender discrimination as 'merely a perceived problem.' This is your charge, Judge, to explain what you meant by what you have said and what you have written."

In his introductory remarks, Biden treated Roberts with a dose of his seriousness and his legal depth that often was obscured by his folksy manner, friendly smile, and long-windedness. And he wrapped it in a sober observation about what was at stake for Roberts and for the people he would serve on the highest court. It was a point he had previously made to Bork. "The Constitution provides for one democratic moment, Judge, one democratic moment before a lifetime of judicial independence," he said. "This is that moment, when the people of the United States are entitled to know as much as they can about the person we are entrusting with safeguarding our future and the future of our children and grandchildren."[18] It was a somber thought, aimed at prodding a guarded witness to be more forthcoming than self-interest might dictate.

In Roberts's opening statement, and in Biden's subsequent effort to draw him out, the nominee—a mild-mannered, mild-looking, soft-spoken man—demonstrated he was unlikely to be soft-soaped into revealing anything damaging to his nomination chances. He laid out a deceptively simple and appealing description of his job, as he saw it to be. "Judges are like umpires," he said. "Umpires don't make the rules, they apply them. The role of an umpire and a judge is critical. They make sure everybody plays by the rules, but it is a limited role. Nobody ever went to a ball game to see the umpire. Judges have to have the humility to recognize that they operate within a system of precedent shaped by other judges equally striving to live up to the judicial oath, and judges have to have the modesty to be open in the decisional process to the considered views of their colleagues on the bench."[19]

The nominee concluded by saying he came to the committee "with no agenda . . . no platform," but with a commitment to "confront every case with an open mind," segueing again into his baseball metaphor: "I will decide every case based on the record, according to the rule of law, without fear or favor; to the best of my ability, and I will remember that it's my job to call balls and strikes, and not to pitch or bat."[20] In other words, no "legislating from the bench" for this job applicant.

In questioning Roberts the next day, Joe Biden, the old baseball (and football) star at Archmere Academy, showed he was no stranger to the national pastime himself. He opened by massaging the witness, telling him he had "hit a home run yesterday" in his opening testimony, and that the conductor and others on Biden's train ride home that night had commented favorably, "Oh, he likes baseball, huh?"

Then he gave the judge a taste of his own medicine: "As you know, in Major League Baseball they have a rule—Rule 2.00 defines the strike zone. It basically says from the shoulders to the knees. And the only question about judges is, 'Do they have good eyesight or not?' They don't get to change the strike zone. They don't get to say that was down around the ankles, you know, and I think it was a strike. They don't get to do that."

Biden continued: "But you are in a very different position as a Supreme Court Justice. As you pointed out, some places of the Constitution define the strike zone—two-thirds of the senators must vote; you must be an American citizen, to the chagrin of Arnold Schwarzenegger, to be president of the United States—I mean born in America to be a president of the United States. . . . [T]he strike zone is set out." But in the case of "unreasonable search and seizure [in the Fourteenth Amendment], what constitutes unreasonable? So, as much as I respect your metaphor, it is not very apt because you get to determine the strike zone. . . . Your strike zone on reasonable or unreasonable may be very different from another judge's view. . . ."

What he was talking about, Biden said, was "the Liberty Clause of the Fourteenth Amendment. It doesn't define it. All the things we debate about here, and the Court debates, the 5–4 decisions, they are almost all on issues that are ennobling phrases in the Constitution that the Founders never set a strike zone for. You get to go back and decide."[21]

From all this, Biden finally got Roberts to say he did agree there was "a right of privacy to be found in the Liberty Clause of the Fourteenth Amendment," and one that "extends to women." Well, what about abortion? Biden asked. Roberts replied: "Well that is in an area where I think I should not respond"—a matter that might come before him for ruling on the Supreme Court. A long dialogue ensued over when a judicial nominee should or should not respond to a question touching on a possible future requirement to rule. Biden kept pitching and Roberts kept fouling the pitches off, until Biden's allotted time to question the witness ran out.[22] In the end, Roberts in this particular at-bat got what amounted to a base on balls from the committee, and was confirmed by a comfortable 78–22 vote of the full Senate.

By the time Biden reached a self-imposed deadline for deciding whether to make a serious second bid for the presidency at the end of 2005, nothing had occurred to dissuade him. In Bush's nomination of yet another conservative federal appeals judge, Samuel Alito, to fill the O'Connor seat that initially was to go to Roberts, Alito had no baseball metaphor to rely on when his confirmation hearing began in the second week of the new year.

Biden in his introductory comments focused, as he had in the Bork nomination hearings, on his concern that the court's existing balance would be disturbed. Alito would be taking the seat of moderate Justice O'Connor, who had been the swing vote in many 5–4 decisions. But after a few stabs at Alito for his membership in a conservative club as a student at Princeton, and a few questions about gender and racial discrimination, Biden subsided and Alito was easily confirmed.

Biden surprised some of his committee colleagues shortly afterward by saying on NBC's *Today* show that he thought Senate hearings no longer produced any useful information and should be scrapped, with nominations going directly to the full Senate for confirmation or rejection.

"The system's kind of broken," he said. "Nominees now,

Democrat and Republican nominees, come before the U.S. Congress and resolve not to let the people know what they think about the important issues," especially on a president's authority to take the country to war. "Just go to the Senate floor and debate the nominee's statements, instead of this game," he said, sounding frustrated.[23]

The man who had succeeded Biden as Judiciary chairman, Republican Arlen Specter of Pennsylvania, noted rather acidly that Biden "never wanted to dispense with the hearings when he was chairman."[24] In fact, some of Biden's most impressive moments had come in his astute and well-informed, if marathonlike, interrogations of judicial nominees.

With his 2008 presidential ambitions still on track, but with Hillary Clinton garnering most of the speculation as a prospective juggernaut in the Democratic Party, Biden in 2006 began to carve out an identifying position for himself on how to extricate the party and the country from the morass of Bush's war in Iraq. In May, he joined a president emeritus of the Council on Foreign Relations, Leslie Gelb, a former *New York Times* foreign policy writer, in authoring a plan to deal with the ethnic divisions in the country, and with the militias that were tearing the country apart.

They proposed a five-part solution that would create a federation within Iraq of autonomous ethnic Shiite, Sunni, and Kurdish regions. The Sunnis would be enticed to participate with a share of oil revenues to sustain their region. The rights of ethnic minorities and women would be guaranteed by the United States. And a gradual U.S. withdrawal would be orchestrated by 2008, with a UN or other international conference assembled to recognize the borders and support the federation. The plan would be modeled after the successful arrangement that had ended ethnic cleansing in Bosnia under UN guidance.

The scheme generally met with mixed reactions, most of them lukewarm, or criticisms of impracticality. It was later undermined by a recommendation of the bipartisan Iraq Study Group that the

United States continue to support a strong central government in Baghdad. But at least it gave Biden an alternative to the pullout that Murtha and others proposed, and to the Bush stay-the-course policy that seemed to promise an endless American presence and involvement in Iraq.

Meanwhile, Bush and his Republican Party were paying an increasingly high political price for their failure to extricate the United States from the Iraq nightmare. In the November congressional elections, the Democrats seized control of both houses of Congress and put Joe Biden back in the chair of the Senate Foreign Relations Committee for the next legislative session in January 2007. In that election result, Larry Rasky said later, "Biden played a role and gave the Democrats a place to hide from the charge that we were the party of cut and run." And politically for him, Rasky said, "Jill was aboard [for another presidential run in 2008]. She really felt now that Joe was right for the job, not just a good candidate, and right for the time."[25]

Jill said later, "Actually, I feel like I was the one who encouraged him to run. I had spoken to the boys the year before, when, on the way home from Nantucket, we were on the ferry, and I said, 'Look what Bush is doing to this country. Look at this war we're into. Dad's the only one who can change things.'" And when she learned that Beau himself might go to Iraq, she said, she felt all the stronger that her husband should run for president again.[26]

In late December, Biden said he would hold more hearings on the war and would oppose a Bush plan to send as many as thirty thousand additional U.S. troops into Iraq in a temporary "surge" to stem escalating sectarian violence and beef up security around the country. At the outset of the war, Biden had argued that Bush had sent insufficient forces into Iraq to deal with the invasion's chaotic aftermath of insurgent strife and ethnic conflict. Now, with violence receding in the wake of achieved ethnic cleansing by the rival sects and with a sudden Sunni "awakening" against al-Qaeda, Biden contended the surge was not needed and would only prolong the time before the new Iraqi government would assume security of the country. "We've tried the military surge option before, and it failed," he said on CNN. "If we try it again, it will fail again."[27]

But neither did Biden align himself with those in his own party who wanted either to pull out or set a firm timetable for U.S. withdrawal from Iraq. When liberal Democrats in Congress in 2007 sought to cut off funding for the war, Biden balked at doing anything that would jeopardize the safety and support of the military in the field. He was particularly irate, his aide Tony Blinken recalled, "at the failure of the Bush administration to provide our troops with proven life-saving technology, especially mine-resistant vehicles." After investigating Pentagon foot-dragging, he said, and having pushed funds through the Senate to accelerate their delivery, Biden refused to use denial of funding as a tool to end the war.[28]

Nevertheless, he remained committed to achieving that objective in a way that would do harm neither to the American forces serving in Iraq nor to the Iraqi people upon whom Bush's disastrous misadventure had been visited. With that goal in mind, Biden confirmed in January that he would indeed make a second bid for his party's presidential nomination in 2008, and this time around would not be like the first. Unlike 1987, he would not to be distracted by events or the political counsel of others. His father's admonition to him as a boy—when knocked down, "Champ, get up!"—would govern Joe Biden's next campaign.

TWENTY-FIVE

BIDEN FOR PRESIDENT AGAIN

ON JANUARY 7, 2007, Joe Biden chose the top Sunday morning television talk show, NBC's *Meet the Press*, to confirm what had been obvious for months, after more than two dozen trips to presidential primary and caucus states. "I am running for president," he said. "I'm going to be Joe Biden, and I'm going to try to be the best Biden I can be. If I can, I've got a shot. If I can't, I lose."[1]

To anyone who had been paying the least bit of attention to his political fortunes, it was clear what he meant. After the crash of his first presidential bid twenty years earlier, Biden was going to run his own way. It was an unveiled rap at the small group of outside political consultants and advisers who had populated his first presidential campaign, and at Joe Biden himself for allowing them to dominate his strategy and other aspects of a frenetic and disorganized effort.

For all that, key loyalists from the previous White House bid were back—all the Biden family members, unofficial family like Ted

Kaufman, and political family like John Marttila and Larry Rasky, plus newcomers, including former John Kerry campaign press secretary David Wade and longtime Democratic foot soldier Anita Dunn of the 2000 Bill Bradley team.

In a sense, the long shot Biden bid was for some a nostalgic running out the string of a near-impossible dream. The candidate himself seemed well adjusted to the realities, consoled that running again was a low-risk opportunity to use his long experience to contribute to the presidential campaign dialogue.

The significance of the promise to be "the best Biden I can be" was not lost, either, on followers and supporters of his candidacy. They all knew, and particularly at home in Delaware, that the highest standard members of the Biden clan could impose on themselves was the one pounded into them by Joe's parents and grandparents—honesty, integrity, pride, hard work. Other families no doubt set the same standards for themselves, but for the Bidens it was almost a religion, which in itself held the family to strict account. In their relations with each other, and in their Catholic faith, the Bidens offered themselves as the ultimate in veracity. When Joe promised anything "as a Biden," as he often did, he meant you could go to the bank with it.

That was one reason why getting out of the 1988 presidential campaign so early—in the wake of those allegations of plagiarism—was so personally hurtful. The later exoneration after a court review of the circumstances eased the pain somewhat, but the notion that he, Joe Biden, could have had his integrity questioned went to the core of the family pride. This time, vowing to be the best Biden he could be carried with it a particular connotation.

Biden was joining three other Democrats who had already officially announced their candidacies: former senator John Edwards of North Carolina, Governor Tom Vilsack of Iowa, and Representative Dennis Kucinich of Ohio. None of them could match Biden's experience. The undeclared Democratic candidates he had to be most concerned about were Senators Hillary Clinton of New York, already the heavy favorite, and Barack Obama of Illinois, a rising star in the party but an unknown political quantity on the national scene.

At the time Biden had decided to run again, there had been some question in his camp whether Hillary Clinton would do so, with uncertainty over the impact her husband, the former president, might have on her chances. "Joe didn't know whether that would keep her on the sidelines," Larry Rasky remembered, "but I don't think that entered into his thinking. The field never really distracted him."[2] As for Obama, Biden had dismissed him in a 2007 television interview on the New England Cable Network, saying "I'd be a little surprised if he actually does run," and observing that Obama was on "everyone's number-two list," referring to the vice presidency.[3]

As chairman of the Senate Foreign Relations Committee and a veteran globe-trotter, Biden was making himself a prominent voice in the Democratic Party with frequent criticism of the war in Iraq. He and Clinton, along with Edwards, had facilitated that war with their votes for the Bush use-of-force resolution in 2002. Obama, however, at that time was in the state senate in Illinois running for a U.S. Senate seat, and he had denounced going into Iraq as "a dumb war" he could not support. Biden by now had also come to see the war as a huge mistake, and he was working hard to position himself as the toughest critic of how it was being waged by Bush.

Two days before his appearance on *Meet the Press*, Biden had charged that "a significant portion of this administration . . . believes Iraq is lost" and that it had "no answer to deal with how badly they have screwed it up." The Republican administration was just hanging on, he said, so that the next president would "be the guy landing helicopters inside the Green Zone, taking people off the roof," in the manner in which the U.S. involvement in Vietnam had ended nearly three decades earlier.[4]

About a week later, Biden joined Republicans Chuck Hagel and Olympia Snowe of Maine and Democrat Carl Levin of Michigan in sponsoring a nonbinding Senate resolution opposing Bush's latest plans to send an additional twenty thousand troops into Iraq to deal with the disintegrating security there. A House version passed, 246–182, but the Senate resolution fell four votes short of imposing cloture and thus was dropped.

After formally declaring his candidacy on January 31, Biden said on ABC News's *Good Morning America* that "I'm not exploring. I'm in, and this is the beginning of a marathon." He told reporters he believed "I can stem the tide of this slide [in Iraq] and restore America's leadership in the world and change our priorities. I will argue that my experience and my track record, both on the foreign and domestic side, put me in a position to be able to do that."[5] But the polls at this juncture showed him distantly trailing both Clinton and Obama.

Joe Biden certainly could be persuasive, no doubt about that. As the Irish in Boston would say, he could talk a dog off a meat wagon. But he did have that penchant for putting his foot in his motormouth, so much so that by now everything he said would be put through a strainer by the gaffe police of the press. And he didn't have long to wait.

On the night before the announcement of his candidacy on morning national television, Biden gave an interview to a reporter for the weekly *New York Observer*. He talked mostly and at length about his differences with Hillary Clinton and John Edwards on the Iraq war. Near the end of the *Observer* story, he was quoted as saying as a throwaway line about the appeal of Obama: "I mean, you got the first mainstream African American, who is articulate and bright and clean and a nice-looking guy. I mean, that's a storybook, man."[6]

Biden went on to express doubt that the American people would elect "a one-term, a guy who has served for four years in the Senate," adding, "I don't recall hearing a word from Barack about a plan or a tactic [for Iraq]." And then he plunged into a long denunciation of another potential foe, saying, "I don't think John Edwards knows what the heck he is talking about."[7]

The brief article by reporter Jason Horowitz relegated the remark about Obama to the eleventh paragraph of a twenty-six-paragraph story, but it did not escape the eyes of alert reporters. Around midnight on the night before Biden's announcement on national television, Larry Rasky was sitting in the bar at the Grand Hyatt in Washington when an account of the *Observer* story popped up on his BlackBerry. The lead on Biden's criticisms of Clinton and Edwards on Iraq, which was why the interview had been scheduled, was lost in the buried words on Obama. When Rasky was awakened early the next morning by a call from Dan Balz of the *Washington Post* asking for a comment on his candidate's description of Obama, the feeling of "here we go again" immediately struck the aide who had lived through the 1987 debacle in Iowa.[8]

The gaffe police, sensing a pattern, were immediately on the case.

"Articulate" and "clean"? A black man? Was Biden being condescending? Or sarcastic? Or just plain insensitive? In his hometown paper, the *Wilmington News Journal*, the next day's story bore the headline: SEN. BIDEN STUMBLES OUT OF GATE IN '08 RACE

Rasky for one knew what was coming. "The day the campaign ended was the day it started," he said later, "because we had a lot of serious fund-raisers looking at us as an alternative to Hillary and Edwards." Until then, he said, "we felt we could be competitive financially." That week *Time* had run a cover picturing four Republicans and four Democrats—Clinton, Edwards, Obama, and Biden—"and we felt leading up to the announcement day that we were positioned to move into the first tier. Having to play defense on that story on announcement day was like '87 all over again—your worst nightmare."[9]

It so happened that Biden had a hearing of Senate Foreign Relations Committee that morning, focused on Iraq. When the *Observer* story was first picked up by the scandal-mongering Drudge Report online, it was billed as a "Biden Shocker" and focused on Biden going after Hillary Clinton and Edwards. But that didn't matter, as the questions on Obama's articulation and cleanliness came pouring in. At first Biden didn't want to talk about it, but he finally said he had been thinking of one of his mother's expressions: "Clean as a whistle, sharp as a tack."[10]

Intentionally, the first day had been planned to focus on Iraq and policy rather than politics. There was no fly-around in Iowa or visit to New Hampshire, but by the time Biden and Rasky headed for New York and the appearance with Jon Stewart on Comedy Central's *The Daily Show*, scheduled well beforehand, Rasky said, "we were already in full-bore damage control."[11] On the show, Biden tried to explain: "What I was attempting to be, and not very artfully, is complimentary. The word that got me in trouble is using 'clean.' I should have said 'fresh.' What I meant was he's got new ideas."[12] But the sharks were circling.

"Anybody who has been around Biden knows he doesn't have a racist bone in his body," Rasky said. Nevertheless, "it was a downer. But Joe is the most resilient son of a bitch in the world, and he'd certainly been hit harder than this," he said. "We were just trying to navigate through this first week, and once we got out of this cycle and things started to settle down, get back in stride and piece things back together.

"The problem, though, was money. The Biden money was there but the big national money deserted us once their suspicions about Biden were realized. The loss of a major fund-raising event to Hillary Clinton in New Jersey drove a stake in our heart and cost us a couple of million dollars. And then when Obama reported the ridiculously large second-quarter numbers [for 2007], that put us in our place and we were scrambling the last six months.

"We always felt John Edwards was in our space, and if we could get into that space, we could have gotten the job done. I always thought Hillary and Obama liked having Edwards where he was, especially Hillary. They never thought Edwards was a real threat. They thought he was a good blocking back to keep Biden or [Connecticut Senator Christopher] Dodd or [New Mexico Governor Bill] Richardson" out of the way if they could get any traction.[13]

Later Biden again said he deeply regretted any offense that might have been taken about the "articulate" and "clean" comments and had so told Obama, who was less than gracious in response. "I didn't take Senator Biden's comments personally," he said, "but obviously they were historically inaccurate. African American presidential candidates like Jesse Jackson, Shirley Chisholm, Carol Moseley Braun, and Al Sharpton gave a voice to many important issues through their campaigns, and no one would call them inarticulate."[14]

Biden also defended his practice of giving long answers to questions. "I think one of the reasons we're in trouble is we reduce the political discussion to sound bites," he said, being the best Biden he could be. "The American public's a lot more sophisticated than we all give them credit for. And on complicated issues, I'm going to give them straight answers. And if it takes more than three minutes, I'm going to do it."[15]

Biden was true to his word as the campaign went on. As the Democratic field grew to eight, and a series of candidate debates unfolded, he soon won a reputation for providing some of the most informed and pertinent answers to questions posed. In this early period, even as Biden failed to show much strength in the polls, he articulately pressed his case that he was the best-qualified candidate to extricate the country from Iraq. After opposing Bush's "surge" of more troops there, he called for repeal of the 2002 authorization for use of force in Iraq, for which he and all other candidates then in Congress except Kucinich had voted. "We gave the president that power to destroy Iraq's weapons of mass destruction and, if necessary, to depose Saddam Hussein," he wrote in an article in the *Boston Globe*. "The weapons of mass destruction were not there. Saddam Hussein is no longer there. The 2002 authorization is no longer relevant to the situation in Iraq."

Biden proposed a "much narrower and achievable mission for our troops in Iraq" limited to digging out terrorists, training Iraqis for their own defense, and removing all American combat forces by early

2008.[16] And he reiterated his plan for decentralizing the country into three ethnic regions. But politically, the call for repeal of the original use-of-force authorization was an unfortunate reminder that he along with the others had signed on to it in the first place.

Like the other Democratic hopefuls, Biden focused on what would be the two traditional and most critical tests of strength in the early voting the next year: the kickoff precinct caucuses in Iowa and then the first-in-the-nation New Hampshire primary, switching back and forth as time and opportunity offered.

Campaigning in New Hampshire in March, Biden stoutly defended his proposal for the ethnic federations in Iraq, denying its critics' characterization as a "partition" and noting that the new Iraqi constitution provided for just such a federalism approach. "I know that there are a number of serious players in the administration who agree with me, at State and Defense," Biden said at one point. "What I also know is there is a very strong pushback from Cheney and from the White House. . . . What I hoped this would do was embolden those within the administration who do not agree with the policy of Cheney and company to stand up and push back, now that they know they have a lot of support in the Senate."

In a talk at Plymouth State University, Biden said his idea "gives us the only possibility of withdrawing our troops without leaving chaos behind." But if it was intended to light a fuse to his presidential bid, there seemed to be little evidence of that in New Hampshire. *Boston Globe* columnist Scot Lehigh called the Biden initiative "a genuinely bold and gutsy proposal" but noted that the Delaware senator was "stuck in the second tier" of candidates, after fellow senators Clinton and Obama, "and far behind in fund-raising."[17]

In the first major Democratic debate of the season, at South Carolina State University in Orangeburg in April, Biden got in the best answer among the eight candidates when moderator Brian Williams of NBC News, taking note that "words have gotten you in trouble in the past," asked him whether he could manage this time around to curb his verbosity. With a straight face, Biden replied, simply, "Yes." Then he flashed his trademark smile. Williams seemed to wait for more, but that was all he got, as others on the panel laughed and the audience applauded.

Biden himself seemed somewhat surprised at the audience's response. In a postdebate interview, he said "I'm beginning to realize what I should have known all along," that outside of Delaware people didn't know he had a sense of humor. "Part of this is reintroducing myself to the Democratic Party. . . . I found it unusual that someone would think, 'God, Biden was humorous.' "[18]

He knew from this revelation that he had a lot of persuasion and exposure to do. At an annual fish fry hosted by Representative James Clyburn, whose endorsement all the candidates in South Carolina sought, Biden remained long after the other contenders had left. "He just left? Are you serious?" Clyburn asked later. "I guess he wants to meet more people." Clyburn said it reminded him of "my younger days; that's what we used to do—make sure you outsit the other guy."[19]

A month later, again in New Hampshire, Biden told the *Boston Globe*'s Scot Lehigh: "As long as this war is the centerpiece issue, I am in the game. I think people understand that I have thought about this and that the answer I have is practical."[20] But with Bush's troop surge appearing to take on some traction, there was little public support behind the Biden approach.

In Iowa in mid-June, twenty years after the gaffe at the state fair that undid his first presidential bid, Biden acknowledged of that effort that "I didn't deserve to be president. I wasn't mature enough."[21] But in small-town talks with Iowans across the state, he argued that his subsequent long experience as a member and chairman of the Senate Foreign Relations Committee had matured and grounded him to meet the central test of the current times. No longer was he making the "new generation" pitch that had been tried and found wanting back in the 1980s. Now that was the Obama message of hope and passion. But Biden still had the personal bounce and enthusiasm of those days, with thinning gray hair now that lent a shade more gravitas to him.

"Right now I am absolutely convinced that the American public and Democratic Party are looking for someone with the breadth and depth of knowledge in foreign affairs and national security policy, as well as the ability to empathize with the circumstances of average, middle-class people," he told a backyard audience in the town of Emmetsburg. And in a thinly veiled reference to the old plagiarism allegations in his first Iowa experience, Biden said, "I am fully prepared to match my character against anyone running. They go after me, they got a fight. . . . To the best of my knowledge, anything that is embarrassing about my past is pretty well public record. Any of you who take a look at my life will not be able to conclude that I am not an honorable man . . . and that's why I wasn't afraid to get back into this thing."[22] His pride was back in full flower as he campaigned as "the best Biden I can be."

Later, during a speech in Waverly, Iowa, Biden felt comfortable enough about the old plagiarism charge to needle himself about it. About to quote Abraham Lincoln, he paused and checked his notes. "I want to make sure I quote it precisely," he said. "I quote everybody these days." A ripple of laughter went through the crowd. But when reporter Nicole Gaudiano of the *Wilmington News Journal* asked listeners in the Iowa crowd about the episode of two decades earlier, she found few who seemed to recall it. "I don't remember that at all," Johanna Foster, a Wartburg College biology professor, said.[23] Missy Owens, the candidate's niece, who as deputy state director for the campaign accompanied him in Iowa, agreed later there was no hangover from the incident. "Surprisingly, not at all," she said. "He was considered the foreign policy expert, and the rest of it fell away. Nobody brought it up."[24]

Running again twenty years later, Biden said in thinking about doing so he wondered "whether or not that would be relitigated, but in the meantime so much had happened and so much of the truth had come out that it wasn't a major factor. Also, the way in which it was looked at, reviewed, considered in hindsight by the press covering me in 2008, was actually gratifying. It was like everybody had taken a deep breath and gone back and said, 'Oh, okay, I get it,' as opposed to the heat of the moment. One, the remainder of my career sort of made a lie of the allegations. Secondly, there had been a lot written in the period from '87 to '92. . . . There was a perspective placed on it."[25]

But at this early stage of the 2008 race, the polls were suggesting strongly that such lapses of memory would not be enough to boost his chances. Still, thirty-six years earlier, he was nowhere in the early polls either, against Cale Boggs in his supposedly foolhardy race for the Senate. Then, though, he was a young man in a hurry with no baggage to slow him down. And in Hillary Clinton and Barack Obama, he was facing two foes with special constituencies behind them, one of gender and the other of race, that could be critical in a large, multicandidate nomination contest. In a sense, history was working for both Clinton and Obama; the election of either of them would be an American milestone.

Biden the second time around was the same freewheeling average Joe, self-confident and unrestrained. In a new-media amalgam debate combining CNN and the equally freewheeling YouTube electronic democracy in July, he had a ready answer for one gun rights interrogator who defiantly brandished a rifle, calling it his "baby." Biden, reiterating his support for strong gun controls, shot back, "If that's his baby, he needs help. I don't know if he's mentally qualified to own that gun."[26] In this and other debates, the other candidates began to associate themselves with Biden's more serious observations, to the point that while he got much less time to talk than either Clinton or Obama, he was accorded a sort of senior status among the pack. At one point, the Biden campaign ran an ad that said "Joe's right," referring to how the other Democrats seemed often to embrace his foreign policy views.

In August, after Biden had been quoted in *Newsweek* as saying he thought Obama was "not yet ready" for the presidency, he was pressed on the comment in the first televised debate for the Iowa caucuses, at Drake University in Des Moines. "I think he can be ready," he said, "but right now I don't believe he is. The presidency is not something that lends itself to on-the-job training."[27]

Biden pressed on with his underfunded Iowa campaign, defending his continued support of better armor for the troops in Iraq while challenging the Bush strategy. Still stagnant in the polls, he plain-talked around the state, often asking Iowans simply to "look me over."[28]

At an Iowa corn boil in Clinton, Iowa, in late August, Biden waited patiently for his turn as Hillary Clinton addressed a large crowd at the Lumber Kings minor-league ballpark. When she had finished and left, much of the crowd left with her, but Biden came down off the pitcher's mound and went into the bleachers, courting those who remained. He continued to be convinced that Delaware retail politics would work in Iowa. "They want you to look them in the eye," he told Gaudiano at one point.[29]

At the same time, Biden was plainly annoyed at being mired in the so-called second or third tier of possible nominees, a view that others with substantial records of public service, such as Senator Christopher Dodd, were obliged to endure. And his lack of money continued to hamper his efforts, as it did Dodd. Although by this time Biden had raised $6.3 million, that amount was a mere snowflake in the blizzard of cash falling on Obama. Later in the campaign, when an important Senate vote threatened to cause Dodd and Biden to miss a major debate in New Hampshire, they split the $2,100 cost of a chartered plane to make it.[30]

In September Biden, joined by Republican Sam Brownback of Kansas, surprisingly won strong Senate approval of a nonbinding resolution calling for international support of a political solution in Iraq, including the federation approach. The bipartisan vote was 75 to 23, including 23 Republicans. At the time, Brownback was seeking the Republican presidential nomination, but he dropped out shortly afterward with slim support similar to Biden's own in the *Des Moines Register* poll. Nevertheless, he showed up in Des Moines in mid-October for a joint news conference with Biden, touting the plan. Neither one of them had gotten much political mileage from the federation scheme, but Biden vowed to press on for the approaching Iowa caucuses.

With his candidacy merely inching up from 3 percent to 5 in the *Register* poll, he gamely said: "My road to success is Iowa. It's the only playing field left out there. The bottom line is that no one in the country knows me. They know Joe Biden if they watch Sunday morning shows or occasionally turn on C-SPAN. But absent that, they don't know much about me at all."[31]

What many Americans did continue to know about, or at least hear and read about Biden, was his penchant for talking himself into a corner. At a meeting with the *Washington Post* editorial board, duly reported thereafter, he left himself open to an allegation of racial insensitivity. Commenting on the low achievement record of District of Columbia schools compared to that of Iowa's, he observed "there is less than one percent of Iowa's population that is African American." His campaign quickly issued a clarification that Biden was talking not about race, but about contrasting socioeconomic advantages in the two places.[32]

Presenting himself as a serious problem-solver, Biden unveiled a health care reform plan that would expand coverage for children and lower the eligibility age for Medicare. Calling on another aspect of his experience, Biden told an Iowa audience: "I know what it is like to be wheeled into an emergency room, unsure if you'll see your family ever again—and I was lucky because I had health insurance." The polls indicated little upward movement.[33]

Unwilling or unable to get much traction in competition with Clinton and Obama and striving to stay on the high road on the Democratic side, Biden used a debate in Philadelphia in early

November to take a slap at one of the Republican hopefuls. He charged that former New York mayor Rudy Giuliani was "genuinely not qualified to be president." He got the best laugh of the night from the audience by saying of the self-styled hero of Ground Zero that "there's only three things he mentions in a sentence: a noun and a verb and 9/11."[34] Giuliani made no reply, but an aide bitingly said, "Rudy rarely reads prepared speeches, and when he does, he isn't prone to ripping off the text from others. And Senator Biden certainly falls into the bucket of those on the stage tonight who have never had executive experience and have never run anything. Wait, I take that back. Senator Biden has never run anything but his mouth."[35]

Biden's gibe at Giuliani appeared to stem from the former mayor's oft-repeated boasts about the sharp decline in New York City crime during his tenure. Biden claimed credit for his COPS program. But some crime experts took issue with its significance. A General Accounting Office report in 2005 said police hired under the COPS program amounted to less than 3 percent of the total, and Jeffrey Fagan, codirector of the Center for Crime, Community and Law at the Columbia Law School observed, "This is akin to a small dose of medicine. Can a small dose cause a big effect? Only if it's a wonder drug."[36]

Nevertheless, the Biden of 2007 did focus as a presidential campaigner more on achievement than did the Biden of 1987. It was undeniable that he had downshifted somewhat from the passionate pitchman he was in his previous presidential run in Iowa. Now he emphasized his credentials and record as a substantive and knowledgeable senior statesman.

And by and large, he was perceptively much more in control of himself and his campaign than he had been the first time around.

In a CNN debate in Las Vegas in late November, Biden demonstrated that self-control tempered by a bit of Irish humor. The moderator, Wolf Blitzer, in his manner of a circus lion-tamer dealing with the eight-candidate field, focused on the three poll front-runners at the time—Hillary Clinton, John Edwards, and Barack Obama—and left the other five standing idly by like sedated jungle cats.

Blitzer opened the debate by inviting Obama to discourse on his criticisms of Clinton, then urged Edwards to chime in with his anti-Clinton gripes. Biden visibly fumed as the selective exchange dragged on. Finally, the moderator decided to let the other candidates have a word, and turned to Biden for his two cents. The chairman of the Senate Foreign Relations Committee expressed horror at being invited to intrude. "Oh, no; no, no, no!" he demurred, eyes raised in feigned surprise. "Don't do it, no!" he exclaimed to Blitzer. "Don't make me speak!"[37]

The heretofore neglected Biden broke into a broad grin. From his vantage point at the far end of the line of candidates, he offered what he thought of the verbal catfight among the front-runners instigated by the theatrical moderator. The American people, Biden declared, "don't give a darn about any of this stuff that's going on up here." He then tried his best to raise important questions about the war in Iraq, education, and crime, concerning all of which he had years of experience in the Senate. But the debate further disintegrated as some of CNN's celebrity reporters on camera urged CNN-selected voters to pose screened questions.

At one point, when one of CNN's female stars introduced a woman voter who expressed concern about Supreme Court appointees, the star interrupted the woman, asking Biden whether he would "require your nominees to support abortion rights." He rebelled, saying he intended to answer "the question of the woman who was there" first, and then would deal with the interloping CNN glamour queen. As Biden lamented Blitzer's time-management favoring the three front-runners, another spear-carrier in the television extravaganza, Congressman Dennis Kucinich of Ohio, joined Biden, saying what he thought of having to play seventh or eighth banana to Clinton, Edwards, and Obama, not to mention ringmaster Blitzer.[38]

For all that, Biden this time around was mostly a happy warrior on the campaign trail. In an interview with the *New York Times* in Iowa City in December, Biden insisted, "This has been the easiest campaign I've ever run in. I haven't had to game anything. For real. I know what I believe, I know what I want to do, and I'm just comfortable saying it, and laying it out there."[39]

After another Iowa speech at Mason City, Biden reflected to a *Boston Globe* reporter that one critical lesson he had learned from his defeat in the early going of the 1988 presidential race was that he had concentrated so much on selling his passion for change that he had failed to convey the substance he brought to the challenge. "I was the Barack Obama!" he declared with as much amusement as sincerity. His campaign consultant at the time, John Marttila, agreed. "The caucus participants [in Iowa] are well-educated people, well-informed," he said. "This is the Biden constituency. It's news consumers."[40]

So Biden piled on the experience this time around, complete with name-dropping of figures he knew as chairman of the Senate Foreign Relations Committee. "I know many of these world leaders—most of them by their first names," he would say. A former Iowa congressman and state party chairman, Dave Nagle, told the *Globe* of Biden's transformation on the stump. "He was prone to appeal to idealism to excess," Nagle said, "relying more on his speaking ability than on his knowledge."[41]

Soon after, a *Philadelphia Inquirer* reporter wrote that Biden in his second presidential try was "speaking his mind, giving Iowa voters full paragraphs of context instead of sound bites, making issues seem clear rather than simple. Whatever happens, Biden has earned respect from voters and pundits as serious and thoughtful, a kind of redemption twenty years after his first presidential campaign of airy rhetoric blew up amid accusations of plagiarism. . . . These days Biden doesn't let much get him down. It's probably the last hurrah for his White House ambitions, and he's enjoying the ride." Biden himself said "it's so much easier this time because I really, genuinely, know what I believe and what I would do as president. I have a comfort zone. There's not any nobility about it, it's just that I'm okay."[42]

Three days after Christmas, Biden's argument that foreign policy, including the unsettled politics in Pakistan, should be at the front of the campaign agenda received an unfortunate boost when former prime minister and current candidate Benazir Bhutto was assassinated. A television ad was running in Iowa at the time that had Biden saying "we don't have to imagine the crises the next president will face." He told reporters he had twice "urged [Prime Minister Pervez] Musharraf to provide better security for Ms. Bhutto and other political leaders."[43] But that development had no perceivable impact on Biden's prospects in Iowa.

In the final week of the Iowa campaign, Biden professed to see a silver lining in the upbeat reaction of the modest crowds that came to hear him. Campaigning as if he were back in Delaware, he continued to collect endorsements from state legislators and local party officials and to count on word-of-mouth to sustain him. At a rally in Waterloo, when county Democratic chairperson Pat Sass endorsed him, he dropped to one knee and kissed her hand. "That's a big deal for me," he told the gathering of about two hundred Iowans.[44] And in Oscaloosa, he told another crowd: "I've been in public life long enough to feel in my fingertips and to understand by the crowds and responses . . . that this is real. There's something genuinely going on."[45] The polls were suggesting otherwise, but he pressed on.

About twenty other Bidens spread out around the state as they had done for years in Delaware. At a rally in Ames his ninety-year-old mother was present and inquired whether he was eating enough. When a voter asked a question about education, he put his arm around his teacher wife and declared he would be an "education president" because he wanted "to continue to sleep with this woman." Meanwhile brother Jimmy and son Hunter would stand below and occasionally point to their watches to remind him to stop talking.

When his brother walked up to the podium in Newton, the candidate joked: "I'll be glad when I've got Secret Service. They can't tell me what to do."[46]

On the day before the caucuses, with his candidacy still barely registering in the Iowa poll, Biden made a final plea for support. "Folks, now everyone is talking about this race being about either experience or change," he said at the Raccoon River Brewing Company in Des Moines. True enough, experience was the ace card for Hillary Clinton and himself, and change for Barack Obama and John Edwards, still in the picture.

"But folks, it's not about change or about experience," Biden said. "It's about action . . . because the next president of the United States is going to have no time. . . . I want all the caucus-goers in this great state to close their eyes and imagine. If their candidate is president of the United States, not in a year but this very instant, are they confident that they have the sure-footedness, the steady-enough hand to know exactly what they would do in Pakistan? To know exactly what they would do, not generically, in Iraq? Exactly what they would do with the totalitarian tilt of Putin? Know exactly what they would do this moment where the relationship with China is fraying in a way that could become dangerous? Are they ready? Are they smarter than their secretary of state?"[47]

Biden's windup was a characteristically embellished version of the ready-from-day-one pitch, with an implied version that he, apparently everybody's choice to be secretary of state, was the one most ready.

Much earlier, he had sought to put the notion of that consolation prize to rest. Asked by his hometown paper's reporter about it, he was emphatic: "Absolutely, positively, inequitably, Shermanesquely, no. I will not be anybody's secretary of state in any circumstance I can think of. And I absolutely can say with certainty I would not be anybody's vice president, period. End of story. I guarantee I will not do it."[48]

On New Year's Day, Marttila's hopes were buoyed, in spite of the dismal polling numbers for Biden, by a surprisingly large turnout of as many as five hundred people at a rally on a national holiday. "I was confused by it," he said later, not appreciating how the star quality of Barack Obama and Hillary Clinton, and their operations, would swell the turnout beyond expectations. "They just occupied too much oxygen," he said later, and the Bidens too allowed their spirits to rise.[49] They spent the final day of the Iowa caucus campaign flying around the state, thanking now modest turnouts in Waterloo, Dubuque, and Davenport, and the candidate was at peace with himself. "One of the liberating things about it is, I'm not spending time trying to figure what to say," he philosophized, not that he ever was at a loss for words.

Before going to a last caucus in Des Moines, Biden recalled how it was Jill—who disliked campaigning and the disruption of family life—who called the children together and told them how they had to "ask Dad to run," after Bush was reelected in 2004, and then told him. Joe remembered how it persuaded him, "because I knew how much Jill would rather our life be one where I was a lawyer, where we were comfortable," and they could travel as they chose.[50]

As the precinct caucuses got under way across Iowa, Biden said he expected to beat expectations and would stay the race through the month of January "no matter what." He compared himself to the football player he had been at Archmere Academy, who thought only about crossing the goal line. The staff too fully expected to do well enough to continue. "We all expected to go on to New Hampshire," Larry Rasky said.[51] But in the end, as Marttila said, Obama and Clinton "occupied too much oxygen," in money and news-media attention, for Biden to get even a toehold in the race.[52]

Biden was taking a catnap when Jill awakened him with the results, and it was clear to both of them that the Biden campaign had ended. He finished a dismal fifth with only 1 percent, behind Obama, Edwards, Clinton, and Governor Bill Richardson of New Mexico. "We all thought we had a shot at third," niece Missy Owens recalled.[53] Biden's strategists had counted on the Iowa caucus turnout to be much lower than a primary but it had become a hybrid, swelling to such primarylike dimensions that his grass-roots campaign had been overwhelmed. Biden told his followers: "Look folks, there's nothing to be sad about tonight." He said he was content he could say exactly what he believed, and that was what he had done.[54]

Meeting privately with his family, Ashley recalled, "Dad again was the one to rise to the occasion to comfort all of us. [I thought] wait a minute, this should be the opposite, and still Dad is saying how proud he was of the family, what an amazing job we did, he had no regrets, and he was so proud of us. It was a measure of his character. He was the candidate [but] instead of us being the ones comforting him, he comforted us."[55] Missy Owens reflected the attitude among the Bidens: "We took it with heads held high. We did it as a family. We got through it because we had each other."[56]

Son Beau reflected afterward that his father's second bid for the presidency was much more satisfying to him than the first: "I saw him enjoy it more. I saw a candidate say exactly where he wanted to lead this country and had incredibly well-conceived specifics on what he wanted to do and see borne out. He didn't run for president in 2008 to see his name on the ballot." Unlike 1987, his elder son said, when "we had so many people involved who had different motives and perspectives . . . in 2008 he was completely in charge of every aspect." Furthermore, he reminded, other Democratic candidates in the primary debates often took their lead from him on foreign policy questions, prefacing their own remarks by saying, "Joe's right."

Above all, Beau said, his father never forgot in either of his two presidential campaigns that family came first. After the second one was over, Beau said, he had an experience that again drove that point home to himself as a public servant, as attorney general of Delaware at the time, as well as a parent.

He was driving up to his home in Wilmington one afternoon. "My kids didn't know I was home yet, and I was on the phone on what

I thought was an important call," Beau recalled. As he pulled into the driveway, his wife lifted one of their kids to see him. He looked at her and signaled by raising the phone that he was in the middle of a call. Suddenly, he said, "I felt on the back of my shoulder my father tapping, figuratively speaking. Never once in my entire life has my dad done that to me or my brother or my sister. He would have done one of two things, which I did promptly. He would end the call, or he'd say, 'Come on, pal,' open the door, and I'd hop on his lap and sit there as he finished the call. Moms and dads wouldn't do that [keep talking], and I of all people shouldn't need a reminder of that, and I needed that little figurative tap on the shoulder. That to me kind of illustrates the kind of dad he is and always has been."[57]

Valerie observed later that while Joe was naturally disappointed that his second presidential bid also failed, "his happiness never rested on being president. It was a job he tried out for and wanted to be hired for, a very important job application," she said. "But it would not define him by being president."[58]

As for the candidate himself, he observed long afterward, "I swore after '87 I'd never lose another race on any terms other than my own. And so when I ran this time, it was the easiest campaign, the most satisfying campaign I'd ever run. I didn't look at any polls, I went out and I said exactly what I thought, I didn't game it, I was convinced I was right about where to take the country. I honestly thought I was the best prepared for the particular problems we were facing to lead the country, and I was completely at ease, and I think that was what came through in the debates."[59]

Finally, it was all over—or so Biden thought and expected. He would not endorse another candidate until the voters around the country had spoken, and then he would be a loyal supporter of the winner. Biden was, after all, also running for reelection to the Senate, for a seventh term, and although the Republicans would offer a rival candidate, it was a foregone conclusion that fellow Delawareans would want their Joe back in Washington representing them again under the next president, hopefully a Democrat.

Confidently, he told his glum caucus-night crowd: "I'll be going back to the Senate as chairman of the Foreign Relations Committee and I will continue to make the case I've been making," including providing the necessary funds to support the American troops waging the war he now opposed. He promised his congressional colleagues "I will be their worst nightmare if they do not."[60]

There was always the possibility, to be sure, that rather than returning to the Senate he might indeed be the next secretary of state or even vice president. But again he told his hometown reporter otherwise. "If we have a Democratic president," he said, "I can have much more influence, I promise you, as chairman of the Foreign Relations Committee than I can as vice president." And he laughed, walking off to rejoin his family and get back to Wilmington.[61]

TWENTY-SIX

A QUESTIONABLE PRIZE

FOR MOST OF the nation's history, the vice presidency had been avoided like a plague by most ambitious politicians. More often than not, especially in the first hundred years of the Republic, the office was regarded as a dead end, a sort of gold watch in retirement. John Adams, the first occupant under George Washington, recognized the reality of its limitations as well as its potential when he wrote to his wife, Abigail: "In this I am nothing, but I may be everything." He lamented that "my country in its wisdom has contrived for me the most insignificant office that ever the invention of man contrived or his imagination conceived."[1] Its only constitutional functions, of presiding over the Senate with the singular power to break a tie vote, and of succeeding to the presidency in the event of the president's death or incapacitation to serve, left its occupant a standby in both the executive and legislative branches.

Adams himself became so contemptuous of the vice presidency that he wrote: "Is not my election to this office, in the scurvy manner in which it was done, a curse rather than a blessing? Is this justice? Is there common sense or decency in this business? Is it not an indelible stain on our Country, Countrymen and Constitution? I assure you I think it so, and nothing but an apprehension of great Mischief, and final failure of the Government from my Refusal . . . prevented me from Spurning it."[2]

The original method of vice presidential selection—awarding the job to the presidential runner-up—was the unanticipated by-product of "double balloting" by "electors" chosen by each state legislature. To obviate concern that if each elector cast a single vote for his state's favorite son, a majority vote would be impossible, the Constitutional Convention settled on having each elector cast two votes, with one required to go to a nonresident of that state.

George Washington of Virginia was the unanimous choice in the first two elections, with another Federalist, John Adams of Massachusetts, the runner-up. But when Adams was elected Washington's successor in 1796, the runner-up was anti-Federalist Thomas Jefferson, who did not share Adams's political agenda. And in 1800, the double balloting failed to avoid a tie between two anti-Federalists, Jefferson and Aaron Burr, with Federalist Adams running third and denied reelection. It took thirty-six ballots and more than a week of political intrigue in the House of Representatives before Jefferson won out, with Burr left a disgruntled vice president.[3]

In any event, something had to be done to assure continuity of policy in the event the death or incapacitation of the president was to elevate a vice president of a different or opposition political faction. In 1804, in time for the next presidential election, the states ratified the Twelfth Amendment, which provided separate election of the president and vice president, enabling politically like-minded candidates to run as a team and eliminating from the equation any reward for the presidential runner-up.

Even this arrangement, however, encountered criticism during the Senate debate on the amendment, on grounds it was likely to produce mediocrity in the vice presidency. The office, Senator William Cocke of Tennessee argued, "will be a sinecure. It will be brought to market and exposed to sale to procure votes for the president. Will the ambitious, aspiring candidate for the presidency, will his friends and favorites promote the election of a man of talents, probity, and popularity for vice president, who may prove his rival? No! They will seek a man of moderate talents, whose ambition is bounded by that office, and whose influence will aid them in electing the president."[4]

That prediction, with some exceptions, was borne out over the next century by the roster of elected vice presidents whose political careers ended there. After runner-up vice presidents Adams and Jefferson, only two other elected, sitting vice presidents, Martin Van Buren and George H. W. Bush, were subsequently elected to the Oval Office in their own right, as opposed to succeeding by the death or resignation of the president.

Richard Nixon as Dwight D. Eisenhower's vice president lost his first presidential bid in 1960 before winning the office as a private citizen eight years later. It was not surprising, therefore, that political figures with their eye on the White House shunned the vice presidential nominations of their party for so long, despite the fact that nine occupants of the vice presidency have succeeded directly to the presidency without benefit of direct election.

In the first fifty-two years of the Republic, every president served out one or two terms, with little interest in or concern about the identity of the vice president. President James Madison's two vice presidents, George Clinton of New York and Elbridge Gerry of Massachusetts, died in office and neither was replaced to finish the unexpired term, there being no vehicle for selecting a replacement. It was a constitutional shortcoming not corrected until ratification of the Twenty-fifth Amendment in 1967, providing for the presidential nomination of a vice president to a vacancy subject to confirmation by a majority of both Houses of Congress.

In 1832 President Andrew Jackson's vice president, John C.

Calhoun, became the first occupant of the office to resign. The insignificance of the office was driven home to him when Senator John Forsyth from Georgia made a remark that Calhoun, presiding but empowered only to break a tie, perceived as a slur against him. Calhoun demanded, "Does the senator allude to me?" Forsyth in rebuke shot back, "By what right does the chair ask that question?" Calhoun, leading the nullification movement in South Carolina, later returned there and was elected to the U.S. Senate while still serving out the last months of his vice presidency.[5]

In 1841, when sixty-eight-year-old President William Henry Harrison delivered an inaugural address for nearly two hours on a cold, windy day in March and died a month later, John Tyler of Virginia became the first vice president elevated to the presidency. But the Constitution said only that in such circumstances "the powers and duties of said office . . . shall devolve on the Vice-President," who would "act as President." So Tyler rushed to be sworn in, lest he be regarded as only a seat-warmer until another president was elected. He finished out Harrison's term but was never seriously considered by the Democrats for a term of his own.[6]

In 1844 the vice presidency had fallen so low in esteem that when Senator Silas Wright of New York was nominated by the Democratic National Convention, he turned it down. The next vice president, Millard Fillmore, like Tyler assumed the presidency when President Zachary Taylor fell victim to overeating and a blazing sun at a Fourth of July celebration at the Washington Monument in 1850 and died. Fillmore, too, serving two years and eight months without a vice president, was denied a term of his own after his accidental presidency.

The very next vice president, William R. King of Alabama, took the oath in March 1853 with President Franklin Pierce but died the next month, leaving the vice presidency unoccupied for nearly seven years. Nobody seemed to care much. And fifteen years later, when Abraham Lincoln was assassinated in the second month of his second term, Andrew Johnson replaced him, leaving the vice presidential chair empty for almost four years. This was so even as Johnson was impeached in the House for removing a cabinet member without Senate approval, though saved from conviction by a single vote in the Senate.[7]

By now, the vice presidency had reached such a low estate that in 1880, when the Republicans nominated for president James A. Garfield of Ohio, a former Union general and now a congressman, the party in an internal squabble made a deposed collector of the corrupt New York Custom House, Chester A. Arthur, his running mate. In mock approval, E. L. Godkin of the *Nation* wrote there was "no place in which his powers of mischief will be so small as in the vice presidency." Besides, he said, with Garfield only forty-eight and in good health, succession was a "too unlikely contingency to be making extraordinary contingency for."[8] But the assassination of Garfield at Union Station in 1881 made a mockery of that judgment, as Arthur was elevated to the Oval Office for more than three years.

In 1900, after President William McKinley's vice president, Garret A. Hobart, died in his third year in office, the Republicans at last offered a running mate of genuine stature on their next ticket. Governor Theodore Roosevelt of New York was persuaded against his own best judgment to accept the nomination for vice president in McKinley's bid for a second term.

McKinley himself left the choice to the convention, but to the chagrin of his campaign manager, kingmaker Mark Hanna, New York party chairman Senator Thomas Platt was determined to get the independent-minded Roosevelt out of Albany. At only forty-one, Roosevelt preferred to seek a second term as governor, and wrote Platt that "I would a great deal rather be anything, say a professor of history, than vice president."[9] But Platt and Senator Matthew Quay, the Pennsylvania party boss, played on Roosevelt's vanity, even as Hanna warned of the consequences of putting the strong-willed New Yorker in the line of presidential succession. When a national committeeman from Wisconsin, Henry Payne, asked a clearly upset Hanna what was the matter, Hanna shot back: "Matter? Why, everybody's gone crazy! What is the matter with all of you? . . . Don't any of you realize there's only one life between that madman and the presidency? . . . What harm can he do as governor of New York compared to the damage he will do as president if McKinley should die?"[10]

Roosevelt finally "reluctantly" informed the convention that "I cannot seem to be bigger than the party" if it really wanted him. When the McKinley-Roosevelt ticket was nominated unanimously, Hanna solemnly told his Ohio friend, "Now it is up to you to live."[11] Whereupon McKinley stayed mainly on his front porch all fall as was still the custom for the head of the ticket, and Roosevelt campaigned as if he were the presidential nominee. They were elected by a larger margin than in McKinley's first-term victory. Less than a year later, Hanna's worst fears were realized, when McKinley was assassinated and "that madman" was in the Oval Office for nearly the next eight years.

Few vice presidents until recently had ever had much good to say about the office. Thomas Marshall, who endured it in silence and ignorance of the seriousness of the long illness of President Woodrow Wilson, said his job was "to ring the White House bell every morning and ask what is the state of the president."[12] From most reports at the time, he never was told by Wilson's doorkeeper—his wife, Edith—that her husband was barely holding on.

Marshall himself described being vice president "like a man in a cataleptic state: he cannot speak; he cannot move; he suffers no pain; and yet he is perfectly conscious of everything that is going on about him."[13] Even Wilson agreed, observing that "the chief embarrassment in discussing this office is, that in explaining how little there is to be said about it, one has evidently said all there is to say."[14]

That general attitude continued to prevail into the 1930s, when Franklin D. Roosevelt's first vice president, former powerful House Speaker John N. "Cactus Jack" Garner of Texas, frustrated by his lack of influence, famously said, in what probably was a sanitized version, that the office was "not worth a bucket of warm spit."[15] Garner in 1940 decided to challenge FDR for the presidential nomination, lost, and was replaced on the Roosevelt ticket with agriculture expert Henry A. Wallace, who soon after their election lost favor among the party bosses. Ironically, the man who succeeded Wallace in the vice presidency, Harry S. Truman, was critical in the eventual rehabilitation and rejuvenation of the office to something that soon became much more than Garner's epithet.

Truman came into the vice presidency as almost an offhand choice by FDR, after a small group of machine politicians urged him to drop Wallace from the party's ticket in 1944 and take Truman. Roosevelt did not know the man well and never confided in him. Truman didn't learn of the ongoing development of the atomic bomb until informed of it by Secretary of War Henry Stimson after FDR's death, at which he said, "the moon, the stars and all the planets" had suddenly fallen on him.[16] Its successful testing shortly afterward forced Truman to make the most momentous decision of his presidency in authorizing its use against Japan, effectively ending World War II.

The Truman experience drove home the imperative, so often ignored in the past, of keeping the vice president much more intimately involved in the workings and secrets of the national administration he might inherit at any moment. Thereafter, that edict was observed to various degrees by most subsequent presidents, though Vice President Richard Nixon remained essentially an outsider in the Eisenhower administration. When a reporter asked Eisenhower at a news conference for an example of "a major idea of his you had adopted" during Nixon's years as his vice president, the president replied, "If you give me a week, I might think of one. I don't remember."[17] At Eisenhower's next news conference, nobody bothered to ask him whether he had.

Eight years later, when Nixon himself reached the Oval Office, he kept his vice president, Spiro Agnew, similarly in the dark, never informing him in advance of Nixon's heralded "opening to China" secret trip to what was then called Peking. Eventually, with Agnew facing the loss of the vice presidency in an investigation of bribe-taking, he was seen in some quarters as a sort of insurance policy for Nixon's presidency, with Nixon himself facing impeachment in the Watergate scandal and cover-up. The thinking was that Congress would not impeach and convict the president if the result would be to put the tainted Agnew in the Oval Office.[18] In any event, Agnew was forced to resign in 1973, only eight months before Nixon met the same fate in the first presidential resignation in U.S. history.

It fell to Agnew's replacement, House Minority Leader Gerald R. Ford, to resuscitate some respect for the presidential office by succeeding Nixon and ending what he famously called "our long national nightmare" of the Watergate fiasco. Ford was the first unelected vice president, having been nominated by Nixon and confirmed by Congress upon Agnew's departure under procedures of the Twenty-fifth Amendment, which had been ratified only six years earlier. Then, as the first unelected president, Ford quickly struck a heavy blow to his chances of being elected in his own right in 1976 by pardoning Nixon, a highly unpopular action at the time.

The Democrat who narrowly defeated Ford, Jimmy Carter, bestowed upon the vice presidency its most significant upgrading up to that time. He gave his running mate, Walter F. Mondale, unprecedented access to presidential meetings in the White House with visiting heads of state, and otherwise conducted an open-door policy toward him. It was a treatment toward vice presidents continued by subsequent Democratic and Republican presidents alike, rendering the office much more desirable to ambitious politicians whose eyes were on the White House.

The office by now had developed politically in both parties as an obvious stepping-stone to the presidential nomination if its occupant so desired, though not a guarantee of election to the Oval Office. Vice Presidents Mondale, George H. W. Bush, and Al Gore all easily captured their parties' presidential nominations, but of the three only Bush got there. His own vice president, Dan Quayle of Indiana, a boyish and immature fellow given to malapropisms, made a weak and futile bid to carry the Republican banner in 1996, after the defeat of the Bush-Quayle bid for reelection in 1992.

In 2000, after President Bill Clinton's vice president, Gore, was easily nominated for president, the Republican presidential nominee, George W. Bush, son of the former president, took an unusual course in selecting his running mate. He assigned his father's secretary of defense, Richard B. Cheney, to lead the search for his own vice presidential choice, then chose Cheney himself amid speculation that Cheney had maneuvered himself into the job. A man with a history of heart disease when he was selected, Cheney declared himself from the start not interested in pursuing the presidential nomination upon completing his vice presidential service.

Instead, upon election in the notorious campaign decided by a split vote of the Republican-controlled U.S. Supreme Court, Cheney concentrated on efforts to strengthen and expand the powers of the presidency—and his own great influence within the George W. Bush administration. As described in Barton Gellman's account of the Cheney vice presidency, *Angler*, he not only deftly maneuvered the vice presidential nomination for himself but went on to shape the office of the vice presidency into an unprecedented power center in its own right.[19]

Cheney championed the theory of the "unitary executive," holding that the Constitution bestowed total power upon the president as commander in chief of the armed forces in wartime. In the process, he embraced and stoutly defended administration legal positions justifying extreme practices in foreign and domestic intelligence surveillance that dismayed civil liberties defenders. One of the most outspoken of them was Joe Biden, who ultimately was moved to label Cheney the most dangerous vice president in the history of the Republic.

In 2008, well before the long and tumultuous Democratic presidential nomination fight was decided between Barack Obama and Hillary Clinton, Biden had already said he was not really interested in giving up the chairmanship of the Senate Foreign Relations Committee for the vice presidency. After thirty-six years in the Senate, he was accustomed to being his own boss and liked it, and had reservations about being in a job subordinate to anybody. But that was before he was actually confronted with the possibility of a chance to reach the White House after all, if only standing in the wings.

TWENTY-SEVEN

AN OFFER HE COULDN'T REFUSE

WELL BEFORE BARACK OBAMA had secured the presidential nomination, the oddsmakers were putting Joe Biden high on the short list to be chosen for vice president. The customary coy response from possible nominees was to be noncommittal or express lack of interest. A notable exception was Governor John Chafee of Rhode Island, later a U.S. senator, who was asked at the 1967 National Governors Conference whether, if offered, he would agree to be the next Republican nominee's running mate. Chafee immediately replied, "Sure."[1] In return for his candor, he wasn't asked.

Biden, however, was almost of the same mind. On a Sunday morning television show, asked about the vice presidential nomination, he said he didn't want the job, but if asked felt he couldn't say no. At the same time, according to Obama chief strategist David Axelrod later, Biden never asked to be considered.[2]

After withdrawing from the Democratic presidential race in

January, Biden mostly stayed on the sidelines during the next six months as Obama and Hillary Clinton battled tenaciously for the nomination. Obama approached him later in January for his help in the remaining primaries, but Biden told him he intended to remain neutral until the voters had decided on the nominee. He endorsed neither Obama nor Clinton but told each of them he would "work my heart out" for the eventual nominee. "If you win," he told Obama, "I'll do anything you ask me to do." To which Obama replied, Biden said, "be careful because I may ask you a lot."[3] In another conversation in February, Biden recalled later, Obama told him, "The only question I have is not whether I want you in this administration. It's which job you'd like best."[4]

Biden meanwhile wasted no time resuming his responsibilities as chairman of the Senate Foreign Relations Committee. In late February, he and fellow committee members John Kerry and Chuck Hagel flew to Afghanistan, India, Turkey, and Pakistan on a fact-finding tour. Afterward, Biden labeled Afghanistan "the forgotten war" and Pakistan "the neglected frontier," calling for a fresh look at the former and more economic aid for the latter. Afghanistan, he said, was "slipping toward failure because it has never been given a priority" as the war in Iraq dragged on.

Much of the press attention to the weeklong trip was focused, however, on a hairy flight the trio endured in a Black Hawk helicopter that had to make an emergency Afghanistan mountainside landing in a blinding snowstorm. A truck convoy had to drive two and a half hours to reach the pilot and the three senators. It was a jolt of reality for Biden after all those months on the wearying but safe campaign trail.[5]

From then until the end of the primaries, Biden and Obama would talk from time to time. "He'd call not so much to ask for advice as to bounce things off me," Biden said. Obama, he said, had already asked him to "play a more prominent" and "deeply involved" role in his campaign if he was nominated, and Biden said he would even travel with him if Obama so chose.[6]

Although most presidential nominees vow that they will choose a running mate based first and foremost on his or her qualifications to be president, questions of what political strengths the vice presidential nominee may bring to the ticket have been dominant much more often than not. Considerations of geographical and philosophical base, age, and foreign policy or legislative experience nearly always intrude. According to Obama's young campaign manager, David Plouffe, Axelrod's partner in their Chicago political consultant firm, once Obama had the nomination in hand such matters obviously were weighed, and obviously "we didn't want to pick someone who could potentially hurt the ticket."[7]

But in Obama's first meeting on the subject with Plouffe, Axelrod, and the campaign's team for vetting prospective choices, Plouffe recalled Obama's telling them: "I am more concerned and interested in how my selection may perform as an actual vice president than whether they will give a boost to the campaign. A boost would be fine, of course. But I'm not sure that person exists, and I don't want it to infect our thinking."[8]

From the outset, Axelrod and Plouffe said later, Hillary Clinton was definitely in the mix and remained there a long while at Obama's repeated insistence. At a closeted meeting between the two primary contenders at the Washington home of Senator Dianne Feinstein shortly after Obama had clinched the nomination, Clinton, according to Plouffe, had "said she only wanted to endure the full vetting process if it was a near certainty she would be picked." But, he said later, "because we had already researched her so thoroughly, this was not a big problem. We were light-years ahead in our vetting of her than of anyone else."

As for her qualifications, Obama told his two top aides that "if his central criterion measured who could be the best VP, she had to be included in that list. She was competent, could help in Congress, would have international bona fides, and had been through all this before, albeit in a different role. He wanted to continue discussing her as we moved forward."[9]

Obama meanwhile set aides onto a rigorous vetting process of prospective running mates, of which there were about twenty listed at the start. Each was asked to submit answers to a very lengthy questionnaire designed to ferret out any problems in each individual's personal and professional backgrounds that could mar his or her acceptance to the voters, or become disruptive of the campaign and its themes. Among those included in addition to Hillary Clinton and Joe Biden were Senator Evan Bayh of Indiana; and Governors Tim Kaine of Virginia, Kathleen Sebelius of Kansas, and Bill Richardson of New Mexico.

Biden, meanwhile, continued to deny interest in either the vice presidency or the State Department if a Democrat were to be elected. "They are the only two things that would be of interest to anybody," he said in April, but "I've made it clear I don't want either of them. . . . I'm satisfied where I am right now."[10] When it became clear that Obama would be the nominee, Biden said, and "when he asked me if I would consider being vetted [as his possible running mate], I said I'd have to think about it."[11]

Although they had served together in the Senate for nearly four years, Obama and Biden did not know each other all that well. Obama was a young newcomer compared with the veteran Biden, and the latter's comment in one of the debates that Obama wasn't "ready" to be president had not been that surprising. "Biden was a creature of the Senate," Axelrod reflected later, "so he gave some deference to years of service and Obama's was rather short. I think there was a natural kind of wariness; I don't think it was hostile. It was sort of, 'Here's a bright, impressive young guy. Let's see what he's got.' "

Later, Axelrod said, "I think Biden made a good adjustment [to his own loss of the Democratic nomination]. I think at first, and he even may have said, 'I thought, of course, I would have been the best president,' but over time as he worked with Obama, got to know him, got to watch him closely, I think he sincerely bought into him as a potential president."[12]

Biden said at one point after having served thirty-six years in the Senate, he wasn't sure being vice president would be better and more influential than continuing as chairman of the Foreign Relations Committee.[13] By the time Obama met again with his aides on the matter, the list had been pared to about half a dozen names, with Hillary Clinton still there. According to Plouffe, Obama was continuing to say "Hillary has a lot of what I am looking for in a VP—smarts, discipline, steadfastness." But, Plouffe remembered him adding, revealingly: "I think Bill may be too big a complication. If I picked her, my concern is that there would be more than two of us in the relationship."[14]

There again was the old "two-for-the-price-of-one" bargain that Democratic presidential nominee Bill Clinton in 1992 had offered voters in Hillary as a brilliant First Spouse, now seen in reverse by Obama as a not very appealing bargain. He doubtless remembered the taste he had of Bill as a Hillary surrogate in the South Carolina primary in which Bill's remarks about Jesse Jackson's previous primary victories seemed to sensitive ears to diminish Obama's own victory there.

According to Axelrod, "Hillary was in the mix longer than anybody realized, in the sense that then senator Obama in all the meetings we had to discuss this issue always came back to, 'What about Hillary?' " In Biden's favor, he said, "to the extent that she had any aspirations for that, or those around her had aspirations for her, Biden was the least objectionable alternative, because he had a very close relationship with Bill and Hillary Clinton; he was very close to them during the Clinton presidency. He maintained neutrality in the primaries after he dropped out of the race and didn't endorse either candidate. I think his paper was pretty good with the Clintons."[15]

In any event, as Obama prepared to go off on a preconvention vacation in Hawaii in early August, during which he would make up his mind, the list was finally shaved to three, and Hillary Clinton was not on it. The survivors were Joe Biden, Evan Bayh, and Tim Kaine. Bayh, according to Plouffe later, was regarded as the safest choice in terms of dependability to stay on message, but was seen as a bit bland. Kaine, the first governor to endorse Obama, shared his pragmatic approach, and they were quite compatible. But neither man could match Biden's foreign policy experience and associations in the Senate, nor his compelling personal story and identification with blue-collar America.

On the down side, Plouffe wrote later, the old plagiarism charges against Biden "would of course be rehashed"; in addition, "he was also known to best even the Senate's standard for windiness" and "was prone to making gaffes." But, the campaign manager wrote, "it was in his DNA and we couldn't expect that to change. . . . Overall, though, we liked what we saw . . . and through our many shared debates we thought he comported himself quite well" and would do the same in the one vice presidential debate in the fall.[16]

"We knew Biden could be somewhat long-winded and had a history of coloring outside the lines a bit," Plouffe also said. "But honestly, that was very appealing to Obama, because he wanted someone to give him the unvarnished truth. What do you need in a vice president? He knows and understands Congress, has great foreign policy and domestic experience. He had the whole package from a VP standpoint." Particularly persuasive, Plouffe said, was respect for Biden among House and Senate members of both parties voiced to Obama vetting agents Caroline Kennedy and Eric Holder.[17]

Before the final choice was made by Obama, Axelrod and Plouffe under cover called on each of the three. Biden was the first, and they met at Valerie's home just across the Pennsylvania line from Delaware. Plouffe wrote later that Biden launched the critical meeting with "a nearly twenty-minute soliloquy" ranging from the Iowa caucuses to his views on Obama, why he didn't want to be vice president and why nevertheless he would be a good choice. "The last thing I should do is VP," Plouffe recalled him saying. "After thirty-six yeas of being the top dog [on two Senate committees for part of the time], it will be hard for me to be number two. . . . But I would be a good soldier and could provide real value, domestically and internationally."

The two top Obama strategists said later they were more amused than concerned about Biden's opening volley, during which, Plouffe recalled, "Ax and I couldn't get a word in edgewise." But they were impressed by his grasp of the issues "and the fact that while he would readily accept the VP slot if offered, he was not pining for it," nor was there any negotiation involved.[18]

The call, of course, was Obama's alone, and upon his return from Hawaii in August, he and Biden met secretly in a hotel room in Minneapolis to discuss the vice presidency. Obama cautioned Biden that if he accepted the nomination and they were elected, he would have to consider the job as a step up, the crowning political achievement of his career, not a diminution.

According to an interview with *Newsweek*, at one point Obama asked him: "Will this job be too small for you?" And Biden replied: "No, not as long as I would really be a confidant." He told Obama, "The good news is, I'm sixty-five and you're not going to have to worry about my positioning myself to be president. The bad news is, I want to be part of the deal." Biden said he told Obama he was "ready to be second fiddle" as long as he was guaranteed to have a private meeting with him as president every week and was present at all key meetings. They agreed, he said, that his job would be as a chief adviser who would help get Obama's agenda through Congress.[19]

Although Biden had voted for the Bush resolution that paved the way for the invasion of Iraq, and Obama had opposed the war at the time as a candidate for the U.S. Senate, they were now in agreement that its conduct had been disastrous, and both were urging a measured withdrawal of American forces. One Republican and mutual friend of both men, Chuck Hagel, recalls he told Obama around this time that from his outsider's vantage point, Obama had only one sensible choice as his running mate, and that was Biden. Obama, he remembered, said Biden was certainly at the top of his list. "I don't mean to imply I had any special role in the decision," Hagel said later.[20]

In the end, Biden's experience in foreign policy was critical in Obama's choice, underscored by his views on the clash between Russia and Georgia at the time. His summons to the breakaway Soviet state to confer with Georgian President Mikheil Saakashvili on the crisis apparently impressed Obama. One of the running mate also-rans, Evan Bayh, said later that Obama had mentioned that fact to him in a phone call informing him he wasn't being picked.[21] (Later, at the Democratic Convention in Denver, Biden met with Georgian Vice Prime Minister Giorgi Baramidze and expressed his support for Georgia's membership in the NATO alliance.)[22]

When Obama finally decided on Biden, he reached him at a dentist's office, where the senator had taken his wife to have a root canal done. Although his selection of Biden was widely regarded as an asset to his ticket, bringing the wide foreign policy experience the presidential nominee himself lacked, many doubts were raised about whether Obama could afford the sort of distracting gaffes that had marred Biden's own bids for the presidential nomination. But in the end, the pluses outweighed the minuses in the nominee's mind. Upon Biden's selection, he was given the Secret Service code name "Celtic" for obvious reasons.

According to Axelrod, Biden's performances in the series of 2007 debates among the Democratic contenders for the party nomination "unquestionably" played a key role in Obama's decision to choose him. "It was Senator Obama's perception," Axelrod said later, "that Biden was consistently very good. He came across as mature, he came across as commonsensical, he came across as strong. And he exhibited the discipline that people questioned whether he would have.

Obviously it all culminated in that one hilarious one-word answer—'Yes'—to whether he could control himself, posed in the South Carolina debate."

Beyond that, Axelrod said, "We were impressed by the way he carried himself in the campaign." He remembered it was Obama around May 2008 "who first surfaced Biden's name as a potential running mate. One of the things he said was, 'We've been in a lot of debates together and I've been impressed with how he handled himself.' "[23]

Biden said later that while the debates in themselves didn't mean much in terms of the eventual outcome, he recognized that they were an important factor in Obama's choosing him. Without any apparent trace of self-praise, he recalled after one of the last debates having commended Obama on the job he had done, and Obama telling him: "Look, don't you know I know you beat me in the debates?" In light of that remark, Biden said, "Do I think it had an impact? I guess it probably did, because he actually mentioned it."[24]

At the same time, Axelrod said of Biden's notorious verbosity: "That dimension of Biden wasn't exactly a secret; it wasn't a skeleton in his closet. It was a quality that had been joked about. It was a very bruted-about idiosyncrasy. Our attitude was that he demonstrated during the campaign that he had the discipline to color within the lines. . . . Everybody's strength is also their weakness, and Joe Biden's strength is that he is real, he's genuine, he's ebullient . . . and he's credible as a result. The flip side of that is that sometimes he goes on some."[25] In other words, his ticket to acceptability to Obama was, once again, "That's Joe."

"At the end of the day, when Obama made the decision," Axelrod concluded, "his judgment was that the pluses far outweighed the negatives. . . . Look, he knows a lot about a lot. He obviously brings a great expertise on foreign policy, he brings great expertise on criminal justice policy, and he brings a really deep understanding of this town and Congress, and how it works and the relationships here. All of that was appealing to Obama, and we knew that for voters there was comfort in having a running mate with those qualities. Obama was plenty of change for everyone. We didn't need a running mate who validated Obama in the nature of change.

"We needed a running mate who balanced the ticket and who brought some gray hair and years of experience here in Washington. Biden brought that, and he came from a working-class Catholic family in a pivotal part of the country and still spoke to that experience. That's who he is; he's a working-class guy, thinks like that, speaks like that, and that was very appealing. So there were some very strong chips on one side of the argument and then there was that, well, you know sometimes he talks a lot. That was a trade-off we were more than willing to make."

Obama himself, in response to submitted questions for this book after the election, said of his choice: "Well, of course I knew Joe as a colleague in the Senate, where I served on his committee. I knew that he was knowledgeable and tough, and I was impressed by his performance in the debates during the primaries. He was probably as consistent and as good as anyone out there, and he handled the questions with a kind of strength and wisdom that impressed me. I also knew that Joe had a really strong connection with middle-class families and with middle America. He was a guy from a working-class background in Scranton, Pennsylvania, and he's never lost touch with that experience.

"Joe was also a seasoned veteran, and someone who had experienced the pressures of a national campaign. He knew what to expect. I learned as a candidate that it takes time to adjust to the pressures of a national campaign, and I felt that whoever I chose as my running mate needed to be someone who was battle-tested and ready to go."

Obama added the mandatory caveat. "In choosing a running mate, I felt a very strong responsibility to pick someone who would be able to step in and serve as president if something happened to me, and Joe certainly was qualified. That was implicit in our discussions, and it was certainly foremost in my mind as I made the decision. . . . There is no question that Joe Biden was prepared."[26]

Integrating Biden into the Obama campaign, Plouffe said later, "was greatly facilitated by Biden's own attitude." "He immediately said, 'You know what? This is your campaign. If I'm doing something I shouldn't be doing, let me know,'" Plouffe recalled, "and he was true to that."[27]

Finally, Axelrod said, "when Biden agreed to run, all he asked for was to be in the room to be a counselor, to be able to bring the value of his experience to some of the major policy decisions that the president was making." That job description squared easily with how Obama also envisioned the vice presidency under him, Axelrod said. And after Cheney, he said, his boss "knew what he didn't want and wasn't looking for. He wasn't looking for a kind of prime minister, he wasn't looking for a vice president/chief of staff. He was looking for a vice president who could carry his message effectively, who could take on assignments effectively, and could offer broad advice and counsel."[28]

The innovative Obama chose to make his selection of Biden known via the Internet that had been an indispensable part of building wide support for his candidacy. He sent text messages and e-mail to his campaign volunteers before introducing Biden at a massive outdoor crowd at the state capitol in Springfield, Illinois. After his new running mate, in shirtsleeves, trotted across the platform to join him, Obama called him "that rare mix" of a man who for decades had "brought change to Washington, but Washington hasn't changed him. He's an expert on foreign policy whose heart and values are firmly rooted in the middle class. He has stared down dictators and spoken out for America's cops and firefighters. He is uniquely suited to be my partner as we work to put our country back on track."[29]

Biden immediately took on the traditional running mate's role as a critic of the opposition party's nominee, linking Republican John McCain with George W. Bush, the embattled lame-duck president.

After describing McCain as "genuinely a friend of mine" and a courageous war hero, Biden said, "you can't change America and end this war in Iraq when you declare—and again these are John's words—'No one has supported President Bush in Iraq more than I have.'" Biden added, "You can't change America when you know your first four years as president will look exactly like the last eight years of George Bush's presidency."[30]

The choice of Biden effectively ended the demands and hopes of Hillary Clinton's supporters that Obama would choose her as his running mate. In any event, when in the heat of their competition in the primaries Obama airily observed that the former first lady was "likeable enough," her selection seemed most unlikely.

At the Democratic National Convention in Denver, an emotional highlight was the introduction of Biden's son Beau, the attorney general of Delaware and an officer in the National Guard soon to be deployed to Iraq.

Joe Biden the family man responded in a familiar way. "Beau, I love you," he said. "I am so proud of you. Proud of the son you are. Proud of the father you've become. And I'm so proud of my son Hunter, my daughter, Ashley, and my wife, Jill, the only one who leaves me breathless and speechless at the same time." Then he saluted Bill and Hillary Clinton, and thanked Delaware. "Since I've never been called a man of few words, let me say this as simply as I can: Yes. Yes, I accept your nomination to run and serve alongside our next president of the United States of America—Barack Obama."

Biden wasted no time giving the convention a taste of the kind of campaigner he would be. "Let me make this pledge to you here and now. For every American who is trying to do the right thing, for all those people in government who are honoring their pledge to uphold the law and respect our Constitution, no longer will the eight most dreaded words in the English language be: 'The vice president's office is on the phone!'"

After the cheers died down, Biden retold his story—of his humble beginnings in Scranton as a stuttering child of supportive parents (introducing his ninety-one-year-old mother standing by) and how they instilled faith, pride, and toughness in him. He spoke of his departed father's instruction to him that was as good a description of his own life's journey as any—"Champ, when you get knocked down, get up! Get up!" And his mother's Irish admonition "when I got knocked down by guys bigger than me. She sent me back out and demanded that I bloody their nose so I could walk down the street the next day." His mother's creed, he said, was "No one is better than you. You are everybody's equal and everyone is equal to you."

He told Obama's story as well, and of his own long and cherished friendship with John McCain, the Republican standard-bearer, despite McCain's support of the soon-to-depart George W. Bush. He spelled out the Democratic case against them in terms that he said were on the minds of middle-class Americans: "Should Mom move in with us now that Dad is gone? Fifty, sixty, seventy dollars to fill up the car? Winter's coming. How we gonna pay the heating bills? Another year and no raise? Did you hear the company may be cutting our health care? Now we owe more on the house than it's worth. How are we going to send the kids to college? How are we gonna be able to retire?" He concluded, "That's the America that George Bush has left us, and that's the future John McCain will give us."[31]

Biden proceeded to chronicle McCain's record of supporting Bush 95 percent of the time, including his tax cuts for corporate America, tax breaks for oil companies, opposition to an increase in the minimum wage, and so on. The speech left no doubt that Biden was ready, willing, and able to play the running mate's traditional attack role, regardless of his friendship with McCain.

Back in Wilmington, the local *News Journal* that had often been a Biden critic pulled out the stops to boost Delaware's favorite son. An oversize cartoon showing him wearing a cape bore the headline: HE'S A POLITICO! HE'S OUR JOE! . . . NO, HE'S SUPER JOE! A full page followed of testimonials from local readers to Biden's reputation as a good citizen. Typical was one from Mary Hartnett of Wilmington, who wrote of an incident thirty-one years earlier, when a boy snatched her purse and started to run away: "Senator Joe Biden was driving by my house. [He] jumped out of the car, not thinking, 'Gee, I'm a senator. Let the cops do this.' [He] ran through a couple of backyards, over a couple of fences and was gaining on the kid, when the kid threw down the pocketbook and continued to run. Biden retrieved my purse and brought it back to me. For thirty years, I have kept this pocketbook. I wrote him . . . and told him I was keeping it so that, one day if he became president, he would sign it for me. He said he would."[32]

Biden's selection as the Democratic vice presidential nominee had one downside. It generated a regurgitation of critical stories about his campaign supporters, including MBNA, and about business involvements of members of his family. The *Washington Post* reported that his brother Jimmy and son Hunter were being sued on allegations of "defrauding a former business partner and an investor of millions of dollars in a hedge fund deal that went sour." The story said Jimmy and Hunter Biden in turn charged the old partner with "misrepresenting his experience in the hedge fund industry and recommending that they hire a lawyer with felony convictions."

The *Post* reported that Anthony Lotito Jr. claimed in court papers that Jimmy Biden had called him in 2006 to arrange for a high-paying job for his nephew because he "was concerned with the impact that Hunter's [previous] lobbying activities might have" on his father's expected 2008 presidential campaign.[33] The Bidens denied the allegations.

The *New York Times* meanwhile reported that Hunter Biden had received consulting fees from MBNA from 2001 to 2005 at a time when his father was working for passage of a bankruptcy protection bill sought by the credit card industry. The story said Biden voted four times for the bill before it finally cleared the Senate in 2005. It also said Biden was one of only five Democrats who voted against requiring credit card companies to give better warnings about the consequences to consumers of paying only minimum required monthly payments.[34] Finally, *USA Today* reported that Biden in the early 2000s had worked to defeat asbestos-damage lawsuits when his older son Beau was working for a Wilmington law firm handling asbestos litigation cases.[35]

None of these stories, however, drew much national attention for two reasons. First, after having served thirty-six years in the Senate and having run twice for the presidency, Biden was regarded as a known quantity. Second, his Republican opponent for the vice presidency was the opposite—a fresh and sparkling personality unknown to most of the nation and thrust into an inquiring spotlight—Governor Sarah Palin of Alaska. Plucked from national obscurity as a long shot gamble by John McCain to rescue his lackluster campaign, Palin became an overnight sensation. Almost at once, all the speculation concerning Biden was whether he could handle her in their scheduled nationally televised debate in the first week of October.

On September 4 Biden's hometown paper, the *Wilmington News Journal*, splashed on its front page a large photo of Governor Palin addressing the Republican National Convention in St. Paul. On the same page, below the fold, this headline ran under a one-column story: NO DUI IN CRASH THAT KILLED BIDEN'S 1ST WIFE, BUT HE'S IMPLIED OTHERWISE. The story was a resurrection of the tragic 1972 accident in which Neilia Biden and infant daughter Naomi were killed and sons Beau and Hunter injured when their car was in a collision with a tractor-trailer.

What brought the story back into the news was the fact that a syndicated television news show, *Inside Edition*, had recently run an old clip of Biden saying the driver of the truck, Curtis C. Dunn, had "stopped to drink instead of drive." His daughter, Pamela Hamill of Glasgow, Delaware, saw and heard the clip and, distressed, phoned the newspaper, telling a reporter, "To see it coming from his mouth,

I just burst into tears." Her father had died in 1999, and she said he "was always there for us. Now we feel we should be there for him because he's not here to defend himself."[36]

The reporter who wrote the story, Rachel Kipp, said later she detected no political motive or animus in the call.[37] The daughter released a statement on behalf of her family saying her father's "unfortunate involvement" in the Biden deaths "was in no way a result of any negligence of our father," and that "he was not at fault and was never charged with any wrongdoing whatsoever." Noting that Biden said the event had "changed the trajectory" of his life, she said it had also changed her father's. "We're not trying to equate Senator Biden's loss to my father's heartache," she wrote. "But we wanted it to be known that our father never forgot that tragic day," and that he "would often say, 'I wonder how the little Biden boys are doing.' "[38]

Jerome Herlihy, the state official who had reviewed the matter at the time of the crash and had reported no evidence that Dunn had been drinking, reiterated to the Wilmington paper, "The rumor about alcohol being involved by either party, especially the truck driver, is incorrect." In any fatality in which driving under the influence of alcohol was involved "there's going to be a charge," he said, and there was none.[39]

But a few days after the September 11, 2001, terrorist attacks and twenty-nine years after the accident, Biden had addressed the students and faculty of his alma mater, the University of Delaware, and referred to the 1972 accident, saying he knew how the families of the 9/11 victims felt. According to a transcript on Biden's archived Senate Web site quoted by the Wilmington newspaper, he said: "It was an errant driver who stopped to drink instead of drive and hit a tractor-trailer, hit my children and my wife and killed them."[40]

Hamill wrote that Biden's equating his grief with that of the families of the 9/11 victims compelled her to write to him at that time "on behalf of our father, explaining the effect [the accident] had on him, as well as our family." Biden replied to her note, writing longhand:

"Dear Ms. Dunn, I apologize for taking so long to acknowledge your thoughtful and heartfelt note. All that I can say is I am sorry for all of us and please know that neither I nor my sons feel any animosity whatsoever. Warmly, Sen. Biden."[41]

However, six years later, in December 2007, Biden repeated the assertion as he was running in the 2008 Iowa presidential caucuses. Speaking at the University of Iowa, he was quoted in the *New York Times:* "A tractor-trailer, a guy who allegedly—and I never pursued it—drank his lunch instead of eating it—broadsided my family."[42] Inasmuch as an investigation right after the accident had exonerated the driver, the question was why, after all this time, Biden would have reiterated the allegation. His campaign press secretary at the time, David Wade, noted that Biden had said "'allegedly—and I never pursued it,' nor did he encourage reporting on it then or at any other time. He has never called it or thought of it as anything but an accident."[43]

The allegation nevertheless refused to die. Hamill said later she had heard CBS anchor Katie Couric repeating it on the air during the 2008 Democratic Convention and later during the inauguration festivities, and she complained to the network. After an exchange of e-mails between Hamill and Couric, on March 24, two months later, CBS Evening News ran a brief clip of the controversy quoting Herlihy repeating that "there was no indication that the truck driver had been drinking" and reporting Biden saying he "fully accepts the Dunn family's word that these rumors were false." The network's report said only that "multiple news outlets, including CBS, have reported Dunn was drunk," offering no explanation or apology for its having aired the allegation when Herlihy had long ago reported it to be unfounded.[44]

The next day, Hamill said, she received a phone call from Biden. In a ten-minute private conversation, she said, "he did redeem himself by doing that," and that "I did feel closure" at last about the unfortunate allegations.[45] Some friends later speculated that the tragedy was so psychologically damaging to Biden when he made those comments in 2001 and again in 2007 that he must have needed some such explanation within himself in order to cope with his great loss.

Biden's campaigning in the fall for the Obama-Biden ticket touched on happier memories of the past. On Labor Day, he returned to Scranton with his mother for a holiday picnic in the backyard of their old house at 2446 North Washington Avenue, hosted by the current owner and occupant, Anne Kearns. "This is where my family values and faith melded," he told some three dozen neighborhood friends. "I learned about neighborhood. It's your first loyalty."[46]

His hostess, the mother of six, had bought the house from Joe's maternal grandfather, Ambrose Finnegan. She told of finding two steel bed frames in an attic room and painting them for her kids there, where young Joe had slept years earlier. On the wall, Mrs. Kearns said, were two statements: the old World War II joke, "Kilroy Was Here," and then "Joe Biden Was Here." She had painted over them, but asked the vice presidential nominee to sign the wall now.

"Can I go up?" Biden asked, and followed the two Kearns sons to the attic, once used by his aunt Gertie Finnegan. "Where was it signed before?" he asked. "You actually want me to sign the wall?" When assured by one of the boys the family would be honored, Biden pulled out a Sharpie pen and scribbled: "I Am Home. Joe Biden, 9-1-08," saying "if my father was here, he'd smack me for writing on the wall."[47]

The Scranton and the Green Ridge neighborhood Biden found were mostly unchanged to the naked eye, but the town had by now seen hard times. Its population had shrunk from about 125,000 to 74,000, and the coal-mining industry that had been its backbone was mostly a memory.

The downtown Lackawanna County Square sported a plaque noting that the first electric streetcar system was built in Scranton in 1886, running from downtown to Green Ridge. There was also a Distinguished Citizens of Lackawanna County Memorial that listed former governor Robert P. Casey and Congressman Joseph M. McDade, but no mention of Joe Biden.

The sentimental journey to Scranton was a positive kickoff for his vice presidential campaign, but inevitably some negative news stories surfaced. The Associated Press sought and got from the Obama campaign Biden's Selective Service record, which showed he had received five student draft deferments, matching the number Vice President Dick Cheney had received. Biden had been given the deferments as a student at the University of Delaware and then at the Syracuse University Law School. After a physical examination in April 1968, Biden received a 1-Y classification, meaning he could be drafted only in the event of a national emergency. A campaign press aide said the classification was the result of Biden's having had asthma as a teenager.[48]

The *New York Times* also took another look at Biden's relations with MBNA, the giant Delaware-based credit card company, and at his comfortable lifestyle, which critics said belied his constant references to himself as a product of the American middle class. The story noted that "he is among the least wealthy members of the millionaires club that is the United States Senate—he and his wife, Jill, a college professor, earn about $250,000 a year." A review of Biden's finances, the story said, "found that when it comes to some of his largest expenses, like the purchase and upkeep of his home and use of Amtrak trains to get around, he has benefited from resources and relationships unavailable to average Americans."

The *Times* account went on: "As a secure incumbent who has rarely faced serious competition during thirty-five years in the Senate, Mr. Biden has been able to dip into his campaign treasury to spend thousands of dollars on home landscaping and some of his Amtrak travel between Wilmington, Delaware, where he lives, and Washington. And the acquisition of his waterfront property a decade ago involved wealthy businessmen and campaign supporters, some of them bankers with an interest in legislation before the Senate, who bought his old house for top dollar, sold four acres at cost, and lent him $500,000 to build his new home."

The *Times* noted "there is nothing to suggest Mr. Biden bent any rules in the sales, purchase, and financing of his homes. Rather, he appears to have benefited at times from the simple fact of who he is: a United States Senator, not just 'Amtrak Joe,' the train-riding everyman that the Obama-Biden campaign has deployed to rally middle-class voters." The story quoted Ronald Tennant, the loan officer who handled the Biden mortgages, as saying "he was a VIP, so he was treated accordingly by the bank." He wasn't given a below-market interest rate, Tennant said, and "we paid particularly close attention to make sure everything came out right."

According to the *Times*, Biden's Senate salary was $165,000, augmented by teaching law courses at Widener University in Wilmington and advances of $225,000 for his bestselling memoir of 2007. Also noted were the facts that he had obtained or refinanced mortgages twenty-nine times since being in the Senate, and at the time owed $730,000 on two home mortgages as well as some other personal loans. The story mentioned, too, the campaign contributions to Biden by MBNA employees and the sale of his home to MBNA executive John Cochran that figured in the refuted smear against him in his 1996 reelection campaign against Ray Clatworthy.[49]

The highlight of the campaign for the vice presidency was supposed to be the nationally televised debate between Biden and the surprise Republican vice presidential nominee, Governor Palin of Alaska. She turned out to be made for TV—perky, folksy, good-looking, good-humored, and surprisingly quick on the draw, seemingly a match for the glib Irishman from Delaware. She was an unanticipated sensation at the Republican National Convention in St. Paul, and in the month leading up to the debate she drew huge crowds, easily eclipsing Biden's speaking events. But Biden, a smoothie with women, and with the home-grown manners of a gentleman for all his brashness, took the challenge in stride in the debate.

In advance, the Obama strategists took the confrontation with the crowd-pleasing, unpredictable Palin seriously. "We knew, and we prepared for the fact that it could be kind of wild and wooly," Axelrod said.[50]

"Debating Sarah Palin was an enormously difficult task with very little upside," Plouffe said, "because people thought Palin would be lucky to get off the stage without a monumental gaffe. But we knew having studied her that she was a damned good debater."

Biden was grilled in predebate practice sessions to focus and stay on Obama's message rather than his own record, which he was used to proclaiming and defending at length. In doing so, Plouffe said later, Biden "internalized" the case for the presidential nominee to the point that "he began not to see the campaign through the prism of his [own] eyes" but of Obama's. "He kind of wore the uniform and said, 'This is what Obama wants to do, this is what Obama's record is.' Candidly, that was a hard thing for him to do. That was the one thing in debate practice we had to keep going back to, and he eventually mastered it."[51]

In mock-debate preparations, Michigan's sharp and attractive governor, Jennifer Granholm, played Palin with hilarious authenticity. "She really had done her homework," Axelrod said, "and had all her speech patterns down. . . . We thought, well, she's really playing this almost farcically. The real Palin couldn't possibly behave this way. But it turned out be a really apt, almost uncanny, channeling of Palin. When Biden finally came face-to-face with the real Governor Palin, there weren't too many surprises. He knew what to expect."

Biden had a mandate to adhere to the campaign message of concern about and service to middle-class voters, and he stuck to the script. "He time and again returned to the plight of the middle class in this economy," Axelrod said. "That was our strategy and that was our desire that he really hone in on that, how the middle class had fared under the Republicans, why we needed to stay there regardless of where she took the thing. And he was unshakable in that debate. He came back to those themes again and again. And that's what we'd seen during the primaries; we knew he had the capacity to turn in a performance like that."[52]

For ninety minutes, the two nominees exchanged predictable policy positions and defended their parties' standard-bearers. From the outset, Palin injected a tone of folksiness beginning with the introductions of the nominees by Gwen Ifill of the PBS *Newshour* show, asking her opponent, "Hey, can I call you Joe?" In Palin's debate prep, according to John Heilemann and Mark Halperin in their postelection book, *Game Change*, she had repeatedly referred to her opponent as Senator O'Biden until her handlers suggested that she just call him Joe. To which, they wrote, she replied, "But I've never met him."[53] Nevertheless, that's what she did. The reply from Biden, customarily never for a loss of words, was inaudible to listeners, and the two thereafter engaged in cordial sparring.

Most of the debate was given over to discussion of the desperate state of the economy, which Biden led off blaming on "excessive deregulation" that let "Wall Street run wild." Palin claimed she and McCain as a "team of mavericks" would set it all right. She blamed "predatory lenders" for the mess, and when asked about the nation's inadequate health care system, she segued into the old Republican allegation that the answer from Obama and Biden would be raising taxes. In an obvious strategy to find safe ground, she told Ifill and Biden, "I may not answer the questions the way that either the moderator or you want to hear, but I'm going to talk straight to the American people and let them hear my track record, too," as a tax cutter in Alaska.

Biden, though, demonstrated his experience in the exercise. What Republicans liked to call "redistribution of wealth" and "class welfare" in Obama's plan to raise taxes only on earners of $250,000 or more a year, he said, was "tax fairness." She defended drilling in the Alaska National Wildlife Refuge as part of McCain's call to tap all available energy resources. She defended the troop surge in Iraq that Biden had derided, and called his plan for measured withdrawal "a white flag of surrender." She reminded viewers that Biden had once said Obama was "not ready" to be president, and dropped in the name of Henry Kissinger as a credential.

Through it all, Biden remained cordial but persistent in reminding viewers of McCain's general support of the George W. Bush agenda and record. He cited his own role in the successful American interventions through collective action in Bosnia and Kosovo. And when Palin borrowed the old Reagan line of "there you go again" from his debate with Jimmy Carter, Biden countered with an updated version of Reagan's "are you better off than you were" in Carter's tenure.

Near the end of the debate, the moderator sought to bring out the candidates' capacity to adjust their views under pressure. She asked Biden and then Palin: "Can you think of a single issue, policy issue in which you were forced to change your view in order to accommodate changed circumstances?"

Biden responded directly and to the point. "Yes, I can," he said. "When I got to the United States Senate and went on the Judiciary Committee as a young lawyer, I was of the view, and had been trained with the view, that the only thing that mattered was whether or not a nominee appointed, suggested, by the president had a judicial temperament, had not committed a crime of moral turpitude, and was—had been a good student. And it didn't take me long—it was hard to change, but it didn't take me long—but it took about five years for me to realize that the ideology of that judge makes a big difference."

The reason he fought the Bork nomination, he said, was because he understood undesirable changes that might well occur with him on the court. He said "it matters what your judicial philosophy is" and "the American people have a right to understand it and to know it."

Palin in a garbled reply said only that she had "quasi-caved in" in some budget fights as a mayor and governor, and, in an incomprehensible explanation added "there were times when I wanted to zero-base budget and cut taxes even more, and I didn't have enough support in order to accomplish that."[54] Whatever that answer meant, it was no response to the question asked.

Biden outshone Palin with his more polished grasp of the issues and a confident debating style, but the Palin faithful seemed satisfied with her boilerplate conservative replies, as convoluted and confusing as they often were. The debate at Washington University in St. Louis sparked more curiosity than substantive discussion, and Biden came out the clear winner in the postdebate polls. United Press International for one reported that 51 percent of voters scored Biden superior to 36 percent for Palin.

Even so, Palin was credited with exceeding expectations. In her predebate appearances, her inexperience had been so conspicuous that an increasingly nervous McCain campaign camp had held her in the political equivalent of house arrest, sharply restricting her exposure to the news media hordes on her scent. When she eventually was given a freer rein, Palin became more a caricature than a serious adversary, though her popularity and appeal remained high with the party's right-wing base as she continued to give voice to its righteous litany.

Through all this time, Biden's reputation for careless remarks was reinforced with every gaffe no matter how trivial. In the debate, the *Wall Street Journal* called him out on several misstatements regarding the Middle East that went uncorrected.[55] A year after his comment that "you can't go to a 7-Eleven or a Dunkin' Donuts unless you have a slight Indian accent," it was still being recycled in videos and articles. His explanation that he had merely been commenting how the middle-class Indian population was growing in Delaware didn't seem to satisfy his critics.

In mid-September, he added to the gaffe list an observation that "Hillary Clinton is as qualified or more qualified than I am to be vice president of the United States—let's get that straight . . . and quite frankly it might have been a better pick than me. I mean that sincerely, she is first-rate."[56] And then there was Biden's remark to Katie Couric of CBS News that "when the stock market crashed, Franklin

Roosevelt got on the television"—three years before he was president and when television was merely in its experimental stage.[57] Worst of all though, may have been when he urged a wheelchair-bound state senator, Chuck Graham, at a rally in Columbia, Missouri, to "Stand up, Chuck! Let 'em see ya!"[58]

At a closed fund-raiser in Seattle late in the campaign, Biden gave the Republicans a line they could chew on when he said, "Mark my words. It will not be six months before the world tests Barack Obama like they did John Kennedy. Watch, we're going to have an international crisis, a generated crisis to test the mettle of this guy."[59] John McCain seized on it to suggest Obama wasn't ready to face such crisis, but Obama dismissed the remark by saying "I think that Joe sometimes engages in rhetorical flourishes."[60]

Later, a prominent political science professor at Delaware State University, Samuel B. Hoff, pointed out there was ample history of new presidents confronting major crises and decisions in their first six months in office, including Truman's momentous use of the atomic bomb and Kennedy's Bay of Pigs invasion, as well as other military challenges in the first years after. Hoff noted: "Of course, the reality is that the [new] president of the United States [will be] taking office at a time of two active wars involving more than two hundred thousand American troops, as well as a collapsing economy and possibly an impending recession. Essentially, the test has already begun and the clock is already ticking."[61]

After the campaign, David Plouffe said the "gonna be tested thing" was "the one big mistake" Biden made "that the McCain campaign thought somehow would be a magic bullet. But what we heard from our research was that people thought, 'Of course he's going to be tested.' It wasn't something that worked, but obviously it caused an unnecessary distraction. Over sixty days he made one mistake," Plouffe said. "I'd take that any day in the week."[62]

In their postelection inside account of the Obama campaign, John Heilemann and Mark Halperin reported that Biden's penchant for loose talking had caused serious strains with Obama, and had led to his exclusion from the highest-level internal campaign conference calls. Upon Biden's comment that he was more qualified to be president than Obama, the authors wrote, "a chill set in between Chicago and the Biden plane. . . . Joe and Obama barely spoke by phone, barely campaigned together."

The authors also wrote that Obama grew "increasingly frustrated with his running mate after Biden let loose with a string of gaffes, including a statement that paying higher taxes amounted to patriotism and criticism of one of the campaign's own ads poking fun at John McCain. But when Biden, at an October fund-raiser in Seattle, made the prediction that Obama would be tested with an international crisis, the Illinois senator had had enough. 'How many times is Biden gonna say something stupid?' he demanded of his advisers on a conference call." According to the book, others on the call "said the candidate was as angry as they had ever heard him." Obama, the authors wrote, told Biden he was hurting the campaign with such remarks and Biden later apologized, healing whatever breach there was.[63]

Upon the book's publication, Biden spokesman Jay Carney declined comment on what he called "rehashed rumors about the campaign" and said the authors had never checked with the campaign on the accuracy of the quotes. But notably he did not deny their accuracy and said only that Obama and Biden had later "worked together very closely and successfully" in the first year of the Obama-Biden administration.[64]

After the election, when asked for this book about his running mate's proclivity for talking so much, President Obama dismissed it. "Well, obviously, that was something that Joe handled himself during the campaign," he said. "One of the things that impressed me most was how he did it with a combination of humor and discipline. He [had] proved during the primary campaign that he could be a very disciplined candidate, so that was not a major concern to me."[65]

In any event, Biden, the devotee of the ad-lib, during the last stages of the campaign eventually was speaking with a teleprompter, and this famously loquacious backslapper was keeping his distance from the reporters traveling with him. "There came a point toward the end when we felt like the goal line was in sight and we were doing a lot fewer press avails, not just Biden to the extent that was true, but also Obama himself," Axelrod explained. "Our thing was just to get to see as many people as possible, do as many rallies as possible."[66] Of course, meeting and talking with reporters on the plane en route obviously would not have taken up any additional time.

Meanwhile, Biden's reputation for exaggeration was fanned by a story in the *London Evening Standard* dated October 1 about remarks to the National Guard Association: "If you want to know where al-Qaeda lives, if you want to know where Bin Laden is, come back to Afghanistan with me, come back to the area where my helicopter was forced down, with a three-star general and three senators at 10,500 feet in the middle of those mountains; I can tell you where they are."

The *Evening Standard* observed, "But the reality was different. Bad weather forced his helicopter to land, [and] while American jets watched overhead, a land transport team was sent to retrieve him and his party." The story said Biden had "never explicitly said terrorists had brought down his helicopter, but many were left with the impression he had been under attack."[67] This latter observation was not the sort that would have passed muster in the *New York Times* and many other American newspapers, but it achieved worldwide circulation via the journalistically permissive Internet.

Biden's freewheeling way with words didn't seem to handicap him in his special mission to court blue-collar America that had been somewhat cool to Obama in key primary states like Ohio and Biden's native Pennsylvania. In mid-October, Bill and Hillary Clinton joined him in Scranton. Crowds whooped it up for Biden, for the former president who had carried Pennsylvania in 1992 and 1996, and for his wife, who had soundly beat Obama in the earlier state primary.

Of the assignment of shoring up the working-class vote, Biden said near the end of the campaign: "My job has primarily been to go into those territories that are pretty tough territories for the Democrats. I think we've done well. I'm not suggesting I'm the reason for it, but I mean, if the polling data's correct we've done very, very well among Catholics. . . . We've done very well in the very areas it was hoped I might be some value added. Again, it might not be because of me. But the fact is, it feels good."[68]

Generally speaking, however, Biden was relatively under wraps as election day approached. Although the Obama camp professed not to be concerned about his shoot-from-the-hip style, reporters covering him got restless over the fact that he hadn't had a real news conference since right after Labor Day or taken questions from voters since his comment that Hillary Clinton would have been a better running mate for Obama.

On election day, Biden of course was back home in Delaware, taking his mother, wife, and daughter to vote at Tatnall School near his home in Greenville, among locals cheering and waving DELAWARE FOR BIDEN signs. Then it was off to Richmond, Virginia, for a final campaign event and then a flight to Obama's hometown of Chicago.

When it was clear they had won, the president-elect introduced his running mate to the crowd. "I want to thank my partner in this journey," he said, "a man who campaigned from his heart and spoke for the men and women he grew up with on the streets of Scranton and rode with on that train home to Delaware, the vice president–elect of the United States, Joe Biden."[69]

For Biden, the destination was not the one he had set out to reach two years earlier on a second, long shot quest for the Oval Office. As in his first such endeavor more than twenty years earlier, he had soon fallen by the wayside. But this time he found himself almost literally on the doorstep of his original quest, with the promise of at least a small share of the glory and the responsibility he had envisioned for so long.

In the weeks between the election and the inauguration of the two new American leaders, Biden spent most weekdays in Chicago conferring with the president-elect and his top staff members. He and Obama discussed his role as vice president in greater detail, cementing the arrangement in which the older man would serve as what he himself called "adviser in chief."

Unlike his Democratic predecessor, Al Gore in the Bill Clinton administration, who initially was assigned heavy responsibilities in the specific fields of science and technology, Biden would be an insider generalist. Among his functions in Chicago was giving Obama his suggestions on cabinet appointments and assessing prospective appointees in terms of Biden's knowledge of, and experience with, many of them.

David Axelrod told the *New York Times* of Biden, "I think his fundamental role is as a trusted counselor and an adviser on a broad range of issues."[70] And although during this period Biden was in direct contact with a number of foreign leaders, he told associates that he would not be the key administration adviser on foreign affairs. That task would go to the new secretary of state, Hillary Clinton, who, with Biden's strong concurrence, Obama had already nominated. Obama said later that Biden had been "involved in persuading her to serve," and was thereafter "in consultation" with her on the assignments given to him by the president.[71]

On Biden's sixty-sixth birthday, Obama at their weekly lunch gave him two Chicago sports caps, one for the White Sox and one for the Bears, and a dozen cupcakes each with a lighted candle, kidding him, "You're twelve years old!" Biden quickly snapped back: "Maybe in dog years!"[72] From all indications, the Obama-Biden team was off to a comfortable start. Joe Biden, soon to step down as the chairman of the Senate Foreign Relations Committee to play second banana on a broader stage, seemed already adjusted to fill the subordinate role with Biden-like relish for a new challenge.

In his first nationally televised interview as vice president–elect in late December, Biden told George Stephanopoulos of ABC News that Obama had already lived up to his commitment to make him a fully informed partner in the new administration. "Every single solitary appointment he has made so far," Biden said, "I have been in the room. The recommendations I have made in most cases coincidentally have been the recommendations that he's picked. Not because I made them, but because we think a lot alike."

Although Obama had already asked Biden to head up a new task force focused on the needs of the middle class, the vice president–elect said it was a job with a limited time frame and he would not be saddled with a specific area of responsibility in the way Al Gore had been under Clinton. And he reiterated that he intended to restore what he saw as balance in the job of vice president after Dick Cheney's controversial eight-year tenure. Reminded that he had called him "probably the most dangerous vice president ever" for the advice he had passed on to Bush, including his belief in the concept of the unitary executive giving all power to a president in time of war, Biden held fast. That idea, he said, "is dead wrong. . . . I think that it caused this administration in adopting that notion to overstep its constitutional bounds, at a minimum to weaken our standing in the world and weaken our security. I stand by that judgment."[73]

At the same time, Biden had already made clear that he had no intention of seeing the vice presidency revert to its earlier stature as a useless fifth wheel. At a minimum, he indicated he intended to reestablish the office to the status it had before Cheney occupied it as a shadowy and fearsome figure of immense power in his own right. No one was going to cast Biden as the "real" president as Cheney often was characterized by Bush's critics. But neither was he going to be a Dan Quayle, whose qualifications for even the minimal requirements of the job were repeatedly questioned and ridiculed during his uneventful four years in waiting. Joe Biden was taking the office with the expectation of being much more than an ornament and the determination to make it so.

But before that would happen, Joe Biden on January 6, 2009, took the oath of office of the U.S. Senate for the seventh time, having been easily reelected in Delaware on the same day he had won the vice presidency. At age sixty-six he was the youngest person ever to do so. In a moment of cordial irony, he was sworn in by Vice President Cheney, president of the Senate, and after the official Senate photo, Biden invited Cheney to join him in another shot with other Biden family members. It was almost as if Biden could not bear to leave the Senate after thirty-six years, and he so told old colleagues who came up to greet him, and later old Delaware friends who had come to the Capitol for the occasion.

Nine days later, he resigned the seat so that his longtime friend and Senate chief of staff, Ted Kaufman, appointed by Delaware Governor Ruth Ann Minner, could be sworn in. Kaufman had agreed to serve only until a special election could be held in 2010 to fill the rest of the unexpired term. The arrangement had all the earmarks of a seat-warming deal, with the expectation that Biden's older son, Beau, the Delaware attorney general serving with his National Guard unit in Iraq, would run for his father's old seat upon his return. But that could wait. Beau made no commitment as he came home on special leave for one of the largest and certainly most historic inauguration days the nation's capital had ever seen.*

---

* In January 2010, Beau Biden surprisingly announced he would not run for his father's former Senate seat, choosing to seek another term as Delaware's attorney general. He cited as a major reason his interest in concluding work he had already begun on legislation protecting against child abuse, a matter close to his father's own concerns.

TWENTY-EIGHT

VICE PRESIDENT BIDEN

AT TWO MINUTES before noon on the clear, crisp morning of January 20, 2009, under the watchful eye of president-elect Barack Obama, Joseph Robinette Biden Jr. of Delaware became the vice president of the United States on the south steps of the U.S. Capitol. He held the Biden family Bible in his left hand and raised his right hand as he repeated the oath of office given him by Justice John Paul Stevens, the senior member of the Supreme Court. Jill Biden stood radiantly at her husband's side with their three grown children, Beau, Hunter, and Ashley, nearby. It was a day and a setting Joe Biden had dreamed about for at least thirty years. If he was not the featured player, nevertheless his brightest smile shone down on the huge crowd gathered in celebration of the inauguration of America's first African American president and his chosen partner in governing the nation.

After the solemn oath-taking by himself and Obama, and the new president's stirring inaugural address, the Bidens joined the Obamas in a traditional luncheon in the Capitol and then walked gaily down Pennsylvania Avenue behind the new president and first lady, Joe and Jill beaming and blowing kisses to celebrants shouting "Joe! Joe!" in the familiar way he was always greeted. Once in a while he would break into a little trot, with Jill, Beau, Hunter, and Ashley keeping up.

Prior to the swearing-in of himself and the new president, Biden had joined Obama at the White House for coffee with the departing George W. Bush and Dick Cheney in the customary if forced cordiality of the traditional meeting. No new vice president had come to the office with a more impressive résumé than the man who had served thirty-six years in the U.S. Senate from Delaware and as chairman of two of its most powerful committees, Judiciary and Foreign Relations, part of the time simultaneously. John Nance Garner, Franklin D. Roosevelt's first vice president, had been Speaker of the House; and John F. Kennedy's, Lyndon B. Johnson, had been Senate majority leader, but neither could claim the longevity of service that Biden boasted. Garner had been in the House "only" thirty years and was speaker for three; Johnson had twenty-four years in Congress, a dozen each in the House and Senate, and had been Senate leader for eight years.

Yet for all of Biden's long service, his wide network of political colleagues and friends in both parties, and long-standing associations with foreign leaders across the globe, he came to the vice presidency as somewhat of a question mark to many. For all his hard-earned and deserved credentials, he continued to be seen by some Americans as an unbridled horse in political pastures. He was a loyal Democrat but not a blindly slavish one; he was a consistent supporter of liberal causes, but not a knee-jerk backer of all liberal legislation. He was a hail-fellow-well-met, but on occasion one with a quick Irish temper. Most of all, there was that compulsion to talk, and at times to exaggerate, and that incident two decades earlier when borrowing the language of another politician without attribution had brought accusations of plagiarism, of which he was later exonerated in a court review.

At the same time, the Joe Biden who took the oath as vice president had already demonstrated, in his second bid for the presidency and as the Democratic vice presidential nominee in 2008, both the ability and the willingness to function with admirable restraint and self-control, convincing Obama not only of his wide range of experience but also of his sagacity. Obama had chosen Biden as an obvious balance of age to his own youth and as a man steeped in foreign policy experience to make up for his own lack of the same. But he also saw in him a like-minded pragmatist who could be an active and trusted partner in governance.

In their preinauguration conversations in Chicago on the role Biden would play, that latter consideration appeared essential and critical. When Obama had first sounded out Biden on his availability, Biden had made clear he had no interest in being an ornament, and Obama had made it similarly clear that he had no interest in or intention of making him one. Biden's own job description as adviser in chief to the president without specific portfolio squared with what Obama said he wanted in his vice president. And as they went about the business of organizing the new administration, Biden arranged his prospective office accordingly.

Biden had said before the inauguration that he hoped to "restore the balance" of the vice presidency in the wake of predecessor Dick Cheney's expansion and insulation of the office as an unprecedented power center in its own right. In the first conversation between Obama and Biden about the vice presidency in that hotel room in Minneapolis, Axelrod said, "that was a stipulated understanding. They both were appalled at the degree to which Cheney executed what really was executive authority in the office. So there was total agreement that the vice president should play a supplementary role and not the primary role."[1]

Later, in office, Obama without reference to Cheney said of Biden's acceptance of the proper boundaries of the vice presidency: "I think that Joe in some ways is the model of what a vice president should be. He is both blunt and supportive. He is willing to take on tough assignments. He never shies away from offering his view, but he also understands that ultimately, as president, I have to make the final decision. In many cases, I think those decisions were better because of his sound advice."[2]

With that in mind, Biden and aides analyzed how the office had been conducted by not only Cheney but previous occupants, including Al Gore, Walter Mondale, George H. W. Bush, and even Dan Quayle. The vice president–elect had direct meetings and consultations with four of the five, excluding only Quayle because Biden regarded his vice presidency as one of ideological focus, caring for the concerns and wants of the conservative base of his party. Biden saw no need or desire to address an ideological focus in his own party. Instead he met and conferred mostly with Gore and Mondale, who had different approaches to the office and the job.[3]

Gore had come to the vice presidency with very strong and specific areas of interest and expertise, conspicuously including the environment, climate change, governmental reorganization (or "reinventing," it was called), and communications technology. In his eight years in the Clinton administration, he had immersed himself in these areas, working with cabinet members also focused on the same concerns, meanwhile by his mere presence fulfilling the office's constitutional role of standby for the president.

Mondale, aware of President Jimmy Carter's inexperience in Washington and particularly in working with Congress, wrote a famous lengthy memo to Carter when the presidential nominee was interviewing prospective running mates, outlining how he proposed to serve if chosen. He suggested he be more in the nature of a general adviser without specific portfolio who would be free to counsel with the president across the board, as circumstances and needs arose. Mondale gave Biden a copy of the memo, and they met three times. Biden, who thought similarly, concluded that this model was better suited to Obama's needs and his own desires and skills. With

Obama's full agreement, Biden proceeded to organize his own vice presidency accordingly.

With the role of the office in governing increasing, Biden expanded his staff to accommodate the greater responsibilities given him by Obama. Generally following the Mondale model, Biden carried over several staff departments from previous vice presidential offices and added a few others tailored to Obama's particular needs. Under Chief of Staff Ron Klain, who had held the same job in the Gore vice presidency, Biden created economic, domestic, and national security councils paralleling those of the president.

Then, as specific assignments came to him from Obama to oversee the economic recovery, monitor the stimulus outlay to states and cities, run a task force to look out for the interests of the middle class, and assume oversight responsibilities on the war in Iraq, support staff in those areas were added. With these assignments, Klain said later, the Biden version became "a bit of a hybrid" of the Mondale and Gore models, including both advisory responsibilities across the board and specific portfolios.[4]

By intent, the Obama and Biden staffs also had a degree of intertwining and overlap. Tom Donilon, a longtime political and foreign policy adviser to Biden going back to his first presidential campaign, became Obama's deputy national security adviser. Tony Blinken, Biden's foreign policy adviser, accompanied Obama on his 2008 trip to Iraq. And Jared Bernstein, an Obama campaign expert on economic policy in 2008, became Biden's chief economics adviser. That arrangement was a distinct difference from the Cheney years, where the Cheney staff, ruled by the iron-fisted David Addington, was an independent entity reputedly often at cross-purposes with the Bush staff.

Biden had always been on the go either at work or at play, or both, and on top of all these challenges he managed in his first months to take on speechmaking and fund-raising for the two Democratic gubernatorial races in New Jersey and Virginia and for several campaigns of Democrats running for congressional seats in 2010. The new vice president had before him the full plate he wanted, and then some, in saying he didn't want to be a ceremonial ornament.

Biden was given an office in the West Wing of the White House, a perk initiated by Carter for Mondale, and ready access to the president, fulfilling his condition in his preelection discussions with Obama that he be "in the room" when all major administration decisions were made. According to Klain, that meant in the first year attending all daily morning intelligence briefings with the president, a weekly one-on-one lunch with him, and conversing directly with Obama three or four times a day. During the first year, Biden was also at the president's side for most television appearances, announcements, swearing-ins, and the like. And, as with Mondale and Gore before him, in addition to his own office in the White House West Wing, Biden acquired staff working space in the adjacent Eisenhower Executive Office Building, in pre-Pentagon days the home of the War Department.

Biden's long service in Congress and particularly as chairman of the Senate Foreign Relations Committee brought him immediately into a hands-on role alongside Obama dealing with the overload of crises inherited from the Bush administration, as well as Obama's own priority-agenda initiatives that he declined to put on the back burner. The obvious items in the first category were the military stalemates in Iraq and Afghanistan and the economic meltdown at home; and in the second was the push to achieve broad health care reform in the new presidency's first year.

Concerning the two wars, a week before his swearing-in Biden had already gotten a jump on Obama's oversight assignment with a trip to Pakistan, Afghanistan, and Iraq. He was mandated by the president-elect to make a baseline assessment of the state of play in the region in advance of Obama's first policy review as chief executive to determine his course of action in Afghanistan. He returned convinced that the new administration needed clarity on its goals, with the prime focus on returning to the pursuit of al-Qaeda and its leadership.

This view was subsequently reflected in the commissioned report of former CIA official and counterterrorism expert Bruce Riedel of the Brookings Institution. The primary goal of disrupting, defeating, and dismantling al-Qaeda in the report substantially followed Biden's recommendation and was a key ingredient in Obama's initial decision in his first weeks as president to send twenty-one thousand more U.S. troops to Afghanistan. Biden's input foreshadowed his foreign policy advisory role that soon would dominate all his other assignments in terms of underscoring the influence of his voice in the administration.

Beyond this initial task as vice president–elect of appraising the challenge in Afghanistan, Biden upon taking office also was assigned in the foreign policy sphere the oversight of nuclear nonproliferation efforts. These included congressional approval of a comprehensive test-ban treaty and monitoring the lockdown of nuclear weapons in Russia and around the world. Finally, he was charged with shepherding the transition from American military to diplomatic and economic involvement in Iraq.[5]

Regarding the economic meltdown at home, Biden was aggressively involved in courting the Republican votes of Senators Susan Collins of Maine and Arlen Specter of Pennsylvania for Obama's $787 billion stimulus package. Upon its passage less than a month after the inauguration, the president made Biden the point man on tracking the implementation of the American Recovery and Reinvestment Act, and on monitoring the federal outlay. The assignment thrust him as "Sheriff Joe" into the midst of competition among states and cities, obliging him to deal directly in an administrative role with the affected cabinet secretaries and governors, mayors, and other local officials across the country.

Often in person and at least weekly by phone, the new vice president set deadlines for the dispensing of the stimulus money, called for regular periodic progress reports on projects undertaken, and generally functioned as a troubleshooter to hear and respond to the needs and complaints of state and local officials. He did so on a bipartisan basis, conferring in the early period with such Republican governors as Haley Barbour of Mississippi, Bobby Jindal of Louisiana, and Mark Sanford of South Carolina on how to get the money flowing quickly to the most needy and deserving places. In the first year, Biden visited more than fifty recovery projects around the country, including a new solar panel manufacturing plant in California, road construction in Maine, a new rail transit facility in Florida, and the extension of broadband connection in rural north Georgia.

In early conversations with Obama, the vice president urged him not to overlook the problems of middle-class families, prompting the president to appoint him to chair a new Middle Class Task Force. Together they established four areas as the pillars of its agenda: child care, elderly care, college student assistance, and savings for retirement, with chief economist Jared Bernstein as the executive director. Biden also pitched in on Capitol Hill in the task to pass health care insurance reform, consulting weekly with Senate Majority Leader Harry Reid, while taking care not to infringe on Senate prerogatives, which some earlier vice presidents had done. Pointedly, he did not ask to attend the Senate Democratic caucus as Cheney previously had attended his party's caucus.[6]

Biden's full load of responsibilities was lightened somewhat by the fact that for the first time in thirty-six years he would be saving three or more hours a day by giving up his Wilmington-Washington commute and living in the Vice President's Residence at the old Naval Observatory on Massachusetts Avenue. It was only a few minutes by chauffeured limousine from his White House office, and even with an early start and a late night he could have breakfast and dinner with Jill and see other family members living nearby with regularity.

In his first year, Klain confirmed, Biden kept up as best he could his customary attendance with his grandchildren's school events and for a change was able to do some social entertaining, inviting in his first months forty or more senators from both parties, sometimes with their spouses, for dinner at the vice presidential residence. Early on, he told his sister, to her surprise, that "my favorite vice president is Dan Quayle," because during his tenure Quayle had built a beautiful flagstone patio and swimming pool that was a magnet for the Biden grandkids and friends.[7]

For perhaps the first time in thirty-six years, Biden also would occasionally partake of a Senate or administration social event. In those first months, Joe and Jill Biden also were seen at various downtown restaurants, getting acquainted with the nation's capital city, which to them had mainly been his workplace all those years.

The change of venue did not affect, however, Biden's practice of remembering friends' celebrations and occasions of sorrow with unannounced personal appearances, though usually with his Secret Service protectors in tow. He also found time weekly, Klain said, to slip up to the Senate gym at the Capitol for a workout with some of his old colleagues. His schedule, when he wasn't out of town or abroad on administration business, was to spend the weekdays in Washington and at least half the weekend at the family home in Wilmington.

Meanwhile, Jill resumed teaching at Northern Virginia Community College in nearby Arlington, with Secret Service agents staying outside her classroom, dressed casually and relatively inconspicuous. If students asked about the vice president, she would acknowledge "he's one of my relatives" and let it go at that, which seemed to satisfy them.[8] She found time as well to continue her interest in boosting the community college system and extending assistance and compassion toward the families of military men and women serving in Iraq and Afghanistan.[9] That work took on personal pertinence when Beau Biden served a tour in Iraq with his National Guard unit.

As Biden assumed his heavy load of responsibilities that assured he would not be dismissed as a mere statue, the journalistic gaffe squad was ever alert for real or perceived missteps by the new vice president.

On inauguration weekend, Jill got the Bidens in a little temporary hot water on the *Oprah Winfrey* television show when she left an impression that her husband may have been offered his choice between the vice presidency and the office of secretary of state. She told her hostess she had observed to him, "If you're secretary of state, you'll be away, we'll never see you. I'll see you at a state dinner once in a while."[10] The White House, anxious to counter a notion that Hillary Clinton was Obama's second choice for that post, quickly stated Biden had been offered the vice presidency only.

In the first days also, reporters did not fail to catch and convey his good-natured but somewhat barbed ribbing of Chief Justice John Roberts at a swearing-in of key members of the White House staff. In administering the oath of office to the president days before, Roberts had garbled the designated words, and now Biden needled him about it. Obama, standing at Biden's side, reached out and put his hand on his vice president's arm with a look of mild disapproval. That was enough to start tongues wagging that "there goes Joe again," or that this could be an early sign of a coolness or even a schism at the top.

David Axelrod, however, dismissed Obama's gesture as an indication that "the president felt some solicitude for the chief justice and didn't want to pile on."[11] The senior adviser told the *Wall Street Journal:* "All of the vice president's insight and experience dwarf any minor gaffe or misstep."[12]

Nevertheless, Biden's public words were closely parsed for any evidence of a mistake or a diversion from lockstep agreement and support of the president. In a speech in early February, speaking about a conversation he had had with Obama, Biden said: "If we do everything right, if we do it with absolute certainty, if we stand up there and we really make the tough decisions, there's still a 30 percent chance we're going to get it wrong."[13]

Biden didn't say, though, what the subject discussed was, so Obama was asked by a *Washington Times* reporter at his first White House news conference: "Since the vice president brought it up, can you tell the American people, sir, what you were talking about? And if not, can you at least reassure them it wasn't the stimulus bill or the bank rescue plan, and if in general you agree with that ratio of success, 30 percent failure, 70 percent success?"

Obama replied, as laughter broke out, "You know, I don't remember exactly what Joe was referring to, [slight pause] not surprisingly. But let me try this out. I think that [what] Joe may have been suggesting . . . is that given the magnitude of the challenges that we have, any single thing that we do is going to be part of the solution, not all of the solution." He went on at greater length, saying, "not everything we do is going to work out exactly as we intended it to work out." He talked about "the need for a recovery package of a certain magnitude," tax cuts, investments, dealing with the financial system, housing, better regulation. He concluded: "Now, those are big, complicated tasks. So I don't know if Joe was referring to that, but I used that as a launching pad to make a general point about these issues. . . . I have no idea. I really don't."[14] In other words, Obama was joining the "That's Joe" chorus.

One of the more prominent platoon leaders of the gaffe squad, Maureen Dowd of the *New York Times*, she of the leaked Biden plagiarism story of twenty years earlier, caught the slur and blew the whistle on Obama for it. She wrote: "Joe is nothing if not loyal. And the president should return that quality, and not leave his lieutenant vulnerable to *Odd Couple* parodies of the sort aired earlier on television's *Saturday Night Live*."[15]

According to a *Newsweek* report later, "Biden felt insulted. Through staffers, Obama apologized, protesting that he had meant no disrespect. But at one of their regularly scheduled weekly lunches, Biden directly raised the incident with the president. The veep said he was trying to be more disciplined about his own remarks, but he asked in return that the president refrain from making fun (and require his staff to do likewise). He made the point that even the impression that the president was dissing him was not only bad for

Biden, but bad for the administration. The conversation cleared the air, according to White House aides who did not want to be identified discussing a private conversation."[16]

The report suggested Biden's continuing sensitivity to the view in the news media that his tongue was his worst enemy and a concern that Obama was buying into it. The president's feeling of solicitation for Justice Roberts attributed to Obama by Axelrod regarding Biden's ribbing didn't appear to be present on this latter occasion. In any event, the newsmagazine's report indicated Biden had no hesitancy letting his boss know how he felt about his own loose tongue. At one point in their first year together, Biden observed that "the thing about our relationship that I like is it's not built on any kind of fluff. We're not like each other's best buddy, have to go to the movies together. It's I know he respects me; he knows I respect him."[17]

This minor dust-up had no notable effect on Biden's new role as a major administration spokesman on both domestic and foreign policy and a more visible figure than many past vice presidents in governing functions.

Barely two weeks in office, Obama sent him to Munich to make the administration's first prominent speech on foreign affairs, at an annual European security conference. In a direct refutation of the Bush unilateralist foreign policy, Biden said, "I come to Europe on behalf of a new administration . . . determined to set a new tone not only in Washington but in America's relations around the world" that would "work in a partnership whenever we can, and alone only when we must." He said his country henceforth would "strive to act preventively, not preemptively" to avoid use of force "to stop crises from occurring before they are in front of us . . . starting with diplomacy."[18]

Of the war triggered by the 9/11 terrorist attacks, Biden said Obama had already ordered "a strategic review of our policy in Afghanistan and Pakistan," which before the year was out would occupy the major portion of the new administration's foreign policy planning and in which Biden would play a significant role. And he reached out to the NATO partners and Russia to give more to the effort to defeat the Taliban and al-Qaeda.[19]

At the close of the comprehensive speech, in referring to the strengthening of the continent's defense through an increased partnership between NATO and the European Union, Biden said: "The United States rejects the notion NATO's gain is Russia's loss, or that Russia's strength is NATO's weakness. The last few years have seen a dangerous drift in relations between Russia and other members of our alliance. It is time, to paraphrase President Obama, to press the reset button and to revisit the many areas where we can and should work together."

Apparently to calm any discomfort among former Soviet Union states caused by that comment, Biden said "we will not recognize any nation having a sphere of influence" and that "sovereign states have the right to make their own decisions and choose their own alliances."[20]

Back home, one of Biden's first prominent domestic assignments was kicking off the new Middle Class Task Force. At the University of Pennsylvania in Philadelphia, he led a three-hour discussion on efforts to create "green" jobs associated with environmental matters in which six cabinet secretaries, both Pennsylvania senators, three U.S. House members, Governor Edward Rendell, and Mayor Michael Nutter all took part. Aiming squarely at a prime Democratic constituency, Biden said "we will measure the success or failure of this administration not merely on whether the economy is technically recovered . . . but on whether the middle class at the end of the day is growing, the middle class is reaping its fair share of growth." He promised that Obama would "restore to the center of our efforts the middle class that has been long forgotten, out of view, and left out of our investments."[21]

In early March Biden was back in Europe for a meeting of the North Atlantic Council in Brussels. He urged the member nations to step up their contributions in Afghanistan and solicited proposals from them on how to go forward, to be included in Obama's reassessment of policy in the region. Biden returned to Washington and was handed a chore most vice presidents abhor but this one seemed to relish. Obama asked him to fill in for him at the annual Gridiron Club dinner, the traditional roast by the elite club of newspaper reporters and columnists, which every president since Grover Cleveland attended in his first White House year. Obama ruffled some feathers among the Fourth Estate by choosing instead to go to Camp David with his family, and Biden the famous marathon talker made the most of the assignment.

He opened by reporting that "Axelrod really wanted me to do this on teleprompter, but I told him I'm much better when I wing it." He added, "I know these evenings run long, so I'm going to be brief—talk about the audacity of hope." He said, "President Obama does send his greeting, though. He can't be here tonight because he's busy getting ready for Easter. He thinks it's about him."

Biden was just getting warmed up. He turned his guns on the gathered press: "I understand these are dark days for the newspaper business, but I hate it when people say newspapers are obsolete. That's totally untrue. I know from firsthand experience. I recently got a puppy, and you can't housebreak a puppy on the Internet." And he didn't spare his predecessor. "I never realized just how much power Dick Cheney had," he said, "until my first day on the job. I walked into my office, and you know how the outgoing president always leaves the incoming president a note on his desk? I opened my drawer, and Dick Cheney had left me Barack Obama's birth certificate!"[22]

In the White House these days, however, it was not all laughs. The somber economic meltdown was taking on greater intensity, and the other dire challenges that George W. Bush had left on Obama's doorstep were demanding emergency attention. They were threatening to clog the presidential agenda at the expense of the drastic changes the new president had promised in his campaign. Reforming health care at home to cover the more than thirty-four million Americans without health insurance topped the list, with ambitious plans for dealing with climate change, education reform, and a host of other items close behind.

Among them was getting American troops out of Iraq, finally begun by Bush on the way out the door, and giving the neglected war in Afghanistan the attention it required and circumstances demanded. Obama had already agreed to dispatch twenty-one thousand more American forces to the beleaguered country, bringing the total to sixty-eight thousand, and commanders in the field were clamoring for more. The review undertaken by Obama gained an intensity in March as Secretary of Defense Robert Gates and the Joint Chiefs of Staff chairman, Admiral Mike Mullen, were now pushing for four thousand more to train Afghan security forces. Late in the month, the president went to Camp David for the weekend to consider his options and returned agreeing to the limited increase for training, not combat, promising to revisit the question of troop strength later in the year. Included was a heavy component of diplomacy, with the dispatch of special emissary Richard Holbrooke to Afghanistan and Pakistan, and Biden continuing to handle his special responsibility of pushing Prime Minister Nouri al-Maliki's government in Iraq to achieve political accommodations with the contentious factions there.

Biden, who as chairman of the Senate Foreign Relations Committee had opposed the troop surge in Iraq, cautioned Obama against going down the path of another Iraq-like buildup that could run into stiff congressional resistance and public war-weariness. He reminded the president of the original imperative of going into Afghanistan—rooting out and destroying al-Qaeda terrorists. What came out of all the discussions was a compromise between what the president's military advisers wanted and what civilians like Biden, more attuned to political sentiment, accepted. The vice president's repeated visits to Iraq and Afghanistan as a senator, another during the transition, and again soon after taking his new office, were said to be particularly influential in Obama's decision to go forward in a limited way.

Joe Biden may have continued to draw snickers from critics who denigrated his colloquial style, but within the Obama circle, he was vindicating the president's appraisal of him as a worker bee within the partnership structure they had agreed upon. As a senator, he had always gauged what was acceptable to public opinion, convinced that any policy involving loss of American lives without public backing would not be sustainable. Basketball fan Obama in an interview with the *New York Times* likened Biden to a player "who does a bunch of things that don't show up in the stat sheet. He gets that extra rebound, takes the charge, makes that extra pass."[23]

The same story reported that Obama "has come to see Mr. Biden as a useful contrarian in the course of decision-making." In a White House usually overpopulated with yes-men regardless of party, Biden was one insider apparently not hesitant to be a no-man when he thought it warranted. As an old political veteran who after thirty-six years as a senator was given to observe that he had never had a boss until he went to work for Obama, his independent streak seemed to be an added credential for a president tolerant of loyal dissent.

Speaking of the existence of "an institutional barrier . . . to truth-telling in front of the president," Obama told the *Times* that "Joe is very good about sometimes articulating what's on other people's minds, or things that they've said in private conversations that people have been less willing to say in public. Joe, in that sense, can help stir the pot."[24]

At the same time, Biden made clear he was a team player, interestingly contrasting himself with Cheney in a CNN interview that was a rarity now for this man who as a senator was always accessible and free with his views and innermost thoughts. With Cheney as vice president, Biden said, "there was a divided government. There was Cheney and his own sort of separate national security agency, and then there was the National Security Agency."[25] In other words, for all his own willingness to speak his mind within the Obama inner circle, he was making no effort to create a power center of his own.

Within that context, the president told *60 Minutes*, "Joe's not afraid to tell me what he thinks. And that's exactly what I need, and exactly what I want."[26]

Joe Biden saying exactly what he thought continued on occasion, however, to feed the clinging public impression that he was more than a bit careless in failing to think through the ramifications of some things he would say. When the first wave of concern over a possible swine flu pandemic hit the country in April 2009, he drew wide public criticism for saying on the *Today* show he would advise members of his own family not to take plane trips during that time. "If you're out in the middle of a field and someone sneezes, that's one thing," he said. "If you're in a closed aircraft or a closed container or closed can or closed classroom, it's a different thing." He went on to say he had told his own family members to avoid subways and not to "go anywhere in a confined place now."[27]

Biden's advice not surprisingly sent a tidal wave of dismay through the transportation industry. Another *New York Times* member of the gaffe police, Gail Collins, observed: "As the White House's unfiltered talking head, Biden is the perfect warning bell to show the White House when they are veering out of control. A kind of mental canary in the governmental mine shaft."[28] The vice president's remark was probably no more than any parent might say to his or her kids, but the general reading was, there goes Joe again. The comments were a mere interlude in his whirling dervish schedule of travel around the country monitoring the economic recovery and stimulus money outlays. But they dominated the news about Joe Biden for days thereafter.

More significant for the Obama administration and the Democratic Party around this time was Biden's key role in persuading longtime Republican Arlen Specter to jump ship suddenly and rejoin the Democratic ranks. The immediate catalyst was an internal poll for Specter showing that his independence and occasionally maverick ways had finally caught up with him. It indicated that he was likely to lose in the next Republican senatorial primary to a younger Pennsylvania conservative, former representative Pat Toomey. But intensive courting by Biden, a fellow Amtrak commuter for years, and by Pennsylvania Governor Ed Rendell finally pushed Specter over the line. Aides reported later that Biden had met Specter six times and telephoned him eight times more in the eventual successful courtship. Specter's defection, and the eventual Senate seating of Democrat Al Franken in a contested Minnesota race, gave Obama a near-veto-proof Senate, further enhancing Biden's stock with the new president.[29]

Through the spring of 2009, Biden also managed to sandwich in trips to Chile, Costa Rica, Lebanon, and the Balkans while focusing at home on overseeing implementation of the economic stimulus package. In mid-May, the vice president submitted his first quarterly progress report on the Recovery and Reinvestment Act. It claimed that in the first seventy-seven days, one hundred fifty thousand jobs had been "created or saved," more than $88 billion had been made available to localities for various projects, including more than three thousand in transportation construction in fifty-two states and territories. Also, Biden reported, 95 percent of working families had begun to get tax credits and unemployment benefits had gone up twenty-five dollars a week. In the next hundred days, he said, six hundred thousand more jobs "are expected to be created or saved" under the Recovery Act,[30] and he continued to spew out other periodic progress reports on such matters as cleaning up lead poison in housing and helping military families hit hard by having to sell their homes.

Also in May, Biden made a sentimental journey back to central New York, delivering the commencement address at Syracuse University, where he had gone to law school. Immediately afterward, he and his entourage sped to Bellevue Elementary School in the middle-class Swarthmore neighborhood, where Neilia had taught grade-schoolers and the young couple had their first home in an apartment on nearby Stinard Avenue. The kids at Bellevue had written him inviting him to visit, and they sat on the floor as he regaled them with the story of his courtship of Neilia.

"Now look, guys, my name is Joe Biden and I used to hang out at this school," he began. "You know why? Because I was in love with the most beautiful teacher I ever saw," and how, while she was busy in teachers' meetings after classes, he would play basketball with the Bellevue kids in their playground behind the school. He told them that while President Obama lived in the White House and as vice president he worked there, it was "the people's house" that belonged to each of them. Any one of them might live there later by working hard, he said.

Asked by a young student whether "when I was a kid did I want to be the vice president," Biden replied: "When I was a kid I wasn't sure what I would be able to be because I used to stutter really badly." He proceeded to give the kids an example of how he used to talk, and how he had conquered the problem. He told them how Obama had succeeded in spite of growing up without a father, who "picked up and left and went back to Africa," and was raised by a single mother and his grandparents. He concluded by telling a little girl in the front row that "Maybe I'm talking to the first woman president."[31]

It was an emotional experience for Biden as he left and strode over to Stinard Avenue, where old neighbors greeted him, some with tears. Pat Wojenski, a young girl befriended by Neilia at Bellevue, identified herself to him in another tearful scene. Later, she planted a tree in Neilia's memory at the entrance to the school, with a placard on a large boulder next to it. Brother-in-law Jack Owens, who had known Neilia from the beginning, accompanied him on the sentimental journey, himself fighting back tears.[32]

Biden also found time during the spring to return to Bosnia, Kosovo, and Serbia, where peacekeeping efforts had been one of his major interests as chairman of the Senate Foreign Relations Committee. As multiethnic sniping threatened to rekindle the deadly divisions in the Balkans, he scolded the Bosnian parliament in

Sarajevo. "God, when will you tire of that rhetoric? This must stop. Let me be clear: Your only real path to a secure and prosperous future is to join Europe. Right now you're off that path." The alternative, he warned, was to "descend into the ethnic chaos that defined your country for a better part of a decade. The choice is yours."[33]

Biden threw cold water on speculation of a second Dayton Peace Conference, saying it would only harden nationalist positions. In Serbia, he pushed for its admission to NATO and the European Union and said it would not be necessary for Serbia to recognize the independence of Kosovo for the United States to support Serbian admission. And in Kosovo, where he received a hero's welcome from flag-waving schoolchildren lining the streets, he called its independence "absolutely irreversible" and "the only viable option for stability in the region."[34]

On Biden's return to Washington, another assignment awaited him as Obama's Joe-of-all-trades. The sudden announcement by Justice David Souter that he was resigning from the Supreme Court gave Obama his first opportunity to make an appointment. He nominated Federal Appellate Court Judge Sonia Sotomayor of New York as the first Hispanic American on the highest bench. Biden, who had chaired the Senate Judiciary Committee through every Supreme Court confirmation hearing since 1987, was called on to help prepare her.

A *Wall Street Journal* editorial issued a scathing diatribe against him entitled "How Joe Biden Wrecked the Judicial Process," alleging his insistence that a nominee be subjected to questioning about his or her judicial philosophy broke new and unwarranted ground.[35] That approach had indeed brought Bork down, to the dismay of conservative stalwarts. Sotomayor, however, artfully dodged all challenges and was confirmed by a 58–31 vote, with nine Republican senators supporting her.

At the beginning of July Biden, in his role as overseer of efforts to advance political and ethnic reconciliation in Iraq, was back there for talks with Prime Minister al-Maliki. Only days after the United

States withdrew most combat troops from Iraqi cities under the security agreement with his government, Biden reassured him that America would remain engaged in Iraq as parliamentary elections approached, while cautioning that support was contingent on progress in achieving that reconciliation.

He also pressured al-Maliki on the eventual enactment of a new elections law, and above all monitored the remaining scheduled withdrawal of U.S. forces to be completed by 2011. And he prodded al-Maliki on a range of outstanding political issues, from power-sharing and oil revenue–sharing among competing factions and regions to boundary disputes, the status of Kirkuk, and integration of Kurdish militia into the Iraqi army, while trying to keep a lid on decreasing but still troublesome violence.[36] Then Biden was back in Ohio in his assigned task of selling and monitoring the dispersal of more funds from the $787 billion stimulus package at the heart of the Recovery and Reinvestment Act.

In late July the vice president made a quick trip to Ukraine and Georgia. It was designed to ease fears there that the Obama administration's determination to "reset" relations with Russia after the deterioration during the Bush years would undermine U.S. support. In Kiev, Biden was received with reserved cordiality as he praised Ukraine on its independence while urging its leaders to work to ease its dependence on Russian energy resources. In Tbilisi, on the other hand, he was given a hero's welcome by flag-waving crowds for his support in the previous year's military confrontation with Russia over the ethnic enclaves of South Ossetia and Abkhazia.

At an elegant reception at the new Georgian presidential palace, President Mikheil Saakashvili reminded Biden of his earlier promise that "I will never, ever abandon you" and then draped a gold medal around his neck, greeting him as "Joe, my dear friend." In an exchange with a roomful of refugee children broadcast on national television, the vice president set aside the reset button on Russia for the moment with a harsh remonstrance.

The Russians, he said "used a pretext to invade your country" in the hope of crushing democracy, and he promised to "make clear to the world, and to the Russians particularly, that we stand with you, and that if they fail to meet their commitments, that it is a problem for them. . . . A lot of you think maybe Russia did what they did and they paid no price. They paid a pretty big price already, diplomatically," he said, among Georgia's neighbors.[37]

On the heels of these visits, Biden gave an interview to his journalistic bête noire, the *Wall Street Journal*, that was also read as a postponement to pressing the reset button with Russia. He observed that it now had "a withering economy" with other severe domestic problems, leading him to conclude that "we vastly underestimate the hand that we hold." He said the United States must be careful, however, not to overplay that hand. "It won't work if we go in and say, 'Hey, you need us, man; belly up to the bar and pay your dues.' It is never smart to embarrass an individual or a country when they're dealing with significant loss of face. My dad used to put it another way: Never put another man in a corner where the only way out is over you." He went on: "It's a very difficult thing to deal with—loss of empire. This country, Russia, is in a very different circumstance than it has been at any time in the last forty years, or longer."[38]

In this little lecture, Biden seemed not to realize that his comments would only further embarrass or trigger resentment in Russia by describing their "loss of face" and "loss of empire." The response in Moscow was swift. According to the next day's *New York Times* report from there, President Dmitry Medvedev's former chief foreign policy adviser, Sergei Prikhodko, said: "The question is, who is shaping the U.S. foreign policy, the president or respectable members of his team?"[39]

Later, the *Los Angeles Times* quoted a former adviser to Russian leader Boris Yeltsin, Andranik Migranyan, on Biden's assessment: "I don't want to be rude, but if he continues these kinds of comments, he will be perceived as a clown and no one will take him seriously"[40]—

the Russian version of "there he goes again." White House press secretary Robert Gibbs defensively commented: "The president and vice president believe Russia will work with us not out of weakness but out of national interest," noting that Obama in his recent visit to Moscow had said "the United States seeks a strong, peaceful, and prosperous Russia, one that will be an even more effective partner in meeting common challenges."[41]

But Joe Biden was a handy guy to have around. When Obama got himself in hot water for saying a cop had "acted stupidly" in arresting Harvard professor Henry Louis Gates Jr. for trying to enter his own home without a key, the president invited Gates and the police officer to the White House for a beer. He also asked Biden, celebrated as his administration's blue-collar guy from Scranton, to join them at a backyard picnic table. Biden, a teetotaler, had a nonalcoholic brew. It made a great photo for the media and the story of "the beer summit" faded.[42]

Such frivolities soon vanished at the White House, as the war in Afghanistan began generating more American casualties, and the president was confronted in late summer with a sober report from the NATO and U.S. commander there. Marine General Stanley McChrystal, an expert in counterinsurgency warfare sent to Afghanistan to appraise and direct the nearly eight-year-long conflict, warned in a sixty-six-page document that without a significant troop increase the war could be lost.

Presented on August 30 to Secretary of Defense Robert Gates and sent to the White House in early September, it said: "Failure to gain the incentive and reverse insurgent momentum in the near term (next twelve months) while American security matures risks an outcome where defeating the insurgency is no longer possible." In a short summary, McChrystal observed, "While the situation is serious, success is still achievable."[43]

The dire report broke into public awareness in the *Washington Post* on September 20, in another of reporter Bob Woodward's exclusives confirming his position as the capital's prime acquirer of high-level leaks. By that time Obama had already set in motion a series of private administration meetings among the president, vice president, the secretaries of defense and state, the chairman of the Joint Chiefs of Staff, the nation's intelligence chief, and White House officials to decide on the Afghanistan war strategy going forward.

Speculation on how many more troops McChrystal was asking for beyond the sixty-eight thousand Americans already there ranged from ten thousand to forty thousand or even more. Obama in March, in authorizing the twenty-one thousand increase, had said at the time he would reassess the situation later, and McChrystal's report put it on the administration's front burner. As the Iraq war had dragged on, Biden had continued to support the troops, but had warned as early as Senate Foreign Relations Committee hearings in July/August 2002 about the waning of public support at home. He had insisted that no war could be long sustained without such backing.

Complicating the whole Afghanistan situation, for Biden particularly, was the cloud that enshrouded Afghanistan President Hamid Karzai, by 2008 awash in rumors of complicity in governmental corruption and drug dealing. On one senatorial visit to the beleaguered country, then senator Biden had become so fed up with Karzai's denials of the corruption there that he stormed out of a dinner he and colleagues were sharing with the Afghan leader.[44] Now, in 2009, Karzai's recent reelection was so marred by allegations of fraud, confirmed by a UN electoral investigation, that a runoff was ordered. The delay complicated Obama's sober deliberations on whether to implement McChrystal's call for more troops to wage a more ambitious counterinsurgency in Afghanistan.

Karzai at first balked at the runoff and then rejected the appeal of his prospective opponent, former foreign minister Abdullah Abdullah, for removal of the elections official who had overseen the fraudulent first election. Adbullah, anticipating the same result in the runoff under the circumstances, finally bowed out in frustration and Karzai after all was declared reelected.

Meanwhile, Obama's war policy review stretched through the fall of 2009, ultimately requiring ten closed-door White House meetings with chief military, diplomatic, and political advisers, with his vice president at his side. Biden tenaciously questioned the basic assumptions of the options presented, keeping in the forefront the mission against al-Qaeda along with the fight against its Taliban sponsors. His concerns and reservations for a time appeared to be working as a brake on a rush to judgment, and on the requested scope of any troop surge.

Unexpectedly, the vice president got a lifeline in his argument for fewer troops and a more concentrated effort against acts of terrorism in the theater. The American ambassador in Kabul, Lieutenant General Karl Eikenberry (ret.), sent memos directly to the president that quickly became public, taking issue with the most ambitious aspects of the McChrystal buildup.

Biden intensified his leading role in Obama's review of the Afghanistan war strategy. "The president was desirous of having someone who would question every premise and assumption that people brought to the table about the policies they were proposing," recalled Tony Blinken, Biden's chief foreign policy adviser. "The vice president, in this review that lasted for almost three months, was probably the most insistent voice in raising questions about premises and assumptions. And that created an extremely intellectually rigorous and serious review and produced a much stronger common foundation for the policy that emerged, because we discarded a lot of the premises and assumptions because they didn't hold up under cross-examination."[45]

As the meetings proceeded with due deliberation, Cheney as the Bush administration's new public voice charged Obama of "dithering" and thus inviting the military defeat of which McChrystal had warned in calling for swift buildup. But the president would not be stampeded, instead encouraging continuing discussion in which Biden played devil's advocate regarding the proposals of the military leadership.

General McChrystal clearly was not pleased at the delay and, inferentially, at Biden's role in it. After a speech to the International Institute for Strategic Studies in London, the general was asked whether he would support a decision by Obama to focus on al-Qaeda with rocket-loaded drones and Special Forces. This was a Biden-favored idea, and McChrystal snapped: "The short, glib answer is no."[46]

Obama was unhappy at the general's discussing the Afghanistan forward strategy while high-level deliberations were going on. The president was in Denmark in a failed effort to win the 2016 Olympic Games for Chicago, and he summoned McChrystal to Air Force One as it sat on the Copenhagen airport tarmac to express his displeasure face-to-face.

On the president's return to Washington, the strategy talks continued, with McChrystal centered on defeating the Taliban and Biden focused on al-Qaeda. Obama was not going to agree to another nation-building marathon similar to Iraq, but he wasn't going to walk away from Afghanistan either. The ultimate compromise was a hybrid plan wherein McChrystal would get additional troops in a more rapid deployment, and a deadline of July 2011 for a start to their withdrawal. The idea was that by that time, it would be clear whether the American goals in Afghanistan could be reached or not.

Throughout the long deliberation, Blinken said later, "the vice president's emphasis was never on the number of troops, which unfortunately became the barometer by which people gauged the topic. The number that was in the original McChrystal plan was leaked in the press—forty thousand—and all of a sudden that became however one looked at the debate. The vice president's approach was totally different. It was, 'Let's get the strategy right, and as long as we come up with the strategy, then we can decide how to resource it with troops but the troops follow the strategy, not the other way around.' So he was never fixated on the number of troops. He was focused on making sure we had the strategy right, and I think the strategy that emerged very much reflects many of the ideas the vice president brought to the table and were shaped in part by the concerns that he raised." It all went back, Blinken said, to Biden's first trip to Iraq as vice president–elect and what he had told Obama, and what they had then reminded others: "that we had to keep our eye on the overall goal, and the overall goal was al-Qaeda. And the strategy that emerged, [that] the president signed off on, did just that."

Biden pointedly reminded the Obama inner circle of the original rationale for American involvement in the country—the imperative response to the 9/11 terrorist attacks hatched there—and that most al-Qaeda operatives had since moved into Pakistan. Yet, he noted, the administration was pouring about thirty dollars into Afghanistan for every dollar put into nuclear-armed Pakistan, and he questioned the priorities.

In Biden playing this role, Blinken said, "it was in the nature of who he is, but also strongly encouraged by the president, [who] immediately recognized [its] importance and utility. It also gave the president space to kind of sit back and have the vice president do some of the hard and pointed questioning and the president could not show his own hand." During breaks in the meetings, Biden frequently huddled with Obama, and in advance of each of the sessions Obama chaired, Biden wrote him a memo raising issues and ideas he might want to pursue.[47]

It was now mid-November and Obama had finally made his decision. He called all his Pentagon and military leaders to the Oval Office and summoned Biden, who was in Nantucket for his family's annual Thanksgiving holiday there, to attend. According to *Newsweek* reporter Jonathan Alter in his book *The Promise: President Obama, Year One:* "As they walked along the portico toward the Oval Office, Biden asked if the new policy of beginning a significant withdrawal in 2011 was a direct presidential order that couldn't be countermanded by the military. Obama said yes."[48]

Inside the Oval Office, Alter wrote, Obama asked Secretary Gates; Admiral Mike Mullen, chairman of the Joint Chiefs of Staff; and General Petraeus in turn whether they signed on to the strategy, and all said they did, including the 2011 timeline. "When Obama talked to McChrystal by teleconference," Alter continued, "Obama couldn't have been clearer in his instructions. 'Do not occupy what you cannot transfer [to the Afghan National Army].'" And later, he wrote: "In a West Wing interview, Biden was adamant. 'In July of 2011 you're going to see a whole lot of people moving out. Bet on it,' Biden said as he wheeled to leave the room, late for lunch with the president. He turned at the door and said once more, 'Bet. On. It.'"[49]

After Obama had made his decision, Biden reviewed and summarized his own role for this book, observing:

> *First of all, I had an unfair advantage because I talked about this with the president more often and more regularly than anybody else. I knew, not just by body language [but] by direct assertion that he agreed with me on [the] basic fundamentals of the policy. So in that sense, I went into the debate in the situation room for those many hours we debated, knowing he agreed with me on the strategy.*
>
> *One, that we were not in Afghanistan for nation-building. We were not going to commit to provide and guarantee resources to build that country for the next ten years. Number two, [we agreed] that the COIN strategy was not the appropriate strategy for signing on indefinitely to a nation-building campaign. Three, I was constantly focused on, and I knew he agreed, that the fundamental reason for us being there was al-Qaeda. The [other] national security interest we had was keeping Pakistan from disintegrating, and radicals gaining control of nuclear weapons, and that securing an Afghan government was in the service of the first two objectives.*
>
> *There was one exchange with a lot of the principals where I asked the question: "If there was no al-Qaeda and Pakistan was stable, would you be making recommendations to put tens of thousands of troops and sending hundreds of billions of dollars to*

beat the Taliban?" And the answer with several of the members was yes. Mine was emphatically no. It's the difference between how Afghanistan is viewed, and some in the administration—and you'd expect this kind of disagreement—believed that it was central, it was a linchpin in being able to work toward stabilizing everything from Iran to the subcontinent. I disagreed with that; the president disagrees with that. So . . . I always argued that the number of troops is much less important than the strategy.

In every memo I wrote to the president, which were multiple, including a long, twenty-page handwritten memo I wrote to him a couple of days before the decision, was focused on making the case: (1) that this is a three-dimensional problem—al-Qaeda, Pakistan, and Afghanistan; (2) that there be a limit on the number of troops so this wouldn't be a constant, creeping escalation whatever troop level was announced; (3) that there be a date at which we would begin the drawdown of America forces with the aim of drawing down all combat forces out, à la Iraq; (4) that it was not necessary to defeat the Taliban because the Taliban was and is part of the fabric of the Pashtun society—20 to 30 percent of it is incorrigible and must be defeated, and the remainder should be integrated into Afghan society; (5) that the return of the ability of the Taliban to overthrow the Afghan government was simply not within their power even if we did not add a single additional troop; (6) that the Taliban was not seeking to establish a new caliphate, they were not an existential threat to the United States of America, any more than the Shining Path in Peru [another terrorist group] is an existential threat to the United States of America . . . ; (7) that al-Qaeda's return to Afghanistan was highly unlikely, for two reasons: even if the Taliban gained political ascendancy, they would not invite al-Qaeda back in. It was against their interest, which was controlling Afghanistan, not an international jihad. And secondly, why would al-Qaeda want to leave the sanctuary in Pakistan, 60 to 120 clicks [kilometers] from

Islamabad, in order to go back to Kandahar? Why would they want to do that?

Those principles are principles that the president bought and agreed with. They were different than the arguments initially made by some of his military advisers. . . . There were others making these arguments too in the national security apparatus, including the Defense Department; it wasn't unified, [but] the intelligence community, some of his chief advisers—they agreed with me.

But the other thing I argued for: If you're going to put troops in, put them in! This wasn't a surge. This wasn't a surge. The way they had the McChrystal report . . . running the troops in over a period of a year and some months, that's no surge, that's a buildup. So here's what the president decided, the way he talked about it, the way he made it clear to the generals; I was in the room with him that Sunday in the Oval Office. What he decided to do was give them the troops they said they needed quicker, so that you could find out, and they agreed, that by December [2010] we'll know whether the concept is working.

The president made it absolutely clear that by December [2010] as we review this, the only purpose of the review will be to determine not whether we put any more troops in—there will not be any more troops in. The only question will be how rapidly we begin to draw down troops in July of '11. And the president said repeatedly to [the generals] in that meeting, and subsequently publicly: "Do not occupy what you cannot transfer. The purpose here is to weaken and/or reintegrate the Taliban to the point, while you're simultaneously training the Afghan national security forces, police and army, so that when we begin to leave gradually, as we are in Iraq, the Afghan national security forces can contain and/or control the Taliban and maintain their own sovereignty. . . ." [At the same time] the United States has to send a very strong message to Kabul. . . . Karzai has to step up to the ball and exert

> *sovereignty. . . . So that at a time certain, [there's] a cap on the troops, that it's not a four-corner strategy, it's not nation-building, and there's a limit to our patience in terms of the Afghans having to take more responsibility.*[50]

Asked whether it was his sense that the generals ultimately bought into this plan or just abided by it, Biden replied, "I think they bought into it, maybe a couple abided, but I think they bought into it. Here's what he said: 'Look, guys, tell me now. If what I'm proposing to you cannot work, don't tell me later. Don't tell me that what I've laid out to you—this time-certain, this thirty thousand troops, et cetera. Tell me now if it can't work. Because if it can't work, then we'll have to go to another plan.' He started off with the chairman of the Joint Chiefs and went around the room and they all said, 'Yes, sir. This can work.' "[51]

Earlier, Blinken had identified Biden's likely role in the critical discussions on Afghanistan. Obama, in choosing Biden, he said, "wanted someone who was not going to be a yes-man. He wanted someone who would say what he thought or believed in private, even if it contradicted what other senior people in the administration officials said or even contradicted the president. The president wanted someone who would be willing if it was necessary to be the skunk at the picnic to keep everyone honest."

Now, Blinken said, "I think that is exactly internally what's happened. And externally, once the decision was made, whether it came out the way Biden supported or didn't support, he would be the number one cheerleader for whatever the decision was. Biden pushed back against the conventional wisdom on some issues, had gotten people to think again, and has put us in a better place because he has demanded intellectual honesty in the way issues were looked at."

As for Obama and Biden as a team, Blinken said, "Their styles could not be more different, but they wind up nine times out of ten at exactly the same place in substance. They're both at heart pragmatists and they're both nonideological. At the end of the day, both of them are trying to solve problems, not saying, 'Here's a solution, let me think how I can justify it?' but, rather, 'What's the solution?' "[52]

As a result of the long policy review, Blinken said at the time, "we're no longer talking about doing nation-building in Afghanistan, we're no longer talking about doing a counterinsurgency in every nook and cranny of the country, we're no longer talking about defeating the Taliban, which Biden argued was neither likely nor necessary." Some elements of the Taliban could be integrated into the Afghan army, he suggested, and so there was no necessity to double its size.

Finally, Biden had offered instead the notion of "proof of concept"—the need to undertake something that in which success or failure of the approach could be demonstrated, and within a circumscribed period of time. And for that to happen, the strategy could not be open-ended. It was for this reason that Obama insisted that the movement of additional troops into Afghanistan be significantly accelerated so that the concept could be more rapidly proved one way or the other.

The idea of setting a time frame that ultimately would involve the withdrawal of American forces immediately drew criticism as simply inviting the Taliban to melt into the populace and wait. Biden argued, however, that such an enemy pullback would give the Afghan government more time to get its house in order, and help break the culture of dependency on American assistance into which it had fallen. In any event, the strategy ultimately agreed upon, with all participants telling Obama they were on board, Blinken argued, "was heavily influenced by concerns and ideas that the vice president brought to the table."[53]

Whether or not Biden was generally perceived as having so influential a role in the plan that emerged, he was recognized by others as having been what Brookings Institution foreign policy critic Michael O'Hanlon called an effective "house skeptic." He said Biden had provided "a comfort level" for the wary Obama, who was sending more military forces to Afghanistan even as he accepted the Nobel Peace Prize with a somewhat jarring defense of the concept of just wars.[54] Former senator Chuck Hagel of Nebraska, now a member of Obama's Intelligence Advisory Board who earlier traveled to Afghanistan with Biden, credited him with stretching out the process. By testing the assumptions of those proposing the surge, Hagel said, Biden contributed to the president's understanding of a complicated challenge by focusing on al-Qaeda and Pakistan.[55]

In the end, Obama approved a slightly watered-down but significantly redirected mission, authorizing thirty thousand more American troops to Afghanistan coupled with a public assurance that the plan would be periodically reviewed. With increasing dissent within the Democratic caucus in Congress obviously in mind, the president assured the country there would be an exit strategy for U.S. troops in Afghanistan, setting July 2011 for a start to withdrawal. In congressional testimony, however, Secretary of Defense Robert Gates conceded that the date might have to be adjusted based on conditions on the ground at the time, and that the actual pullout could take two or three years to achieve. The result was a quick-fix solution, giving the generals what they said they needed, at the same time addressing Biden's concern that the commitment not be open-ended.

Biden may have been perceived by critics as having "lost" on Obama's acceptance of the eventual troop surge and its size, but the al-Qaeda threat and Pakistan remained in the forefront of the administration's concern, as Biden wanted. And once Obama disclosed the details of the new approach, the vice president signed on to and worked to sell it, and to encourage NATO countries to provide more troops. In doing so, he delivered on his prenomination pledge to give Obama his honest assessment during the analysis period, then fall into the ranks behind the president's decision once made, with considerable satisfaction that his voice had been heard and significantly acted upon by the president.

The day after Obama outlined his plan, Biden told CBS News,

"All along, you may recall, I'd been arguing the strategy is more important than the numbers. And the president laid out the strategy. This is a regional issue; number one priority—al-Qaeda; number two, Pakistan; number three, giving the Karzai government a fighting chance to be able to sustain itself. The existential threat to the United States remains in the mountains of Pakistan. That's where we have to keep the focus."[56]

Obama, asked later for this book what he thought of critics who might conclude from the outcome of the lengthy deliberations on Afghanistan strategy that Biden had "lost," replied: "I don't think anyone who was party to the very, very exhaustive discussions we had would say that. Joe was enormously helpful in guiding those discussions. The decision that ultimately emerged was a synthesis of some of the advice that he gave me, along with the advice that Secretary Gates and Generals Petraeus and McChrystal offered. I think we arrived at exactly the right answer, and we would [not] have gotten there as quickly, or at all, without each of their contributions. The vice president played a vital role in that process."[57]

Intruding sadly on Biden's whirlwind first year as vice president was the death after a year-long struggle with cancer of Senator Ted Kennedy, the man whose early endorsement of Obama's presidential candidacy had provided a major spark toward his nomination, and who had then embraced the Obama-Biden ticket. In remarks at a pre-funeral memorial at the Kennedy Library, Biden told of how the late senator had campaigned for him in his first Senate race as a twenty-nine-year-old long shot and said it was unlikely he would have been elected and risen to the vice presidency had that not happened.

At the time the car accident killed his first wife and infant daughter, Biden said, Kennedy had called him at the hospital every day and dispatched Boston's best specialists to treat his injured sons, and then was "the prod who convinced me to go to the Senate" when he was about to turn down the seat. On arrival, Biden recalled, Kennedy took him to the Senate gym for a workout, to meet his new colleagues there, and introduced him to two Senate lions, Democrat Warren Magnuson and Republican Jacob Javits—both "stark naked!" And he told of the time he was convalescing after two brain aneurysms when Kennedy took the train to Wilmington, alone and unannounced, for a visit and six hours later caught the Amtrak back to Washington.

With a mixture of pathos and humor, Biden held the Kennedy family and friends entranced as he said Ted Kennedy's death did not mark "the end of the Kennedy era" because there remained a new generation with "more talent, more commitment, more grit, more grace than any family I've ever seen." More than any monuments, he said, addressing the next generation of Kennedys present, "they left us you." He concluded with a line from one of his Senate friend's best-remembered speech lines: "The dream still lives." There were many wet eyes as Biden left the microphone.[58]

Soon afterward, Biden experienced another bittersweet moment when, as president of the Senate, it was his duty to swear in the new Democratic senator from Massachusetts. Paul Kirk, Kennedy's long-time Senate aide and close friend, was appointed as the interim senator until a special election in 2010, under arrangement by the state legislature. Thirty-six years earlier, when young Joe Biden was running for the Senate himself, he had called on Kirk for advice on how the Kennedy political machine worked. Kirk put him onto the Kennedy family neighborhood teas that were instrumental in John Kennedy's first campaign for Congress. With Kennedy political adviser Matt Reese pitching in, the Biden family took a page from that Kennedy playbook with its own marathon neighborhood coffees that were so instrumental in Joe's entry into the Senate.[59]

Now, after more than a year as vice president of the United States, Joe Biden was showing no signs of slowing down in his self-described role as Obama's "fireman in chief," battling political, diplomatic, and even military blazes wherever they flared up. With difficult midterm elections approaching that would imperil the Democratic majority in Congress, Biden also had a full schedule of campaigning ahead. He continued to demonstrate a remarkable energy and durability for a man of sixty-seven carrying such a heavy load of responsibility in an administration laden with challenges inherited and taken on by a president committed to sweeping change at home.

The commitment extended as well to helping to alter the course of American foreign policy. Biden carried on his assignment of monitoring progress in Iraq toward a stable government in Baghdad through the approaching parliamentary elections. And as the administration's watchdog on nonproliferation of nuclear weapons, he defended an Obama call for maintaining the efficiency of the U.S. nuclear infrastructure even as the president was pursuing a new nuclear arms treaty with Russia.

In a major speech at the National Defense University in Washington in late February 2010, Biden said the updating "is not only consistent with our nonproliferation agenda, it is essential to it." The speech came as Obama moved toward the new treaty, and the vice president said the investment "allows us to pursue deep reductions without compromising our security."[60] About a month later, Obama was able to announce such reductions after tough and contentious negotiations with Russian President Dmitry Medvedev.

Meanwhile, Biden was off on another Middle East mission to encourage movement on the stalemated talks for a two-state solution to the Israeli-Palestinian conflict. The construction of new housing for Jews in disputed West Bank territory remained a major bone of contention. And Israeli Prime Minister Benjamin Netanyahu, under pressure from Washington, had earlier announced a ten-month partial freeze on such settlements, excluding Jerusalem. But upon his arrival there, Biden unexpectedly ran into a hornet's nest.

He had just completed a rousing speech in Jerusalem and, in Netanyahu's presence, had vowed "absolute, total, unvarnished commitment to Israeli security" when the Israeli interior ministry announced that sixteen hundred new housing units would be built in East Jerusalem. The ministry said the decision had been three years in the making, had nothing to do with Biden's arrival, and, indeed, that the chagrined Netanyahu had only just been informed of the announcement himself.

Biden immediately condemned the decision in scathing terms, calling it "precisely the kind of step that undermines the trust we need right now," and that it "runs counter to the constructive discussions that I've had here in Israel. We must build an atmosphere to support negotiations, not complicate them," he lectured. "Unilateral action taken by either party cannot prejudge the outcome of negotiations on permanent status issues."[61]

The timing of the new construction plan could not have been worse. It came on the heels of a report by Obama's special intermediary in the Israeli-Palestinian dispute, former Senate majority leader George Mitchell, that the two sides had just agreed to four months of talks through intermediaries designed to break the stalemate. A Palestinian spokesman called the new development on West Bank construction "a dangerous decision that will torpedo the negotiations and sentence the American efforts to complete failure."[62]

Biden learned the news of the new settlements as he and his wife, Jill, were being driven to have dinner with Netanyahu and his wife. The vice president thereupon delayed his arrival as a demonstration of his disapproval. The next day Biden went on to Ramallah in the Palestinian Territory. There he told Authority President Mahmoud Abbas that the latest Israeli decision "undermined that very trust, the trust that we need right now in order to . . . have profitable negotiations," and was "why I immediately condemned the action."[63]

Biden repeated the condemnation the following day in a speech at Tel Aviv University, though first reiterating at length his commitment to Israel. He said his criticism came "at the request of President Obama," which drew applause, adding that "sometimes only a friend can deliver the hardest truth."[64] As a man often criticized for undisciplined speech, Biden comported himself otherwise in this particularly sensitive diplomatic situation, though some commentators commented that it had brought him humiliation. The Obama team quickly backed him up. Secretary of State Clinton chimed in with a telephone rebuke of Netanyahu and his government for the ill-timed announcement.

Biden got back to Washington in time to join in Obama's all-out drive to bring a successful end to the long and rancorous debate over health care reform. Thereafter, with the first major health care reform since the enactment of Medicare in hand, it fell to the vice president to introduce Obama at a wildly celebratory White House signing of the bill, before most of the Democratic faithful who had brought it about and a national television audience.

"History is made when a leader steps up, stays true to his values, and charts a fundamentally different course for the country," the vice president said. "History is made when a leader's passion—passion—is matched with principle to set a new course. Well, ladies and gentlemen, Mr. President, you are that leader." Where earlier presidents since Theodore Roosevelt had failed, Biden said, Obama had turned "the right of every American to have access to decent health care into reality for the first time in American history."

The vice president went on at length, as was his custom, in what was an eloquent introduction, concluding with an embrace of his boss. Unhappily for Biden, however, a sensitive microphone picked him up whispering to Obama: "This is a big fucking deal!" The official White House transcript reported only that Obama had replied, "Thank you, Joe," and that the vice president had responded: "Good to be with you, Mr. President."[65] But there went Joe again, letting his tongue wag loose, this time sotto voce. The news media, and especially the tabloid press, were not going to give him a pass on it.

The *New York Daily News* tabloid front page blared: CURSE OF JOE BIDEN.[66] And on the television networks and cable outlets, it was reported that McKay Hatch, seventeen, a student at South Pasadena High School and founder of a No Cussing Club, took him to task.

"This is a huge deal," the young Californian said, insisting that Biden "needs to be a good role model for kids to use clean and appropriate language." Hatch mailed Biden a club T-shirt and a penalty jar to hold fines for violations.[67]

The day after his latest verbal gaffe, Biden, at a fund-raising event in a private home in Baltimore, told the small gathering that at that morning's staff briefing, Obama had said, "You know what the best thing about yesterday was? Joe's comment." He said Obama told him he had tried to have a T-shirt made bearing it but couldn't get it done in time. Biden said he wisecracked: "If you thought it was so good, why didn't you say it?"

In a more serious vein, Biden added: "The thing that this portends the most is that this country is capable of handling complex, ideological[ly] divisive, consequential issues. If we were unable to move the ball on this issue, not only in the political sense might we be dead . . . but in terms of being able to deal with other major issues on our plate, we would have been done. Absolutely done. What happened two days ago represents fundamental change in an incredibly, incredibly difficult political environment. . . . Reports of the demise of the Democratic Party in November are premature."[68]

There was no indication, however, of the demise of Joe Biden's reputation as a verbal loose cannon. A few days later, when Jill Biden herself introduced Obama at Northern Virginia Community College in Alexandria where she teaches, he responded thus, "Thank you, Dr. Biden, for that outstanding introduction—and for putting up with Joe." The audience laughed heartily, needing no explanation.[69]

For all the needling, however, Obama continued in his second presidential year to rely on his vice president, especially in the realm of foreign policy. As the troop surge in Afghanistan sought by General McChrystal and agreed to by the president went forward, Biden kept a skeptical eye on developments, while remaining committed to the plan laid out in the fall, especially the July 2011 start date for withdrawal, for which Biden was a principal advocate.

In late June, as McChrystal was preparing for his major military push against the Taliban stronghold of Kandahar, he placed an urgent phone call from Afghanistan to the vice president. Biden was returning in early evening from an event in Illinois aboard Air Force Two and was startled by the general apologizing to him for a profile about McChrystal that was to appear in *Rolling Stone* magazine. The writer had reported on banter between McChrystal and staff aides critical of Obama and mocking Biden and other members of the president's decision-making national security team. One aide intimated that Obama had been intimidated by McChrystal; another was quoted calling National Security Advisor James Jones "a clown"; and the general himself was cited on one occasion to have said when Biden was mentioned, "Are you asking me about Vice President Biden? Who's that?" to which an aide was said to have chimed in: "Biden? Did you say bite me?"[70]

Biden had neither known about nor read the story. When Biden hung up, he called the president at the White House, who also had heard nothing of it. Obama immediately contacted press secretary Robert Gibbs to get the article and distribute it to his key advisers on war policy, and he had several of them summoned to help him assess the damage. Biden joined the group the next morning and the outlook was grim for McChrystal's future. The president had Secretary of Defense Robert Gates order the general to return to Washington posthaste to meet with Obama the following morning.[71]

Meanwhile, Biden spoke with Obama before the morning presidential briefing on national security matters and then joined him, as usual, for it. Already a consensus was building that it made sense to change command of the war in Afghanistan, both to respond to McChrystal's violation of protocol and arguably his insubordination, and to assure cohesion of the war decision-making team. Obama quickly raised the name of General Petraeus to replace McChrystal in Afghanistan. As author of the army's manual on counterinsurgency, Petraeus had been in the middle of the previous fall's weeks-long strategy sessions on Afghanistan policy from which the eventual plan had emerged. Biden, according to the participants, became the prime advocate of Petraeus, who was the obvious and perfect fit.

The next morning, the general, in his meeting with the president, apologized and offered no excuses. Obama fired him on the spot, and McChrystal left the White House grounds with dispatch. Later, the president's national security meeting ensued without the cashiered general. Early that afternoon, Obama held a televised news conference in which he praised McChrystal for his long military service and announced he would be replaced as the top commander in Afghanistan by Petraeus, the nation's best-known and acclaimed active military leader.[72]

With that one astute stroke, Obama had kept in place the Afghanistan plan hammered out with exhaustive effort the previous year. The post-surge progress would still be reviewed at year's end, and the July 2011 timetable for a start in U.S. troop withdrawal also remained in place, but with fresh observations from Obama that he was committed to a start only.

"We didn't say that we'd be switching off the lights and closing the door behind us," he pointedly said. "We said that we'd begin a transition phase in which the Afghan government is taking on more and more responsibility."[73] Biden for once had no comment, but aides affirmed his agreement with the president.

As noted earlier, in the formative policy debates the previous fall, Biden and McChrystal were known to be in disagreement over the troop surge and the prime enemy in the war. McChrystal insisted it was the Taliban; Biden said it was still al-Qaeda, and he never saw the defeat and destruction of the Taliban as essential, since the American mission was not nation-building as it was under President Bush in Iraq—hence Biden's insistence on a timetable for troop withdrawal.

Such sentiments were muted, however, by the time the Afghan strategy had been hammered out. When Obama called on each member of the national security team to speak up with any doubts about it, none—neither McChrystal nor Biden—did so. Now, in June, Obama declared that with the errant general gone he would insist on "unity of purpose" in keeping the strategy on track under Petraeus, the wide consensus choice to take over the reins of the war.

Some Republican senators in Congress retained suspicions about Biden's commitment. Just prior to the McChrystal fiasco, Alter's book came out in which the vice president was quoted as saying of Afghanistan: "In July of 2011, you're going to see a whole lot of people moving out. Bet on it." Now, Senator Lindsey Graham of South Carolina said Obama should tell him "to shut up," and Senator Christopher Bond of Missouri agreed with a comment that the president should have "a muzzle put on the vice president when it comes to this war."[74] Graham subsequently backed off somewhat, saying on the Fox television network that "I said it tongue-in-cheek." But he added that if the withdrawal timetable was still the policy, he should say so, and if not, "he should stop saying it."[75]

As Obama's second year in the White House proceeded, it was clear that this issue of the time frame for starting the U.S. troop exit from Afghanistan would remain a subject of partisan dispute. Liberal Democrats like Congresswoman Barbara Lee insisted it must remain firm. But Republicans and at least one Democrat, Senator Dianne Feinstein of California, argued that if Petraeus, the new commander in Afghanistan, were to say along the way he needed more time to do the job, Obama should grant it to him.[76]

As Biden continued his heavy lifting on both domestic and foreign fronts—and those occasional inevitable verbal gaffes—Obama was asked for this book how in his mind his vice president had handled himself so far and had accepted his subordinate government role. He replied: "In terms of his role, he has been a great, great partner. When we met to discuss the possibility of him joining the ticket, he said that all he wanted was to be a good counselor, to be in the room when the decisions were made, to offer his opinion and to lend the weight and value of his experience to me as we work through these very difficult issues. He's done that over and over and over again in a very productive way, and taken on some very tough assignments. Joe has done a remarkable job overseeing the Recovery Act. He's also been a principal emissary to Iraq during a very difficult period and has played a tremendously productive role over there. He's got wisdom, he's got experience, he's blunt, and he's loyal—all the qualities you could ask for in a governing partner and in a vice president."[77]

When Biden was chosen by Obama and elected with him, it was assumed that the vice presidency would be his last elected office. Were Obama to run for and win a second presidential term and Biden with him, his vice president would be seventy-four years old in 2016 if he were to seek the presidency again; it would make him the oldest American ever to run for the office. Ronald Reagan was seventy-three when he sought his second term.

Biden's unwaning vigor and enthusiasm, however, permitted some supporters to entertain the possibility that he would go after it a third time after all. Senator Ted Kaufman of Delaware, his interim replacement in the Senate and before that his longtime Senate chief of staff and confidant, said in late March 2009: "I can't believe that he won't think about it."[78] But on that subject, the new vice president remained uncommonly silent, while not dismissing it, either. And so Joe Biden continued to serve in the shadow of the American presidency, the standby to the incumbent, but not quite declaring an end to a half-century dream of the improbable optimist from the blue-collar, middle-class streets of Scranton, Pennsylvania.

EPILOGUE

ATTAINMENT OF THE office of the vice presidency did not quite satisfy Joe Biden's ambition of reaching the White House, but it did go a long way toward delivering him a considerable degree of public esteem and respect that had often eluded him along the way. Nevertheless, his reputation for long-windedness and a certain penchant for hyperbole continued to hang over him, and with it a view, particularly among conservative critics, that he lacked intellectual heft. His colloquial style sometimes came off in the manner of a stand-up comic, exchanging banter with his audience. And when he wanted to emphasize a point, he would often say: "I'm serious; I'm not joking, I'm not exaggerating," as if acknowledging he sometimes was given to both. But diminution of his substance was an attitude both unjust and personally grating to him, and it was capped by an impression among some Americans that he was, with his occasional exaggerations, a less-than-honorable man—the most unsavory and disputatious allegation of all.

For Biden, whose twin pillars of personal belief and strength were faith and family, this characterization assailed the heart of his being. Nothing was more sacred to him than his church and nothing more cherished than his blood family. It was in his devotion to both that, in upholding or defending a position or a promise, he often gave his listeners "my word as a Biden," meaning it could be depended upon beyond any doubt.

Nothing so aroused his resentment than any implication to the contrary. That was why, more even than his failed presidential campaign in 1987, Biden was shattered by the charges then that he had committed plagiarism—lied!—in what he had said at that Iowa State Fair speech, essentially borrowed from British politician Neil Kinnock. That allegation in a most basic way challenged his very honor and that of a family in which his parents had unfailingly preached telling the truth at whatever cost.

So had the prior allegation of plagiarizing one law school paper, which he long after determinedly denied to the point of seeking and getting a judicial exoneration. Biden struggled against the accusations even as he endured a life-threatening illness. In his two major tasks as chairman of the Senate Judiciary Committee, conducting the confirmation hearings of Supreme Court nominees Robert Bork and Clarence Thomas, he bent over backward to demonstrate his fairness and his honesty. He made an impressive case for both in the first, if arguably less persuasively in the second.

In other Senate responsibilities, Biden also committed himself to honorable pursuits: civil rights; public safety against crime and domestic abuse of women at home; against injustice and genocide in the Balkans and elsewhere abroad; and in the quest for peace in various global trouble spots. In all this, he would frequently give "my word as a Biden" as steadfastly reliable currency to friends and adversaries alike. The phrase could and sometimes did take on a Pollyannaish ring, but he offered it always as a sort of personal IOU, with his cherished family reputation as collateral.

In doing so, regrettably he sometimes shaded or stretched facts, out of a natural enthusiasm or desire to be more convincing. When he did so often enough to leave a suspect pattern of behavior, the ever alert news media as well as political critics were quick to seize upon it, and regurgitate similar previous excesses, intentional or otherwise. So it was not long before the loquaciousness and the freewheeling recitations became wedded in a public impression that Joe Biden was a careless—even irresponsible—and self-promoting windbag. That reading did not at all square with the man who put his family name, pride, and honor behind all he said.

Tony Blinken, Biden's chief foreign policy adviser, who has traveled widely abroad with him since 2002, offers this contrary view: "I was struck very early on by how the image that certain people had of Biden didn't correspond with the reality. What you see is what you get is exactly right. He is exactly the same behind the closed doors of his office as he is on TV. But because he presents the image of someone who is very facile with words and very voluble, I think people would be surprised that he really does meet the definition of an intellectual, in the sense that he's a voracious consumer of books and ideas."

Biden, Blinken says, "would constantly be reading and inquiring and sucking in ideas. What struck me about his intellectual process was that he absolutely has no intellectual blinders on. He was more likely to read a book by neoconservatives to understand their thinking as he was [to read] a book by someone who was traditionally liberal," and often seek them out for lengthy discussion. Blinken cited in this regard such conservatives as William Kristol, Robert Kagan, and David Brooks.

When Biden was still in the Senate, his foreign policy adviser says, "you would go into his office and his desk was piled high with books. That pile would be constantly refreshed, and he actually had read them. He'd ask you about them, or the books would be marked up."

Concerning Biden's windiness, Blinken says: "His greatest

strength is his greatest weakness. He does say exactly what he thinks, and he says it sometimes at length. But he thinks, and what you're getting, even if it's unfiltered, is the product of a real thought process and a truly superior mind; a guy who really time and time again connects the dots and sees the relationship from seemingly disparate issues. But the way it's expressed sometimes, in a sort of torrent, doesn't seem at times to be self-edited. He creates that impression. I think he is portrayed by the large public as a guy who is unscripted, and that's a very good thing. They don't necessarily want a politician who reads talking points and can't say anything more than whatever the pollster told him. They know with Joe Biden—good, bad, and indifferent—that's just not him.

"Sometimes," Blinken says, "it gets him into trouble when he says something he shouldn't or he goes on too long. But I think on balance it's a profound asset, because at a time of very scripted and buttoned-downed politicians, he's a guy who says exactly what he thinks, and people like that." In dealing with leaders abroad, Biden's foreign policy adviser says, "I've seen them be extremely taken with his candor and forthrightness and his no-BS approach. That is incredibly refreshing in diplomatic exchanges where people are sometimes used to not saying much to each other. He is very direct, very clear, resulting in meetings that foreign leaders typically hadn't had with American presidents, vice presidents, and secretaries of state. There are occasions where he has talked a lot, but on balance they know they're going to get the unvarnished view from Biden. He doesn't pull punches, and nine times out of ten that's good in diplomacy."[1]

On the same point, another frequent Biden traveling companion, former Republican senator Chuck Hagel, says: "I think anytime someone talks a lot, he puts himself in a position where he can be subjected to intense criticism and analysis. That's just the nature of politics and the nature of leadership, and Joe talks about as much as any senator I've ever been around. Joe likes to talk, he likes people, he likes to engage. He can't help himself. And when you do that, you run the risk to some extent of trivializing your seriousness of purpose, the seriousness of your identity.

"Joe does talk too much, and he knows he talks too much. But that doesn't at all take away from the depth of his knowledge and his great experience and judgment.

"Here's a very honest, open, direct person," Hagel says. "If you ask him a question, he will tell you; even if you don't ask him a question, he will tell you. This is the way he is, and it seems to me we should put a premium on that with our elected officials. Rather than having to pry everything out of him, this guy is about as forthcoming a politician as I've ever been around. And when you're that way, you're going to get yourself in trouble."[2]

Another Republican, former Senate majority leader Bob Dole, in character makes a gag of Biden's talkiness: "He heats up pretty fast, but I've never seen him in a real attack on anybody. He's got a short fuse and a long speaking ability to go with it. But he knows it. When Joe used to get up to speak, we all figured we had at least thirty or forty-five minutes to go to the office, get a haircut. Not that we didn't want to hear what he said. We'd probably heard it before."[3]

In many public assessments of Joe Biden over the years, the overwhelmingly major crisis in his life was the tragic and heartrending loss of his beloved wife, Neilia, and infant daughter, Naomi, in that highway accident only weeks after his upset election to the Senate in 1972. His personal struggle to recover from it while committing himself to the physical and emotional recoveries of his two sons became political lore, along with his daily commuting from a job he at first had no heart to embrace. Somehow, with the support of old friends in Delaware and new ones in the Senate, and eventually the saving love and devotion of his second wife, Jill, Biden recovered and set on a new course again sustained by faith and family.

In the next years, through fealty to both and a more focused commitment to his public life in the Senate, Biden prospered in public support and political possibilities, finally seeking the presidency fifteen years after that first crushing life crisis. Then, in 1987, came the second personal crisis, less obviously shattering but perhaps almost as debilitating to the man whose existence had been anchored by family identity, pride, and honor. In full public view, he said something that drove a knife into public acceptance of his "word as a Biden" and cast a shadow of doubt or even disbelief over that personal currency.

In the midst of it, he struggled in the Bork and later in the Thomas confirmation hearings to resurrect his good name and trustworthiness. As he persevered in chairing the Bork confirmation testimony, Jill Biden immediately recognized what was at stake, as her husband wrote in his memoir at the time—of reporters "starting to see the emergence of *a pattern . . . a character flaw*. . . . [T]he alarm bells went off for Jill right away. They were questioning the one thing . . . I would never be able to defend with words alone. 'Of all the things to attack you on,' she said, almost in tears. 'Your integrity?' "[4] And, when he withdrew from the presidential race in the midst of the Bork hearings, she told him: "You have to win this thing!"[5]

A year after Biden had done so with the denial to Bork of a seat on the Supreme Court, he was still laboring to counter the blot of the plagiarism accusation on his record. His old Senate friend, Republican Bill Cohen of Maine, told the *National Journal:* "This notion of plagiarism hurt him deeply. I think the most important thing for him to achieve is the restoration of his credibility . . . to work hard and be a good senator, and it'll come back. He has got to redeem himself in his own mind. I think he's doing that."[6]

Twenty years later, when Joe Biden sought the presidency a second time, he ran a much more disciplined and constructive campaign, elevating the level of the Democratic debates and demonstrating his maturity and grasp of issues to the point that the vice presidential nomination and then the office itself came to him. Along the way, what were known as the Biden gaffes surfaced again from time to time, keeping that unhappy personality identification alive in the news media and the minds of many Americans, but not sufficiently to bar his political comeback and the election of the man who chose him.

In Joe Biden's first year and a half as vice president, taken fully and effectively into the confidence of President Barack Obama, he served more visibly in his new supporting role than any predecessor on many fronts. To David Plouffe, the 2008 Obama-Biden campaign manager, Biden running for president a second time in 2008 was "a way to vindicate himself" after all the questions of his integrity. "He's a very prideful person," Plouffe said, "and the way he got out of the first campaign did not marry up with the Joe Biden he saw in the mirror. In 2008, there was a redemption. He excelled in [primary] debates and out on the campaign trail in Iowa despite the outcome, and in the vice presidential campaign. I think he's completely freed up now. This is a guy just coming to work every day, calling it as he sees it, with no self-interest."[7]

When his beloved mother, Jean, passed away at age ninety-two on January 8, 2010, just short of the first anniversary of his inauguration as vice president, Biden observed that his family had learned from his parents that "you are defined by your sense of honor."[8] And in that manner, their son had by now restored the bankability of his "word as a Biden" with millions of Americans who have come to accept, as did Delawareans over the years, the positive warts-and-all judgment expressed in the approving accommodation: "That's Joe."

ACKNOWLEDGMENTS

A biography of a public man, particularly one who has been on the national scene for nearly four decades, compels a writer to cast a wide net. But the fact that Joe Biden spent thirty-six years in the U.S. Senate, and was domiciled all that time in the tiny state of Delaware, rendered the task more manageable than might be expected. In more than a year's time spent exploring and examining this particular life of accomplishment and some disappointments, the undertaking of identifying, locating, and listening to central players in the subject's intensely active pursuits, and to many others with perspectives on him, has been enriched by their generosity of time and memory.

Foremost were the recollections and reflections of Vice President Biden; his wife, Dr. Jill Biden; their children, Beau, Hunter, and Ashley Biden; his sister, Valerie Owens, along with her husband, Jack Owens, and their daughter, Missy Owens; and his brother James Biden. All provided invaluable facts and insights into the motivations, ambitions, and even the foibles that over the years have made Joe Biden both an inspirational and a controversial major figure on the American political stage. None has been more helpful than his best friend, sister Valerie. Her combination of enthusiasm, political savvy, Irish warmth, and openness made her an essential and cordial gatekeeper to the Biden inner circle and outer circles as well.

As a group, the Bidens not only certified the remarkably strong family ties binding them but also an uncommon loyalty tempered with candor toward the man who rose from commonplace beginnings in Scranton, Pennsylvania, and Wilmington, Delaware, to a career of influence in domestic matters and foreign affairs.

What might be called his other family—in politics and particularly in the U.S. Senate—has offered sharp and detailed recollections of him in the personal and political crises through which he has passed, from his earliest days in Scranton and Wilmington to the Senate and his elevation to the vice presidency. Among those who have been the most helpful in this chronicle with their reminiscences and generous with their time are:

From the Senate: Senators Evan Bayh, Tom Carper, Orrin Hatch, Daniel Inouye, Ted Kaufman, Paul Kirk, Patrick Leahy, Richard Lugar, Jack Reed; former senators Birch Bayh, William Cohen, Bob Dole, Chuck Hagel;

Former Senate committee staff aides: on Foreign Relations, William Bader, Michael Haltzel, Brian McKeon; on Judiciary, Mark Gitenstein especially, Victoria Nourse, Chris Putala; Senate librarian Brian McLaughlin; Senate Judiciary Committee librarian Charles Papermeister; Senate historian Donald Ritchie, especially for his oral history interviews with Senate aides William Hildenbrand, Nordy Hoffman, Dorothye Scott, and Frank Valeo; civil rights and legal rights activist Ralph Neas;

Biden staff aides in Delaware: Kaufman, Norma Long, John DiEleuterio, Vince D'Anna;

Obama-Biden administration officials: President Barack Obama (in written responses), David Axelrod, Ron Klain, Antony Blinken, Jay Carney, Brian McKeon; also David Plouffe, Obama-Biden campaign manager, 2008;

In Scranton: Tom Bell, Larry Orr, Jim Kennedy, *Scranton Times* librarian Brian Felton;

At Archmere Academy: headmaster Father Joseph McLaughlin, Coach John Walsh, teammate Mike Fay; at the University of Delaware: Fay, Lou Bartosheshky, Tom Lewis, Professor Jeffrey Raffel, Fred Sears, Professors James Soles and Raymond Wolters, David Walsh; Professor Sam Hoff at Delaware State University;

In Syracuse: Bill Brodsky, Joe Covino, Kevin Coyne, John Fahey, Clayton Hale, Don MacNaughton, William Kissel, Professors Thomas Maroney and William Banks of the Syracuse Law School, Jane Fahey Suddaby, Pat Cowin Wojenski;

In Wilmington: Mayor James Baker, David Bakerian of the Delaware Bankers Association, Sidney and Carol Balik, Bebe Coker, Connie Cooper of the Delaware Historical Society, Bert DiClemente, Rich Heffron;

In the Democratic Party of Delaware: especially John Daniello and Henry Topel, Joe Farley, Pete and Karen Peterson, John Reda, Sonia Sloan; in the Republican Party of Delaware: former governor Pete du Pont, Basil Battaglia, Jane Brady, John Burris, Dave Crossan;

Longtime Biden political associates: William Daley, John Marttila, Larry Rasky, Patrick Caddell; other Biden family friends and associates: Pat Evans, Maureen Masterson Greco, Roger Harrison, Bobbi Greene McCarthy, Bob Osgood;

In the press corps: David Ledford, editor of the *Wilmington News Journal;* Anne Haslam, librarian; reporters Nicole Gaudiano, Rachel Kipp, Ron Williams; John Taylor, former editorial page editor; Tom Eldred of the *Delaware State News;* Sean Kirst of the *Syracuse Post-Standard;* Perry Bacon of the *Washington Post;* David Yepsen, former political columnist of the *Des Moines Register;* Curtis Wilkie, former political reporter of the Wilmington newspapers and retired national political correspondent of the *Boston Globe.* Also, over most of the year of writing, my great friend John Mashek, also retired from the *Boston Globe,* offered regular encouragement and impressions.

Finally, my thanks to my agent, David Black; to my editor at William Morrow, Henry Ferris; to his assistant, Danny Goldstein, for shepherding through this project with counsel and dispatch; and to my wife, Marion Elizabeth Rodgers, for her encouraging support through the process.

Washington, D.C., and Bethany Beach, Delaware
June 28, 2010

NOTES

CHAPTER 1: SCRANTON

1. Personals, *Scranton Times*, November 20, 1942.
2. Ibid.
3. Ibid.
4. *Philadelphia Inquirer*, November 20, 1942.
5. Ibid.
6. *Wilmington News Journal*, September 9, 2002.
7. Interview with Vice President Biden, December 21, 2009, Wilmington, Del.
8. Joe Biden, *Promises to Keep: On Life and Politics* (New York: Random House, 2007), 17; Richard Ben Cramer, *What It Takes: The Way to the White House* (New York: Vintage Books, 1993), 261.
9. Biden, *Promises to Keep*, 16; Cramer, *What It Takes*, 263.
10. Cramer, *What It Takes*, 263–64.
11. Biden, *Promises to Keep*, 17.
12. Interview with Vice President Biden.
13. Biden, *Promises to Keep*, 22.
14. Interview with Tommy Bell and Larry Orr, May 4, 2009, Scranton, Penn.
15. Ibid.
16. Ibid.
17. Interview with Jim Kennedy, May 4, 2009, Scranton, Penn.

18. Ibid.
19. Interview with Tommy Bell and Larry Orr.
20. Ibid.
21. Ibid.
22. Interview with Valerie Biden Owens, September 24, 2009, Wilmington, Del.
23. Interview with Valerie Biden Owens, January 26, 2009, Wilmington, Del.
24. Ibid.
25. Biden, *Promises to Keep*, xiii.
26. Ibid., 4.
27. Interview with Tommy Bell and Larry Orr.
28. Ibid.

CHAPTER 2: WILMINGTON

1. Biden, *Promises to Keep*, 6–7.
2. Interview with Valerie Biden Owens, January 26, 2009.
3. Interview with Vice President Biden.
4. Biden, *Promises to Keep*, 9–11.
5. Interview with Tom Lewis, August 3, 2009, Bethany Beach, Del.
6. Biden, *Promises to Keep*, 3–4.
7. Cramer, *What It Takes*, 304.
8. Interview with Vice President Biden.
9. Interview with Valerie Biden Owens, January 26, 2009.
10. Interview with John Walsh, July 24, 2009, Claymont, Del.
11. Ibid.
12. Interview with Lou Bartosheshky, July 24, 2009, Newark, Del.
13. Interview with Tom Lewis.
14. Interview with Mike Fay, July 24, 2009, Claymont, Del.
15. Ibid.
16. Interview with Vice President Biden.
17. Telephone interview with Maureen Masterson Greco, October 30, 2009.
18. Interview with Vice President Biden.
19. Interview with Lou Bartosheshky.
20. Interview with David Walsh, November 20, 2008, Wilmington, Del.
21. Interview with Mike Fay.
22. *Wilmington News*, September 22, 1987.
23. Interview with David Walsh, November 20, 2008.
24. Interview with Valerie Owens, January 26, 2009.
25. Interview with Fred Sears, January 8, 2009, Wilmington, Del.
26. Biden, *Promises to Keep*, 23.
27. Interview with David Walsh, November 20, 2008.

28. Interview with Fr. Joseph McLaughlin, November 19, 2008, Claymont, Del.
29. Interview with Valerie Biden Owens, January 26, 2009.

CHAPTER 3: BUILDING A DREAM

1. Biden, *Promises to Keep*, 25, 27.
2. Ibid., 45–46.
3. Interview with Vice President Biden.
4. Interview with Fred Sears.
5. Interview with Valerie Biden Owens, September 24, 2009.
6. Interview with Fred Sears.
7. Interview with Valerie Biden Owens, January 26, 2009.
8. Interview with Fred Sears.
9. Biden, *Promises to Keep*, 27.
10. Interview with Tom Lewis.
11. Biden, *Promises to Keep*, 27.
12. Interview with Fred Sears.
13. Biden, *Promises to Keep*, 28.
14. Interview with Fred Sears.
15. Biden, *Promises to Keep*, 28–31.
16. Interview with Valerie Biden Owens, January 26, 2009.
17. Telephone interview with Bobbie Greene McCarthy, October 21, 2009.
18. Ibid.
19. Sean Kirst, *Syracuse Post-Standard*, August 24, 2008.
20. Biden, *Promises to Keep*, 31–33.
21. Ibid., 34.
22. Interview with John DiEleuterio, May 21, 2009, Wilmington, Del.
23. Interview with Vice President Biden.
24. Interview with Jack Owens, September 24, 2009, Wilmington, Del.
25. Telephone interview with Roger Harrison, November 15, 2009.
26. Telephone interview with William Kissell, November 15, 2009.
27. Interview with Clayton Hale, November 19, 2009, Syracuse, N.Y.
28. Biden, *Promises to Keep*, 34–35.
29. Telephone interview with Bob Osgood, December 2, 2009; telephone interview with Don MacNaughton, December 3, 2009.
30. Interview with Jack Owens.
31. Telephone interview with Roger Harrison.
32. Interview with Jack Owens.
33. Interview with Clayton Hale.
34. Biden, *Promises to Keep*, 36.
35. Ibid., 39; *Wilmington News*, September 22, 1987.

36. Interview with John Covino, November 17, 2009, Syracuse, N.Y.
37. Telephone interview with William Kissel.
38. Interview with Clayton Hale.
39. Interview with John Covino.
40. Telephone interview with Bobbie Greene McCarthy.
41. Interview with Joseph Fahey, November 17, 2009, Syracuse, N.Y.
42. Interview with Jane Fahey Suddaby, November 17, 2009, Syracuse, N.Y.
43. Interview with Kevin Coyne, November 17, 2009, Syracuse, N.Y.
44. Interview with Pat Cowin Wojenski, November 18, 2009, Syracuse, N.Y.
45. Interview with Kevin Coyne.
46. Biden, *Promises to Keep*, 38.
47. Interview with Thomas Maroney, November 18, 2009, Syracuse, N.Y.
48. Interview with Vice President Biden.
49. Telephone interview with Bill Brodsky, January 15, 2010.

CHAPTER 4: THE MAKING OF A POLITICIAN

1. *Wilmington News Journal* Archive, 1968.
2. Andreas George Schneider, *Delaware: The Politics of Urban Unrest, July 1967–January 1969*, Woodrow Wilson School of Public and International Affairs Scholars Program, June 1970 (revised June 1971).
3. Interview with Mayor James Baker, January 27, 2009, Wilmington, Del.
4. *Wilmington News Journal* Archive, 1968.
5. Schneider, *Politics of Urban Unrest*.
6. *Wilmington News Journal* Archive, 1968.
7. Interview with Bert DiClemente, November 19, 2008, Wilmington, Del.
8. Interview with Senator Ted Kaufman, January 26, 2009, Wilmington, Del.
9. Interview with Valerie Biden Owens, January 26, 2009.
10. Interview with Vice President Biden.
11. Interview with William Quillen, January 8, 2009, Wilmington, Del.
12. Biden, *Promises to Keep*, 38.
13. Interview with Vice President Biden.
14. Biden, *Promises to Keep*, 52–55.
15. Interview with Valerie Biden Owens, January 26, 2009.
16. Telephone interview with Bobbie Greene McCarthy.
17. Interview with Vice President Biden.
18. Ibid.
19. Ibid.
20. Interview with Bert DiClemente.

21. Interview with Sid Balick, January 9, 2009, Wilmington, Del.
22. Carl Leubsdorf, "Lifelong Ambition Led Joe Biden to Senate, White House Aspirations," *Dallas Morning News*, August 23, 2008 (reprint of 1987 profile).
23. Biden, *Promises to Keep*, 48.
24. Interview with David Walsh, November 20, 2008.
25. Interview with Mike Fay.
26. Interview with John Daniello, November 20, 2008, Wilmington, Del.
27. Biden, *Promises to Keep*, 50.
28. *Wilmington Journal*, August 10, 1970.
29. *Wilmington Journal*, October 29, 1970.
30. Interview with John Daniello.
31. Interview with Valerie Biden Owens, January 26, 2009.
32. Interview with John Daniello.
33. Ibid.
34. *Wilmington News*, January 3, 1973.
35. *Wilmington Journal*, November 11, 1970.
36. *Wilmington Journal*, January 20, 1971.
37. Biden, *Promises to Keep*, 51.
38. *Wilmington Journal*, February 10, 1971.
39. Interview with Vince D'Anna, May 5, 2009, Newark, Del.
40. *Wilmington News*, March 23, 1972.
41. Interview with Vince D'Anna.
42. Interview with Henry Topel, December 17, 2008, Wilmington, Del.
43. "An Army of Youth; a Teenage Memory"; interview with David Topel, January 8, 2009, Wilmington, Del.
44. Interview with Henry Topel.
45. Interview with David Topel.
46. Biden, *Promises to Keep*, 53–54, 57, 58.
47. Interview with John DiEleuterio.
48. Biden, *Promises to Keep*, 52–55.
49. Ibid., 59.
50. Interview with Vince D'Anna.
51. Biden, *Promises to Keep*, 59–60.
52. Interview with Valerie Biden Owens, January 26, 2009.
53. *Wilmington News*, September 27, 1971.
54. Interview with former Delaware governor Pierre "Pete" du Pont IV, May 21, 2009, Wilmington, Del.
55. "William F. Hildenbrand, Secretary of the Senate, 1981–1985," by Senate historian Donald Ritchie, Oral History Interviews, Senate Historical Office, Washington, D.C., 96.
56. *Wilmington News*, September 27, 1971.
57. Interview with Jim Kennedy.

58. Telephone interview with Sonia Sloan, May 22, 2009.
59. Telephone interview with Patrick Caddell, July 23, 2009.
60. Interview with Valerie Biden Owens, September 24, 2009.
61. Interview with Valerie Biden Owens, October 6, 2009.

CHAPTER 5: DAVID AND GOLIATH

1. Biden, *Promises to Keep*, 64–65.
2. Interview with Mayor James Baker.
3. *Wilmington Journal*, March 27, 1972.
4. Biden, *Promises to Keep*, 65.
5. Interview with John Marttila, October 23, 2009, Washington, D.C.
6. Ibid., 65–66.
7. Interview with Henry Topel, January 9, 2009, Wilmington, Del.
8. Interview with Rich Heffron, October 13, 2008, Wilmington, Del.
9. *Wilmington Journal*, June 26, 1972.
10. *Wilmington Journal*, August 24, 1972.
11. Ibid.
12. Interview with Jimmy Biden, September 24, 2009, Wilmington, Del.
13. Interview with Valerie Biden Owens, September 24, 2009.
14. Telephone interview with Patrick Caddell, October 8, 2009.
15. Interview with Jimmy Biden.
16. Joe Biden, "Public Financing of Elections: Legislative Proposals and Constitutional Questions," *Northwestern University Law Review* 69, no. 1 (March–April 1974), 1.
17. Interview with Jimmy Biden; Biden, *Promises to Keep*, 70–71.
18. Biden, "Public Financing of Elections," 1.
19. "F. Nordy Hoffmann, Senate Sergeant at Arms, 1975–1981," by Senate historian Donald Ritchie, Oral History Interviews, Senate Historical Office, Washington, D.C., 187–88.
20. Interview with Jimmy Biden.
21. "F. Nordy Hoffmann, Senate Sergeant at Arms, 1975–1981," 196–97.
22. Interview with Valerie Biden Owens, September 24, 2009.
23. Interview with John Marttila.
24. Telephone interview with Bobbie Greene McCarthy.
25. *Wilmington News*, September 26, 1972.
26. *Wilmington News*, October 3, 1972.
27. *Wilmington Journal*, October 26, 1972.
28. Interview with Jack Owens.
29. Interview with John Marttila.
30. *Wilmington News*, October 2, 1972.
31. Biden campaign poster, office of Senator Ted Kaufman, Wilmington, Del.
32. Interview with Senator Ted Kaufman.

33. *Wilmington News*, October 23, 1972.
34. Biden, *Promises to Keep*, 68.
35. Celia Cohen, *Only in Delaware: Politics and Politicians in the First State* (Newark, Del.: Grapevine Publishing, 2002), 205.
36. *Wilmington News*, November 2, 1972.
37. Biden, *Promises to Keep*, 69.
38. Interview with Valerie Biden Owens, January 26, 2009.
39. Celia Cohen, *Only in Delaware*, 204.
40. Biden speech at Democratic fund-raiser, Alexandria, Va., October 8, 2009.
41. *Wilmington News*, November 8, 1972.
42. Biden, *Promises to Keep*, 74.
43. Interview with Vice President Biden.
44. Interview with Valerie Biden Owens, January 26, 2009.
45. Interview with Senator Ted Kaufman.
46. Interview with James Soles, December 17, 2008, Newark, Del.
47. "William F. Hildenbrand, Secretary of the Senate, 1981–1985," by Senate historian Donald Ritchie, Oral History Interviews, Senate Historical Office, Washington, D.C., 96–97.
48. Ibid., 98.
49. Interview with Valerie Biden Owens, January 26, 2009.
50. Interview with Henry Topel, December 17, 2008.
52. *Wilmington News*, November 9, 1972.

CHAPTER 6: THE DREAM SHATTERED

1. *Wilmington News*, January 3, 1973.
2. Biden, *Promises to Keep*, 79.
3. Ibid., 79–80.
4. *Wilmington Journal*, December 19, 1972.
5. *Wilmington Journal*, December 21, 1972.
6. Telephone interview with Bobbie Greene McCarthy.
7. Biden, *Promises to Keep*, 84.
8. *Wilmington Journal*, December 20, 1972.
9. Biden, *Promises to Keep*, 80–81.
10. Telephone interview with Roger Harrison.
11. Biden, *Promises to Keep*, 81.
12. Telephone interview with Patrick Caddell, July 27, 2009.
13. Interview with David Walsh, November 19, 2008, Wilmington, Del.
14. Interview with Valerie Biden Owens, January 26, 2009.
15. "Dorothye G. Scott, Administrative Assistant to the Senate Democratic Secretary and to the Secretary of the Senate (1945–1977)," by Senate historian Donald Richie, Oral History Interviews, Senate Historical Office, Washington, D.C., 251.

16. *Wilmington News*, January 6, 1973.
17. Interview with Vice President Biden.
18. Interview with Valerie Biden Owens, January 26, 2009.
19. Interview with Beau Biden, November 16, 2009, Wilmington, Del.
20. Biden, *Promises to Keep*, 104.

CHAPTER 7: A FRESH START

1. Biden speech to Delaware supporters, Washington, D.C., January 6, 2006.
2. Biden, *Promises to Keep*, 76.
3. Ibid., 91.
4. Ibid., 92.
5. *Congressional Record* 155 (January 15, 2009), S405.
6. Ibid.
7. Biden, *Promises to Keep*, 87.
8. Interview with Hunter Biden, July 7, 2009, Washington, D.C.
9. Interview with Beau Biden.
10. Interview with Jimmy Biden, September 24, 2009; interview with Jack Owens.
11. Interview with former senator Birch Bayh, March 4, 2009, Washington, D.C.
12. Interview with William Bader, September 3, 2009, Washington, D.C.
13. Interview with Senator Patrick Leahy, November 5, 2009, Washington, D.C.
14. Interview with Jimmy Biden.
15. Interview with Valerie Biden Owens, September 24, 2009.
16. Interview with Jimmy Biden.
17. *Wilmington Journal*, January 9, 1973.
18. *Wilmington Journal*, February 1, 1973.
19. *Washington Post*, August 27, 2008.
20. Interview with Jim Kennedy.
21. *New York Times*, October 24, 2008.
22. *Wilmington Journal*, March 5, 1973.
23. Joe Biden letter to John Covino, April 26, 1973.
24. *Wilmington Journal*, January 3, 1974.
25. "Francis R. Valeo, Secretary of the Senate, 1966–1977," by Senate historian Donald Ritchie, Oral History Interviews, Senate Historical Office, Washington, D.C., 860–61.
26. *Wilmington Journal*, June 11, 1973.
27. Interview with Valerie Biden Owens, September 24, 2009.
28. Biden, *Promises to Keep*, 94–94; phone interview with Roger Harrison.
29. Biden, *Promises to Keep*, 94.
30. *Wilmington Journal*, July 26, 1973.

31. Biden, *Promises to Keep*, 94.
32. *Wilmington Journal*, May 24, 1973.
33. *Wilmington News*, May 25, 1973.
34. *Wilmington Journal*, August 9, 1973.
35. *Washington Star-News*, December 11, 1973.

CHAPTER 8: FINDING HIMSELF, AND JILL

1. *Weekly Post*, January 24, 1974.
2. *Wilmington News*, March 23, 1974.
3. "Death and the All-American Boy," *Washingtonian* (September 1974), 87–90.
4. *Wilmington Journal*, April 10, 1974.
5. *Wilmington News*, April 10, 1974.
6. *Wilmington News*, August 7, 1974.
7. *County Post*, August 14, 1974.
8. *Wilmington Journal*, September 9, 1974.
9. *TV News: The People Paper*, September 26, 1974, 45.
10. Ibid.
11. Biden, *Promises to Keep*, 138.
12. Ibid., 83–84.
13. Ibid., 100–102.
14. Interview with Jill Biden, January 20, 2010, Washington, D.C.
15. Ibid.
16. Biden, *Promises to Keep*, 114.
17. Interview with Beau Biden.
18. Biden, *Promises to Keep*, 115.
19. Interview with Jill Biden.
20. Associated Press, July 22, 1975.
21. *Wilmington News*, February 23, 1976.
22. Jules Witcover, *Marathon: The Pursuit of the Presidency, 1972–1976* (New York: Viking Press, 1977), 212–14.
23. *Wilmington Journal*, February 23, 1976.
24. Ted Sorensen, *Counselor: A Life at the Edge of History* (New York: Harper, 2008), 491.
25. Ibid., 499–500.
26. *Wilmington News*, February 18, 1977.
27. *Wilmington Journal*, April 24, 1977.
28. Interview with Jill Biden.
29. Biden, *Promises to Keep*, 116.
30. Interview with Hunter Biden, July 8, 2009, Washington, D.C.
31. *Washington Post*, June 9, 1987.
32. Interview with Beau Biden.
33. Ibid.

CHAPTER 9: CIVIL RIGHTS, JIMMY CARTER, AND REELECTION

1. *Wilmington Journal*, June 28, 1974.
2. *TV News: The People Paper*, September 26, 1974, 11.
3. *Congressional Quarterly* 1975 legislative chronology, 667.
4. Biden, *Promises to Keep*, 125.
5. *TV News: The People Paper*, 44.
6. Ibid.
7. Ibid., 45.
8. *Wilmington News*, December 18, 1974.
9. *Wilmington News Journal*, May 22, 1977.
10. *Wilmington News Journal*, June 19, 1977.
11. Ibid.
12. *Wilmington News Journal*, June 21, 1977.
13. *Wilmington News Journal*, October 12, 1977.
14. *Wilmington News Journal*, November 22, 1977.
15. Biden, *Promises to Keep*, 126.
16. Ibid., 126–27.
17. Interview with Jeffrey Raffel, May 20, 2009, Newark, Del.
18. Interview with Vice President Biden.
19. *Wilmington Journal*, October 27, 1978.
20. *Congressional Quarterly*, February 7, 1997, 223.

CHAPTER 10: MOVING ON TO THE NATIONAL STAGE

1. Biden, *Promises to Keep*, 136, 147.
2. Interview with William Bader, September 3, 2009, Washington, D.C.
3. *Wilmington Journal*, August 12, 1980; August 24, 1980.
4. *Wilmington Journal*, August 13, 1980.
5. *Washingtonian* (December 1985), 91–92.
6. *Wilmington Journal*, December 7, 1980.
7. *Wilmington Journal*, December 7, 1982.
8. *Washingtonian* (December 1985), 88.
9. Biden, *Promises to Keep*, 141.
10. Ibid., 142.
11. Interview with Patrick Caddell, September 26, 2009, Washington, D.C.
12. Biden, *Promises to Keep*, 142–43.
13. Interview with John Marttila.
14. Interview with Vice President Biden.
15. *Washingtonian* (December 1985), 91.
16. Interview with Patrick Caddell.
17. *Wilmington Journal*, June 9, 1984.
18. *Wilmington Journal*, November 7, 1984.
19. Interview with John Burris, November 20, 2008, Wilmington, Del.

20. Interview with former Delaware governor Pierre "Pete" du Pont IV.
21. Interview with Joseph Farley, May 20, 2009, Dover, Del.
22. Interview with Vince D'Anna.

CHAPTER 11: JOE BIDEN FOR PRESIDENT

1. *Washingtonian* (December 1985), 93.
2. Ibid.
3. Interview with William Bader.
4. *New Republic*, September 1, 1986.
5. *Wall Street Journal*, February 1985.
6. *Congressional Quarterly*, February 7, 1985.
7. *Wilmington Journal*, June 26, 1985.
8. Jack W. Germond and Jules Witcover, *Whose Broad Stripes and Bright Stars?: The Trivial Pursuit of the Presidency, 1988* (New York: Warner Books, 1989), 44.
9. Ibid., 44–45.
10. *New York Times*, December 19, 1985.
11. Ibid.
12. *Wilmington Journal*, April 29, 1986.
13. Interview with Ashley Biden, July 24, 2009, Wilmington, Del.
14. *Washingtonian*, February 22, 1986.
15. Senate Committee on the Judiciary, *Nomination of Justice William Hubbs Rehnquist: Hearings before the Committee on the Judiciary on the nomination of Justice William Hubbs Rehnquist to be Chief Justice of the United States*, 98th Cong., 2nd sess., 137–38.
16. Senate Committee on the Judiciary, *Nomination of William H. Rehnquist to be Chief Justice of the United States: Report from the Committee on the Judiciary, United States Senate, together with additional, minority, and supplemental views* (Washington: U.S. Government Printing Office, 1986), 66.
17. Ibid., 67–68.
18. Mark Gitenstein, *Matters of Principle: An Insider's Account of America's Rejection of Robert Bork's Nomination to the Supreme Court* (New York: Simon and Schuster, 1992), 24; *Wilmington Journal*, August 3, 1987.
19. *New Republic*, September 9, 1986.
20. Ibid.; *Current Biography*, January 1987, 13.
21. *Congressional Quarterly* 45 (1986), 226.
22. *Wilmington News Journal*, October 15, 1986.
23. *New Republic*, September 1, 1986.
24. *Wilmington News Journal*, October 15, 1986.
25. *Washington Post*, April 5, 1987.
26. Biden, *Promises to Keep*, 149.

27. Paul Taylor, *See How They Run: Electing the President in an Age of Mediaocracy* (New York: Knopf, 1990), 93.
28. Biden, *Promises to Keep*, 150.
29. Ibid., 156–58.
30. Ibid., 160.
31. *Washington Post*, March 3, 1986.
32. Interview with William Bader.
33. Biden, *Promises to Keep*, 161.
34. *Wilmington News Journal*, June 10, 1987.

CHAPTER 12: A JUDICIAL INTRUSION

1. Biden, *Promises to Keep*, 164–65.
2. Ibid., 165.
3. Gitenstein, *Matters of Principle*, 26.
4. Ibid., 30.
5. Biden, *Promises to Keep*, 167.
6. Ibid.
7. Ibid., 168.
8. Michael Pertschuk and Wendy Schaetzel, *The People Rising: The Campaign Against the Bork Nomination* (New York: Thunder's Mouth Press, 1989), 33.
9. Ibid., 34.
10. Biden, *Promises to Keep*, 169.
11. Gitenstein, *Matters of Principle*, 57.
12. *Wilmington News Journal*, July 7, 1987.
13. *Congressional Record* 133 (July 1, 1987), S9188–89.
14. Biden, *Promises to Keep*, 169–70.
15. Gitenstein, *Matters of Principle*, 56.
16. Ibid., 276–78.
17. Biden, *Promises to Keep*, 170.
18. Gitenstein, *Matters of Principle*, 61.
19. Pertschuk and Schaetzel, *The People Rising*, 231–34.
20. Gitenstein, *Matters of Principle*, 64.
21. Ibid.
22. Ibid.
23. *Wilmington News Journal*, July 11, 1987.
24. *New York Times*, July 16, 1987.
25. Gitenstein, *Matters of Principle*, 66.
26. Biden, *Promises to Keep*, 172.
27. Interview with Mark Gitenstein, August 7, 2009, Washington, D.C.
28. Biden, *Promises to Keep*, 172.
29. Ibid., 172–73.
30. Ibid., 174.

31. Ibid., 178.
32. *Wilmington News Journal*, August 3, 1987.
33. Biden, *Promises to Keep*, 182.
34. Ibid., 183.
35. Gitenstein, *Matters of Principle*, 192.
36. Ibid., 204, 208.
37. Ibid., 100.

CHAPTER 13: DEBACLE IN IOWA

1. Biden, *Promises to Keep*, 184–86.
2. Germond and Witcover, *Whose Broad Stripes and Bright Stars?* 231–32.
3. Biden, *Promises to Keep*, 190.
4. Germond and Witcover, *Whose Broad Stripes and Bright Stars?* 232.
5. Paul Taylor, *See How They Run*, 92–93.
6. Ibid., 95.
7. Germond and Witcover, *Whose Broad Stripes and Bright Stars?* 238.
8. Paul Taylor, *See How They Run*, 84.
9. Germond and Witcover, *Whose Broad Stripes and Bright Stars?* 242.
10. Cramer, *What It Takes*, 707.
11. Germond and Witcover, *Whose Broad Stripes and Bright Stars?* 242.
12. Ibid., 241.
13. Ibid., 234.
14. *Washington Post*, September 17, 1987.
15. *Washington Post*, September 18, 1987.
16. *Washington Post*, September 17. 1987.
17. Germond and Witcover, *Whose Broad Stripes and Bright Stars?* 235.
18. Biden, *Promises to Keep*, 191.
19. Interview with John Marttila.
20. Interview with Vice President Biden.

CHAPTER 14: CONCENTRATING ON BORK

1. Biden, *Promises to Keep*, 191.
2. Senate Committee on the Judiciary, *Nomination of Robert H. Bork to be Associate Justice of the Supreme Court of the United States: Hearings before the Committee on the Judiciary*, 100th Cong., 1st sess., Sept. 15–30, 1987, 47–51.
3. Ibid., 51.
4. *Washington Post*, September 15, 1987.
5. Senate Committee on the Judiciary, *Nomination of Robert H. Bork*, 95–96.
6. Ibid., 97.
7. Ibid., 103–04.
8. Ibid., 112.

9. Ibid., 114–15.
10. Ibid., 116.
11. Ibid., 117.
12. Biden, *Promises to Keep*, 197.
13. Senate Committee on the Judiciary, *Nomination of Robert H. Bork*, 194–95.
14. Gitenstein, *Matters of Principle*, 250–51.
15. Biden, *Promises to Keep*, 201.
16. Ibid.
17. *Wilmington News Journal*, September 18, 1987.
18. Biden, *Promises to Keep*, 201.
19. *Washington Post*, September 18, 1987.
20. Ibid.
21. Cramer, *What It Takes*, 649.
22. *Washington Post*, September 18, 1987.
23. Cramer, *What It Takes*, 652–53.
24. *Washington Post*, September 18, 1987.
25. Senate Committee on the Judiciary, *Nomination of Robert H. Bork*, 320.
26. Ibid., 320–24.
27. Ibid., 325.
28. Ibid., 326.
29. Ibid., 328.
30. Ibid., 329.
31. Gitenstein, *Matters of Principle*, 259–60.
32. Ibid., 260.
33. Biden, *Promises to Keep*, 204.
34. Germond and Witcover, *Whose Broad Stripes and Bright Stars?* 237.
35. Gitenstein, *Matters of Principle*, 263.
36. Ibid., 264.
37. Interview with John Marttila.
38. Biden, *Promises to Keep*, 204–05.
39. Interview with Beau Biden.
40. Interview with Hunter Biden, July 8, 2009.
41. Biden, *Promises to Keep*, 205.
42. Senate Committee on the Judiciary, *Nomination of Robert H. Bork*, 2099.
43. Ibid., 2100.
44. Ibid.
45. *New York Times*, *Washington Post*, September 24, 1987.
46. Biden, *Promises to Keep*, 206.
47. Interview with Jill Biden.
48. Germond and Witcover, *Whose Broad Stripes and Bright Stars?* 238.
49. Interview with Tom Lewis.

50. Senate Committee on the Judiciary, *Nomination of Robert H. Bork*, 2157.
51. Ibid., 2186.
52. Ibid., 2188.
53. Biden, *Promises to Keep*, 207.

CHAPTER 15: A COSTLY VICTORY

1. Interview with Larry Rasky, September 29, 2009, Washington, D.C.
2. Gitenstein, *Matters of Principle*, 268.
3. Ibid., 268–69.
4. Ibid., 269.
5. Senate Committee on the Judiciary, *Nomination of Robert H. Bork to be Associate Justice of the Supreme Court of the United States: Hearings before the Committee on the Judiciary*, 100th Cong., 1st sess., 3893.
6. Gitenstein, *Matters of Principle*, 288–89.
7. Ibid., 293.
8. Ibid., 291.
9. Ibid., 293.
10. Ibid., 294–95.
11. Robert Bork, *The Tempting of America: The Political Seduction of the Law* (New York: Free Press, 1990), 311–14.
12. Gitenstein, *Matters of Principle*, 308.
13. Biden, *Promises to Keep*, 210.
14. Bork, *The Tempting of America*, 307–08.
15. Arlen Specter, *A Passion for Truth: From Finding JFK's Single Bullet to Questioning Anita Hill to Impeaching Clinton* (New York: William Morrow, 2000), 329.
16. Interview with Vice President Biden.
17. Interview with Jill Biden.
18. Bork, *The Tempting of America*, 316, 319–20.
19. Gitenstein, *Matters of Principle*, 314.
20. Ibid., 315.
21. Biden, *Promises to Keep*, 211–13.
22. Interview with Vice President Biden.
23. Biden, *Promises to Keep*, 212–13.
24. Gitenstein, *Matters of Principle*, 316.
25. Ibid., 317–19.
26. *Legal Times* (Dec. 1987/ Jan. 1988).
27. Interview with Mark Gitenstein, February 20, 2009, Washington, D.C.
28. Gitenstein, *Matters of Principle*, 268.

CHAPTER 16: DOWN BUT NOT OUT

1. *Washington Post*, January 1, 1988.

2. *Wilmington News Journal*, January 10, 1988.
3. Ibid.
4. Biden, *Promises to Keep*, 215–16.
5. Ibid., 217–18.
6. Ibid., 221–22.
7. *Washington Post*, January 12, 1989.
8. *Wilmington News Journal*, June 12, 1989.
9. Biden, *Promises to Keep*, 233.
10. Ibid.
11. *New York Times*, September 8, 1989.
12. Interview with Valerie Biden Owens, September 24, 2009.
13. Interview with Beau Biden.
14. Interview with Vice President Biden.
15. Joseph R. Biden and John B. Ritch III, "The War Power at a Constitutional Impasse: A 'Joint Decision' Solution," *Georgetown Law Journal* (December 1988): 367–412.
16. *Congressional Record*, 105th Cong., 2nd sess. (July 30, 1998), S9444.
17. Ibid., S94446.
18. *Wilmington News Journal*, September 3, 1989.
19. *Wilmington News Journal*, October 14, 1989.
20. *National Journal*, October 14, 1989.
21. *Washington Post*, January 12, 1989.
22. *Congressional Quarterly*, October 14, 1989, 2707.
23. *Wilmington News Journal*, May 28, 1989.
24. *Wilmington News Journal*, May 17, 1990.
25. Ibid.
26. Interview with M. Jane Brady, January 26, 2009, Wilmington, Del.
27. *Wilmington News Journal*, July 31, 1990.
28. *Wilmington News Journal*, October 1, 1990.
29. Interview with M. Jane Brady.
30. *Wilmington News Journal*, November 30, 1990.

CHAPTER 17: HOLDING OUT FOR PEACE

1. Senate Committee on Foreign Relations, *U.S. Policy in the Persian Gulf: Hearings before the Committee on Foreign Relations*, 101st Cong., 2nd sess., September 5, 20, and October 17, 1990, 19–20.
2. Ibid., 99–103.
3. Senate Committee on Foreign Relations, *U.S. Policy in the Persian Gulf: Hearings before the Committee on Foreign Relations*, 101st Cong, 2nd sess., December 4–5, 12–13, 1990, 27–29.
4. Ibid., 96–97.
5. Senate Committee on Foreign Relations, *U.S. Policy in the Persian*

Gulf: Hearings before the Committee on Foreign Relations, 102nd Cong., 1st sess., January 8, 1991.
6. Ibid.
7. Ibid., 1–3.
8. Ibid., 201–02.
9. Ibid., 6–7.
10. *Wilmington News Journal*, February 17, 1991.
11. *Wilmington News Journal*, February 18, 1991.
12. *Wilmington News Journal*, January 29, 1991.

CHAPTER 18: CLARENCE AND ANITA

1. Senate Committee on the Judiciary, *Nomination of Judge Clarence Thomas to be Associate Justice of the Supreme Court of the United States: Hearings before the Committee on the Judiciary*, 102nd Cong., 1st sess., 2–3.
2. Ibid., 3.
3. Ibid., 4.
4. Ibid., 116.
5. Ibid., 127.
6. Ibid., 6.
7. Ibid., 268.
8. Jane Mayer and Jill Abramson, *Strange Justice: The Selling of Clarence Thomas* (Boston: Houghton Mifflin, 1994), 218.
9. Senate Committee on the Judiciary, *Nomination of Judge Clarence Thomas*, 222.
10. Ibid., 222–23.
11. Interview with Senator Patrick Leahy.
12. Mayer and Abramson, *Strange Justice*, 231–35.
13. Ibid., 237.
14. Ibid., 245.
15. Specter, *A Passion for Truth*, 347.
16. Ibid., 349.
17. Ibid., 350.
18. Senate Committee on the Judiciary meeting, October 7, 1991.
19. Mayer and Abramson, *Strange Justice*, 262.
20. *New York Times*, October 11, 1991.
21. Mayer and Abramson, *Strange Justice*, 268.
22. Ibid., 269.
23. Ibid., 271.
24. Anita Hill, *Speaking Truth to Power* (New York: Doubleday, 1997), 155–56.
25. Ibid., 321–24.

26. Ibid., 327–29.
27. Ibid., 337.
28. John C. Danforth, *Resurrection: The Confirmation of Clarence Thomas* (New York: Viking Press, 1994), 80–84.

CHAPTER 19: IN SEARCH OF TRUTH AND FAIRNESS

1. Senate Committee on the Judiciary, *Nomination of Judge Clarence Thomas to be Associate Justice of the Supreme Court of the United States: Hearings before the Committee on the Judiciary*, 102nd Cong., 1st sess., part 4, 1–3.
2. Ibid., 5.
3. Ibid., 6–10.
4. Ibid., 27.
5. Ibid., 57–58.
6. Specter, *A Passion for Truth*, 378–79.
7. Senate Committee on the Judiciary, *Nomination of Judge Clarence Thomas*, part 4, 136–37.
8. Ibid., 157–58.
9. Hill, *Speaking Truth to Power*, 202.
10. Ibid., 161.
11. Ibid., 188–89.
12. Ibid., 254–55.
13. Mayer and Abramson, *Strange Justice*, 337–38.
14. Senate Committee on the Judiciary, *Nomination of Judge Clarence Thomas*, 266–67.
15. Mayer and Abramson, *Strange Justice*, 340.
16. Telephone interview with Senator Orrin Hatch, September 28, 2009.
17. Senate Committee on the Judiciary, *Nomination of Judge Clarence Thomas*, 439.
18. Ibid., 452–54, 460, 479, 484.
19. Ibid., 520, 531.
20. Mayer and Abramson, *Strange Justice*, 343.
21. Florence Graves, Alicia Patterson Foundation, *Revisiting the Thomas–Hill Hearings*, 1994, 4.
22. Specter, *A Passion for Truth*, 385.
23. Graves, *Revisiting the Thomas-Hill Hearings*, 5.
24. Interview with Vice President Biden.
25. *Congressional Record* 137 (October 15, 1991), S26304.
26. Danforth, *Resurrection*, 196.
27. Graves, *Revisiting the Thomas-Hill Hearings*, 6.
28. Interview with Vince D'Anna.
29. *Congressional Record* 138 (June 22, 1992), S8853–67.

30. Periscope, *Newsweek*, January 23, 2006, 7.
31. Interview with Professor Raymond Wolters, May 19, 2009, Dover, Del.
32. Interview with former Delaware governor Pierre "Pete" du Pont IV.

CHAPTER 20: COZY CORPORATE CAPITAL OF AMERICA

1. Larry Nagengast, *Pierre S. du Pont IV, Governor of Delaware, 1977–1985*, Oral History Series (Dover: Delaware Heritage Commission, 2006), 39.
2. Interview with former Delaware governor Pierre "Pete" du Pont IV.
3. Interview with Joseph Farley.
4. Interview with David Bakerian, April 30, 2009, Dover, Del.
5. *Wilmington News Journal*, October 30, 1996.
6. Ibid.
7. *Wilmington News Journal*, November 8, 1996.
8. Interview with Dave Crossan, May 1, 1996, Dover, Del.
9. "The Senator from MBNA," *American Spectator*, 1998; National Review Online, 2008.
10. Patrick Healy and Michael Luo, "A Senate Stalwart Who Bounced Back," *New York Times*, August 24, 2008.
11. Cohen, *Only in Delaware*, 278.
12. Interview with Vice President Biden.
13. Interview with David Bakerian.
14. Ibid.
15. Cohen, *Only in Delaware*, 2.
16. Interview with Senator Tom Carper, February 24, 2009, Washington, D.C.
17. Interview with Joseph Farley.
18. Cohen, *Only in Delaware*, 7.
19. Interview with David Bakerian.
20. Interview with Joseph Farley.

CHAPTER 21: FIGHTING CRIME AND ABUSES OF POWER

1. Germond and Witcover, *Whose Broad Stripes and Bright Stars?* 5.
2. Biden, *Promises to Keep*, 238; interview with Hunter Biden, July 8, 2009.
3. Interview with Vice President Biden.
4. Interview with Beau Biden.
5. Interview with Chris Putala, June 29, 2009, Washington, D.C.
6. Biden, *Promises to Keep*, 239.
7. Interview with Victoria Nourse, July 16, 2009, Washington, D.C.
8. Ibid.
9. *New Republic*, September 24, 2008.

10. Interview with Ashley Biden.
11. Ibid.
12. Biden, *Promises to Keep*, 240; interview with Victoria Nourse.
13. Interview with Vice President Biden.
14. Interview with Victoria Nourse.
15. Biden, *Promises to Keep*, 240–41.
16. Interview with Victoria Nourse.
17. Biden, *Promises to Keep*, 242.
18. Interview with Jill Biden.
19. *New Republic*, September 24, 2008.
20. Ibid.
21. Biden, *Promises to Keep*, 245–46.
22. Interview with Vice President Biden.
23. Telephone interview with Senator Orrin Hatch, September 28, 2009.
24. *Wilmington News Journal*, September 26, 1994.
25. Interview with Victoria Nourse.
26. Interview with Chris Putala.
27. Victoria Nourse, *The Accidental Feminist in Transcending the Boundaries of Law* (New York: Routledge, 2010).
28. Interview with Professor Raymond Wolters.

CHAPTER 22: SENATE GLOBE-TROTTER

1. Biden, *Promises to Keep*, 132.
2. Ibid., 131–32.
3. Ibid., 145.
4. Ibid.
5. *Wilmington News Journal*, August 3, 1986.
6. *Wilmington News Journal*, April 23, 1979.
7. *Wilmington News Journal*, June 18, 1981.
8. Interview with Mark Gitenstein, June 27, 2009, Washington, D.C.
9. Senate Judiciary Committee report, May 3, 2000.
10. Interview with former senator William Cohen, April 7, 2009, Washington, D.C.
11. Ibid.
12. Biden, *Promises to Keep*, 248–52.
13. Ibid., 262.
14. Ibid., 263.
15. Ibid., 266.
16. *To Stand Against Aggression; Milošević, the Bosnian Republic and the Conscience of the West: A Report of the Subcommittee on European Affairs of the Senate Foreign Relations Committee* (Washington, D.C.: U.S. Government Printing Office, April 1993), 21.

17. Biden, *Promises to Keep*, 260–67.
18. *Washington Post*, October 7, 2008.
19. Biden, *Promises to Keep*, 276.
20. Ibid.
21. *Wilmington News Journal*, February 20, 1994.
22. Biden, *Promises to Keep*, 281–83.
23. *Washington Post*, October 7, 2008.
24. Biden, *Promises to Keep*, 284–89.
25. Interview with Michael Haltzel, August 11, 2009, Washington, D.C.
26. Ibid.
27. Senate Committee on Foreign Relations, *Meeting the Challenges of a Post–Cold War World: NATO Enlargement and U.S.-Russia Relations: A report to the Committee on Foreign Relations, United States Senate* (Washington, D.C.: U.S. Government Printing Office, 1997), v–vi.
28. Ibid., vi.
29. Ibid., 1–2.
30. *Roll Call*, May 12, 1997.
31. Senate Committee on Foreign Relations, *The Debate on NATO Enlargement: Hearings before the Committee on Foreign Relations*, 105th Cong., 1st sess., October 7, 9, 22, 28, 30, and November 5, 1997, 5–6.
32. *New York Times*, January 25, 2001.
33. *National Journal*, February 10, 2001.
34. Senate Committee on Foreign Relations, *Progress in the Balkans: Kosovo, Serbia, and Bosnia and Herzegovina: A Report to the Committee on Foreign Relations, United States Senate* (Washington, D.C.: U.S. Government Printing Office, February 2001), v.
35. Interview with former senator Chuck Hagel, June 8, 2009, Washington, D.C.

CHAPTER 23: WARS OF NECESSITY AND CHOICE

1. Biden, *Promises to Keep*, 298.
2. Ibid.
3. Ibid., 299–300.
4. Ibid., 300.
5. *Delaware Capitol Review*, January 14, 2002.
6. Biden, *Promises to Keep*, 303.
7. *Delaware Capitol Review*, January 14, 2002.
8. Ibid.
9. Biden, *Promises to Keep*, 304.
10. Ibid, 307.
11. Ibid, 308–11.
12. *Wilmington News Journal*, September 20, 2001.

13. Senate Committee on Foreign Relations, *The International Campaign Against Terrorism: Hearing before the Committee on Foreign Relations, United States Senate*, 107th Cong., 1st sess., October 25, 2001, 1–2.
14. Biden, *Promises to Keep*, 321.
15. *New York Times*, January 13, 2002.
16. Biden, *Promises to Keep*, 325.
17. Ibid.
18. Ibid., 329–30.
19. Ibid., 331.
20. *New York Times*, July 31, 2002.
21. Senate Committee on Foreign Relations, *Hearings to Examine Threats, Responses, and Regional Considerations Surrounding Iraq: Hearings before the Committee on Foreign Relations, United States Senate*, 107th Cong., 2nd sess., July 31 and August 1, 2002, 1–3.
22. Interview with Antony Blinken, June 3, 2009, Washington, D.C.
23. Biden, *Promises to Keep*, 332.
24. *Washington Times*, August 5, 2002.
25. Biden, *Promises to Keep*, 336.
26. Trent Lott, *Herding Cats: A Life in Politics* (New York: Regan Books, 2005), 238–41.
27. Biden, *Promises to Keep*, 339–40.
28. Interview with Michael Haltzel.
29. Biden, *Promises to Keep*, 339–40.
30. *Delaware State News*, October 9, 2002.
31. Interview with Vice President Biden.
32. *Washington Post*, March 10, 2003.
33. Interview with Vice President Biden.
34. *National Journal*, March 11, 2003.
35. Biden, *Promises to Keep*, 343–44.
36. Ibid., 345.
37. Ibid., 347.
38. *Congressional Record* 149 (June 27, 2003), S8827.
39. Ibid., S8830.
40. *New York Times*, October 1, 2003.
41. *Washington Post*, November 9, 2003.

CHAPTER 24: REASSESSING A QUAGMIRE

1. CNN, August 12, 2003.
2. Interview with Larry Rasky.
3. MSNBC, March 17, 2004.
4. Biden, *Promises to Keep*, 342.
5. Senate Committee on Foreign Relations, *The Nomination of Hon.*

John D. Negroponte to be U.S. Ambassador to Iraq: Hearing before the Committee on Foreign Relations, 108th Cong., 2nd sess., April 27, 2004, 8.
6. Ibid., 9.
7. Ibid., June 22, 2004, 108–729.
8. Democratic National Convention, Boston, July 29, 2004.
9. *Wilmington News Journal*, October 2, 2004.
10. Senate Committee on Foreign Relations, *Accelerating U.S. Assistance to Iraq: Hearing before the Committee on Foreign Relations*, 108th Cong., 2nd sess., September 15, 2004, 8–9.
11. *Rolling Stone*, February 2, 2005.
12. Interview with Larry Rasky.
13. Interview with John Marttila.
14. *Face the Nation*, CBS, June 19, 2005.
15. Ibid.
16. *Washington Post*, November 22, 2005.
17. *Nation*, August 29, 2005.
18. *Confirmation Hearing on the Nomination of John G. Roberts Jr. to be Chief Justice of the United States: Hearing before the Committee on the Judiciary*, 109th Cong., 1st sess., September 12–15, 2005, 16–18.
19. Ibid., 55.
20. Ibid., 56.
21. Ibid., 185.
22. Ibid., 187–94.
23. *Today*, NBC, January 1, 2006.
24. Ibid.
25. Interview with Larry Rasky.
26. Interview with Jill Biden.
27. CNN, December 29, 2006.
28. Interview with Antony Blinken, June 3, 2009.

CHAPTER 25: BIDEN FOR PRESIDENT AGAIN

1. *New York Times*, January 8, 2007.
2. Interview with Larry Rasky.
3. Confirmed in phone interview with Chet Curtis, New England Cable Network, October 5, 2009.
4. *Washington Post*, January 5, 2007.
5. *New York Times*, January 31, 2007.
6. *New York Observer*, January 31, 2007.
7. Ibid.
8. Interview with Larry Rasky.
9. Ibid.
10. *Wilmington News Journal*, February 1, 2007.

11. Interview with Larry Rasky.
12. *The Daily Show*, January 31, 2007.
13. Interview with Larry Rasky.
14. *Washington Post*, February 1, 2007.
15. Ibid.
16. *Boston Globe*, February 27, 2007.
17. *Boston Globe*, March 10, 2007.
18. *Wilmington News Journal*, April 27, 2007.
19. *Wilmington News Journal*, April 30, 2007; *Boston Globe*, May 15, 2007.
20. *Los Angeles Times*, June 18, 2007.
21. Ibid.
22. *Boston Globe*, July 25, 2007.
23. *Wilmington News Journal*, August 19, 2007.
24. Interview with Missy Owens, October 29, 2009, Washington, D.C.
25. Interview with Vice President Biden.
26. ABC News, August 23, 2007.
27. *Philadelphia Inquirer*, August 27, 2007.
28. *Washington Post*, October 25, 2007.
29. *Wilmington News Journal*, August 26, 2007.
30. Interview with Nicole Gaudiano, October 23, 2009, Washington, D.C.
31. *Wilmington News Journal*, October 22, 2007.
32. *Washington Post*, October 25, 2007.
33. *Modern Healthcare*, October 29, 2007.
34. *New York Times*, November 1, 2007.
35. Ibid.
36. *St. Petersburg Times*, November 5, 2007.
37. Jules Witcover, "The Invisible Joe Biden," Tribune Media Services, November 21, 2007.
38. Ibid.
39. *New York Times*, December 14, 2007.
40. *Boston Globe*, December 12, 2007.
41. Ibid.
42. *Philadelphia Inquirer*, December 12, 2007.
43. *Washington Post*, December 28, 2007.
44. *Wilmington News Journal*, January 3, 2008.
45. Ibid.
46. *Wilmington News Journal*, January 4, 2008.
47. *New York Times*, January 2, 2008.
48. Interview of Joe Biden by Nicole Gaudiano, August 26, 2007, Clinton, Iowa.

49. Interview with John Marttila.
50. *Wilmington News Journal*, January 3, 2008.
51. *Wilmington News Journal*, November 5, 2008.
52. Interview with John Marttila.
53. Interview with Missy Owens.
54. *Wilmington News Journal*, January 6, 2008.
55. Interview with Ashley Biden.
56. Interview with Missy Owens.
57. Interview with Beau Biden.
58. Interview with Valerie Biden Owens, September 24, 2009.
59. Interview with Vice President Biden.
60. *Wilmington News Journal*, January 4, 2008.
61. *Wilmington News Journal*, January 5, 2008.

CHAPTER 26: A QUESTIONABLE PRIZE

1. Jules Witcover, *Crapshoot: Rolling the Dice on the Vice Presidency* (New York: Crown Publishers, 1992), 18.
2. Donald Young, *American Roulette: The History and Dilemma of the Vice Presidency* (New York: Holt, Reinhart, and Winston, 1965), 9.
3. Ibid., 14.
4. Witcover, *Crapshoot*, xiii.
5. Young, *American Roulette*, 33.
6. Ibid., 43–46.
7. Eugene H. Roseboom, *A History of Presidential Elections*, 2nd ed. (New York: Macmillan, 1964), 210.
8. Young, *American Roulette*, 94.
9. Witcover, *Crapshoot*, 54.
10. Young, *American Roulette*, 116.
11. Witcover, *Crapshoot*, 55.
12. Ibid., 59.
13. Ibid.
14. Ibid.
15. Ibid., 8–9.
16. Merle Miller, *Plain Speaking: An Oral Biography of Harry S. Truman* (New York: Berkley Books, 1974), 193.
17. Witcover, *Crapshoot*, 128–39.
18. Jules Witcover, *Very Strange Bedfellows: The Short and Unhappy Marriage of Richard Nixon and Spiro Agnew* (New York: Public Affairs, 2007), 275–89.
19. Barton Gellman, *Angler: The Cheney Vice Presidency* (New York: Penguin Press, 2008), 1–30.

CHAPTER 27: AN OFFER HE COULDN'T REFUSE

1. Aboard USS *Independence*, October 11, 1967.
2. Interview with David Axelrod, September 16, 2009, Washington, D.C.
3. *New Yorker*, October 13, 2008.
4. Ibid.
5. *Wilmington News Journal*, February 27, 2008.
6. *Washington Times*, May 30, 2008.
7. David Plouffe, *The Audacity to Win: The Inside Story and Lessons of Barack Obama's Historic Victory* (New York: Viking, 2009), 284.
8. Ibid.
9. Ibid., 284–85.
10. *Wilmington News Journal*, April 12, 2008.
11. *New Yorker*, October 13, 2008.
12. Interview with David Axelrod.
13. *Wilmington News Journal*, November 5, 2008.
14. Plouffe, *The Audacity to Win*, 287.
15. Interview with David Axelrod.
16. Plouffe, *The Audacity to Win*, 288–90.
17. Interview with David Plouffe, December 7, 2009, Washington, D.C.
18. Plouffe, *The Audacity to Win*, 290–91.
19. *Newsweek*, November 13, 2008.
20. Interview with former senator Chuck Hagel, June 8, 2009.
21. Interview with Senator Evan Bayh, February 25, 2009, Washington, D.C.
22. *Wilmington News Journal*, November 20, 2008.
23. Interview with David Axelrod.
24. Interview with Vice President Biden.
25. Interview with David Axelrod.
26. E-mail response from President Barack Obama, March 1, 2010.
27. Interview with David Plouffe.
28. Interview with David Axelrod.
29. *Boston Globe*, August 24, 2008.
30. Ibid.
31. Democratic National Convention, Denver, Colorado, August 27, 2008.
32. *Wilmington Journal*, August 29, 2008.
33. Ibid.; *Washington Post*, August 24, 2008.
34. *New York Times*, August 25, 2008.
35. *USA Today*, August 27, 2008.
36. *Wilmington News Journal*, September 4, 2008.
37. Telephone interview with Rachel Kipp, October 27, 2009.
38. *Wilmington News Journal*, September 4, 2008.

39. Ibid.
40. Ibid.
41. Ibid.
42. *New York Times*, December 14, 2007.
43. *Wilmington News Journal*, September 4, 2008.
44. CBS Evening News, March 4, 2008.
45. Telephone interview with Pamela Hamill, October 31, 2009.
46. *Scranton Times*, September 2, 2008.
47. Ibid.
48. *Newsday*, Associated Press, September 1, 2008.
49. *New York Times*, September 2, 2008.
50. Interview with David Axelrod.
51. Interview with David Plouffe.
52. Interview with David Axelrod.
53. John Heilemann and Mark Halperin, *Game Change: Obama and the Clintons, McCain and Palin, and the Race of a Lifetime* (New York: HarperCollins, 2010), 405.
54. Vice presidential debate, St. Louis, October 2, 2008.
55. *Wall Street Journal*, October 10, 2008.
56. *New York Daily News*, September 11, 2008.
57. CBS Evening News, September 24, 2008.
58. *London Spectator*, November 8, 2008.
59. *New York Daily News*, October 21, 2008.
60. *Time*, October 10, 2008.
61. Interview with Professor Samuel Hoff, May 5, 2009, Dover, Del.; *Delaware State News*, November 10, 2008.
62. Interview with David Plouffe.
63. Heilemann and Halperin, *Game Change*, 413–14, 419.
64. *Politico*, January 10, 2010.
65. E-mail response from President Obama.
66. Interview with David Axelrod.
67. *London Evening Standard*, October 1, 2008.
68. *Wilmington News Journal*, November 5, 2008.
69. Ibid.
70. *New York Times*, November 26, 2008.
71. E-mail response from President Obama.
72. *Wilmington News Journal*, November 20, 2008.
73. ABC News *This Week*, December 21, 2008

CHAPTER 28: VICE PRESIDENT BIDEN

1. Interview with David Axelrod.
2. E-mail response from President Obama.
3. Interview with Ron Klain, October 1, 2009, Washington, D.C.

4. Ibid.
5. Interview with Antony Blinken, December 18, 2009, Washington, D.C.
6. Interview with Ron Klain, December 23, 2009, Washington, D.C.
7. Interview with Valerie Biden Owens, September 24, 2009.
8. Interview with Jill Biden.
9. Interview with Ron Klain, December 23, 2009.
10. *The Oprah Winfrey Show*, January 20, 2009.
11. Interview with David Axelrod.
12. *Wall Street Journal*, January 30, 2009.
13. *Wall Street Journal*, February 6, 2009.
14. White House Press Office, February 9, 2009.
15. *New York Times*, February 15, 2009.
16. Newsweek.com, October 10, 2009.
17. Interview with Vice President Biden.
18. White House Press Office, February 7, 2009.
19. *Los Angeles Times*, February 8, 2009.
20. *Washington Post*, February 28, 2009.
21. *Washington Post*, March 11, 2009.
22. Gridiron Club Dinner, March 21, 2009.
23. Mark Leibovich, "Speaking Freely, Biden Finds Influential Role," *New York Times*, March 28, 2009.
24. Ibid.
25. CNN, April 4, 2009.
26. *60 Minutes*, April 26, 2009.
27. *Today*, April 30, 2009.
28. *New York Times*, May 2, 2009.
29. *Washington Post*, April 29, 2009.
30. Office of the Vice President, May 13, 2009.
31. Video by Stephen J. Pallone, Syracuse.com, *Syracuse Post-Standard*, May 10, 2009.
32. Interview with Pat Cowin Wojenski, May 10, 2009, Syracuse, N.Y.
33. Associated Press, *Boston Globe*, May 20, 2009.
34. *Balkan Watch*, June 1, 2009.
35. *Wall Street Journal*, May 22, 2009.
36. *New York Times*, July 23, 2009.
37. *Washington Post*, July 24, 2009.
38. *Wall Street Journal*, July 25, 2009.
39. *New York Times*, July 26, 2009.
40. *Los Angeles Times*, August 8, 2009.
41. *New York Times*, July 26, 2009.
42. *New York Times*, July 31, 2009.
43. *Washington Post*, September 20, 2009.

44. Newsweek.com, October 10, 2009.
45. Interview with Antony Blinken, December 18, 2009.
46. Jonathan Alter, *The Promise: President Obama, Year One* (New York: Simon and Schuster, 2010), 377.
47. Interview with Antony Blinken, December 18, 2009.
48. Alter, *The Promise*, 389.
49. Ibid., 393.
50. Interview with Vice President Biden.
51. Ibid.
52. Interview with Antony Blinken, December 18, 2009.
53. Ibid.
54. Telephone interview with Michael O'Hanlon, December 7, 2009.
55. Telephone interview with former senator Chuck Hagel, December 22, 2009.
56. CBS News, December 2, 2009.
57. E-mail response from President Obama.
58. YouTube video, August 28, 2009.
59. Telephone interview with Senator Paul Kirk, November 2, 2009, Washington, D.C.
60. *Washington Post*, February 19, 2010.
61. White House Press Office, March 9, 2010; *New York Times*, March 10, 2010.
62. *New York Times*, March 10, 2010.
63. White House Press Office, March 10, 2010.
64. White House Press Office, March 11, 2010.
65. White House Press Office, March 23, 2010.
66. *New York Daily News*, March 23, 2010.
67. *Los Angeles Times*, March 24, 2010.
68. Office of the Vice President, pool report, March 24, 2010.
69. CBS News, March 30, 2010; *Washington Post*, March 31, 2010.
70. "The Runaway General," *Rolling Stone* 1108/1109 (June 25, 2010).
71. *Washington Post*, June 23, 2010.
72. CNN, June 24, 2010.
73. *Washington Post*, June 25, 2010.
74. Ibid.
75. Fox News, June 27, 2010.
76. Ibid.
77. E-mail response from President Obama.
78. *New York Times*, March 29, 2009.

EPILOGUE

1. Interview with Antony Blinken, June 3, 2009.
2. Interview with former senator Chuck Hagel, June 8, 2009.

3. Interview with former Senate majority leader Bob Dole, June 16, 2009, Washington, D.C.
4. Biden, *Promises to Keep*, 191.
5. Ibid., 206.
6. *National Journal*, October 14, 1989.
7. Interview with David Plouffe.
8. Statement of Joe Biden, January 8, 2010.

BIBLIOGRAPHY

Alter, Jonathan. *The Promise: President Obama, Year One* (New York: Simon and Schuster, 2010).

Biden, Joseph R., II. *Promises to Keep: On Life and Politics* (New York: Random House, 2007).

Bork, Robert H. *The Tempting of America: The Political Seduction of the Law* (New York: Free Press, 1990).

Bronner, Ethan. *Battle for Justice: How the Bork Nomination Shook America* (New York: W. W. Norton, 1989).

Cohen, Celia. *Only in Delaware: Politics and Politicians in the First State* (Newark, Del.: Grapevine Publishing, 2002).

Cramer, Richard Ben. *What It Takes: The Way to the White House* (New York: Vintage Books, 1993).

Danforth, John C. *Resurrection: The Confirmation of Clarence Thomas* (New York: Viking Press, 1994).

Gellman, Barton. *Angler: The Cheney Vice Presidency* (New York: Penguin Press, 2008).

Germond, Jack W., and Jules Witcover. *Whose Broad Stripes and Bright Stars?: The Trivial Pursuit of the Presidency 1988* (New York: Warner Books, 1989).

Gitenstein, Mark. *Matters of Principle: An Insider's Account of America's*

Rejection of Robert Bork's Nomination to the Supreme Court (New York: Simon and Schuster, 1992).

Heilemann, John, and Mark Halperin. *Game Change: Obama and the Clintons, McCain and Palin, and the Race of a Lifetime* (New York: HarperCollins, 2010).

Hill, Anita. *Speaking Truth to Power* (New York: Doubleday, 1997).

Lott, Trent. *Herding Cats: A Life in Politics* (New York: Regan Books, 2005).

Mayer, Jane, and Jill Abramson. *Strange Justice: The Selling of Clarence Thomas* (Boston: Houghton Mifflin, 1994).

McGuigan, Patrick B., and Dawn M. Weyrich. *Ninth Justice: The Fight for Bork* (Washington, D.C.: Free Congress Research and Education Foundation, 1990).

Meese, Edwin, III. *With Reagan: The Inside Story* (Washington, D.C.: Regnery Gateway, 1992).

Miller, Merle. *Plain Speaking: An Oral Biography of Harry S. Truman* (New York: Berkley Books, 1974).

Nagengast, Larry. *Pierre S. du Pont IV, Governor of Delaware, 1977–1985.* Oral History Series (Dover: Delaware Heritage Commission, 2006).

Nourse, Victoria. *The Accidental Feminist in Transcending the Boundaries of Law* (New York: Routledge, 2010).

Pertschuk, Michael, and Wendy Schaetzel. *The People Rising: The Campaign Against the Bork Nomination* (New York: Thunder's Mouth Press, 1989).

Plouffe, David. *The Audacity to Win: The Inside Story and Lessons of Barack Obama's Historic Victory* (New York: Viking, 2009).

Roseboom, Eugene H. *A History of Presidential Elections*, 2nd ed. (New York: Macmillan, 1964).

Sorensen, Ted. *Counselor: A Life at the Edge of History* (New York: Harper, 2008).

Specter, Arlen. *A Passion for Truth: From Finding JFK's Single Bullet to Questioning Anita Hill to Impeaching Clinton* (New York: William Morrow, 2000).

Taylor, Paul. *See How They Run: Electing the President in an Age of Mediaocracy* (New York: Knopf, 1990).

U.S. Congress. Senate. Committee on Foreign Relations. *U.S. Policy in the Persian Gulf: Hearings before the Committee on Foreign Relations*, 101st Cong., 2nd sess., September 5, 20, and October 17, 1990.

———. *U.S. Policy in the Persian Gulf: Hearings before the Committee on Foreign Relations*, 101st Cong, 2nd sess., December 4–5, 12–13, 1990.

———. *U.S. Policy in the Persian Gulf: Hearings before the Committee on Foreign Relations*, 102nd Cong., 1st sess., January 8, 1991.

———. *To Stand Against Aggression; Milošević, the Bosnian Republic and the Conscience of the West: A Report of the Subcommittee on European Affairs of the Senate Foreign Relations Committee.* Washington, D.C: U.S. Government Printing Office, 1993.

———. *Meeting the Challenges of a Post–Cold War World: NATO Enlargement and U.S.-Russia Relations: A report to the Committee on Foreign Relations, United States Senate.* Washington, D.C.: U.S. Government Printing Office, May 1997.

———. *The Debate on NATO Enlargement: Hearings before the Committee on Foreign Relations*, 105th Cong., 1st sess., October 7, 9, 22, 28, 30, and November 5, 1997.

———. *Progress in the Balkans: Kosovo, Serbia, and Bosnia and Herzegovina: A Report to the Committee on Foreign Relations, United States Senate.* Washington, D.C.: U.S. Government Printing Office, February 2001.

———.*The International Campaign Against Terrorism: Hearing before the Committee on Foreign Relations, United States Senate*, 107th Cong., 1st sess., October 25, 2001

———. *Hearings to Examine Threats, Responses, and Regional Considerations Surrounding Iraq: Hearings before the Committee on Foreign Relations, United States Senate*, 107th Cong., 2nd sess., July 31, and August 1, 2002.

———. *The January 27 UNMOVIC and IAEA Reports to the U.N. Security Council on Inspections in Iraq: Hearing before the Committee on Foreign Relations*, 108th Cong., 1st sess., January 30, 2003.

———. *The Future of Iraq: Hearing before the Committee on Foreign Relations*, 108th Cong., 1st sess., February 11, 2003.

———. *Iraq: Meeting the Challenge, Sharing the Burden, Staying the Course: A Trip Report to Members of the Committee on Foreign Relations.* Washington, D.C.: U.S. Government Printing Office, July 2003.

———. *The Iraq Transition—Obstacles and Opportunities (Part III): Hearing before the Committee on Foreign Relations*, 108th Cong., 2nd sess., April 22, 2004.

———. *The Nomination of Hon. John D. Negroponte to be U.S. Ambassador to Iraq: Hearing before the Committee on Foreign Relations*, 108th Cong., 2nd sess., April 27, 2004.

———. *Accelerating U.S. Assistance to Iraq: Hearing before the Committee on Foreign Relations*, 108th Cong., 2nd sess., September 15, 2004.

U.S. Congress. Senate. Committee on the Judiciary. *Nomination of Justice William Hubbs Rehnquist: Hearings before the Committee on the Judiciary on the nomination of Justice William Hubbs Rehnquist to be Chief Justice of the United States*, 98th Cong., 2nd sess., July 29–31, and August 1, 1986.

———. *Nomination of Robert H. Bork to be Associate Justice of the Supreme Court of the United States: Hearings before the Committee on the Judiciary*, 100th Cong., 1st sess., September 15–30, 1987.

———. *The Constitutional Roles of Congress and the President in Declaring and Waging War: Hearing before the Committee on the Judiciary*, 102nd Cong., 1st sess., January 8, 1991.

———. *Nomination of Judge Clarence Thomas to be Associate Justice of the Supreme Court of the United States: Hearings before the Committee on the Judiciary*, 102nd Cong., 1st sess., September 10–13, 16–17, 19–20, and October 11–13, 1991.

———. *Confirmation Hearing on the Nomination of John G. Roberts Jr. to be Chief Justice of the United States: Hearing before the Committee on the Judiciary*, 109th Cong., 1st sess., September 12–15, 2005.

———. *Confirmation Hearing on the Nomination of Samuel A. Alito Jr. to Be an Associate Justice of the Supreme Court of the United States*, 109th Cong., 2nd sess., January 9–13, 2006.

Wilmington News Journal Archives, Wilmington, Delaware, 1969–88.

Witcover, Jules. *Crapshoot: Rolling the Dice on the Vice Presidency* (New York: Crown Publishers, 1992).

———. *Marathon: The Pursuit of the Presidency, 1972–1976* (New York: Viking Press, 1977).

———. *Very Strange Bedfellows: The Short and Unhappy Marriage of Richard Nixon and Spiro Agnew* (New York: Public Affairs, 2007).

Young, Donald. *American Roulette: The History and Dilemma of the Vice Presidency* (New York: Holt, Reinhart, and Winston, 1965).

INDEX